Lecture Notes in Mathematics

A collection of informal reports and seminars
Edited by A. Dold, Heidelberg and B. Eckmann, Zürich

70

Proceedings of the Summer School in Logic Leeds, 1967

N. A. T. O. Advanced Study Institute
Meeting of the Association for Symbolic Logic

Edited by M. H. Löb
The University of Leeds
Department of Mathematics

1968

Springer-Verlag Berlin · Heidelberg · New York

Library of Congress Catalog Card Number 68-56951 Printed in Germany. Title No. 3676

A Summer School in Logic was held in Leeds from August 7 - 23, 1967 under the auspices of N.A.T.O. and the Association for Symbolic Logic.

The programme consisted of lecture courses given by S. Feferman, A. Lévy, M. Morley, G. H. Müller, D. Rödding. Invited addresses were presented by W. Craig, G. N. Crossley, P. Geach, G. Takeuti and E. Wette. In addition ten contributed papers were read.

The Summer School was financially supported by N.A.T.O. and to a smaller extent by the British Logic Colloquium.

The present volume contains lecture notes of courses and lectures which were made available by contributors.

M. H. Löb.

Contents

Lectures on Proof Theory

by

Solomon Feferman[1]

§1. Introduction

1(a) Proof theory provides technical tools for a penetrating analysis of derivability in various formal systems. The main impulse for its development came from Hilbert's program [Hl] to establish the consistency of formalizations of mathematics by finitist (combinatorial) means. It was found somewhat later that proof-theoretic methods could also be used to obtain model-theoretic results efficiently when applied to logically complete systems.

Several distinct approaches have been developed in proof theory. The results obtained by these different means overlap to a considerable extent. However, none of these has so far turned out to be most comprehensive, nor is there any indication yet that this will be the case. These lectures present an extension in one direction of the approach due to Gentzen [Gn1]. To my mind this provides the most accessible and elegant development of the material in the common ground, and the most suitable for the further applications treated here.[2]

It is typical of the different approaches that the formal systems $\mathcal{F}$ about which one wants information are not to begin with suitable for a direct analysis. The system $\mathcal{F}$ is transformed or reduced to another system $\mathcal{F}'$ which is suitable, or $\mathcal{F}'$ may simply be a convenient intermediary for a further transformation or reduction to another $\mathcal{F}''$, etc., until we reach a system where a thorough-going

---

1) We wish to thank the organizers of the Leeds Institute in Logic for having provided the opportunity to present these lectures, and for their great helpfulness during the course of the Institute and subsequently with the publication.
Most of the new methods and results which form a part of the material presented here were developed in research carried out during 1966-67 under grant DA-ARO-(D)-31-124-G655 at Stanford University. The preparation of these lectures for publication was supported by grant NSF-GP-6982 at the Massachusetts Institute of Technology.

2) The other principal approaches originated with Hilbert's ε-calculus [H,B], Herbrand's thesis [Hr], and Gödel's interpretation by constructive functionals [Gd2]; cf. Kreisel [Kr3] for an up-to-date survey and bibliography of the entire subject.

analysis can be carried out. (In practice, only one or two modifications is necessary.) In fact, a large part of what is distinctive about each approach is the manner in which this reduction is accomplished.

Consider, for example, any of the usual systems $\mathfrak{F}$ of first-order predicate calculus for deriving the valid formulas ϕ of the given language. The first step in Gentzen's approach is to transform this into a system $\mathfrak{F}'$ for deriving sequents, namely pairs (Γ, Δ) of finite sequences of formulas $\Gamma = \langle \phi_1, \ldots, \phi_n \rangle$, $\Delta = \langle \Psi_1, \ldots, \Psi_m \rangle$. Writing $(\Gamma \supset \Delta)$ for (Γ, Δ), the relation between $\mathfrak{F}$ and $\mathfrak{F}'$ is given by: $(\Gamma \supset \Delta)$ is derivable in $\mathfrak{F}'$ if and only if $(\phi_1 \wedge \ldots \wedge \phi_n) \rightarrow (\Psi_1 \vee \ldots \vee \Psi_m)$ is derivable in $\mathfrak{F}$. The advantage of this transformation is that it is now possible to give rules (for passing from given sequents to new sequents) which isolate the role of each logical connective.[3] However, one is not yet in a position to survey the possible derivations of a given sequent in any informative way. For, the rule of modus ponens in $\mathfrak{F}$ is replaced by a rule called cut in $\mathfrak{F}'$,

$$\frac{\Gamma \supset \Delta, \phi \qquad \Gamma', \phi \supset \Delta'}{\Gamma, \Gamma' \supset \Delta, \Delta'}$$

which gives conclusions leaving no trace of certain components of the argument. The second step in Gentzen's approach is to show that the cut rule is superfluous with respect to the other rules of $\mathfrak{F}'$. That is, let $\mathfrak{F}''$ be obtained from $\mathfrak{F}'$ by omitting the cut-rule. Then if a sequent $(\Gamma \supset \Delta)$ is derivable in $\mathfrak{F}'$, it is already derivable in $\mathfrak{F}''$; this is Gentzen's Hauptsatz or cut-elimination theorem. Every derivation in $\mathfrak{F}''$ has a direct character enjoying what is known as the subformula property, i.e. that each formula of a hypothesis of a rule is a subformula (in a slightly extended sense of the word) of some formula of the conclusion. Now one can proceed to study derivations in $\mathfrak{F}''$ to obtain various results about derivability in $\mathfrak{F}$. We mention particularly Craig's interpolation

---

[3] The reader unfamiliar with such rules may wish at this point to glance at the list of G-rules in §2(c) below.

theorem [Cal]: if $(\emptyset \to \Psi)$ is derivable in $\mathcal{F}$ then there is a formula θ whose basic symbols are common to $\emptyset, \Psi$ such that $(\emptyset \to \theta)$ and $(\theta \to \Psi)$ are derivable in $\mathcal{F}$. (A precise statement of this will be given in §4.) Each of the transformations described above is obtained by a combinatorial argument; one can go on to obtain combinatorial consistency proofs of some formal systems representing part of elementary number theory.[4)]

Gentzen's approach is extended here to the first-order predicate calculus in a language $\mathcal{L}_{HC}$ permitting infinitely long expressions built up using countably infinite conjunctions and disjunctions. This is applied to extend certain results of the model theory of the usual finitary language $\mathcal{L}_{HF}$ to $\mathcal{L}_{HC}$. It is also used to provide a unified framework for constructive consistency proofs of various non-elementary formal systems (which cannot be treated by finitist means).

The following is a more detailed summary of the work. §2 deals with various syntactic and semantic preliminaries for $\mathcal{L}_{HC}$. It turns out to be particularly convenient for the applications to take this as a many-sorted language from the outset. The system of Gentzen rules is presented in §2(c) and a proof of its completeness is sketched. §2(d) introduces a spectrum of sublanguages $\mathcal{L}_A$ of $\mathcal{L}_{HC}$, derived from certain generalizations of recursion theory; this includes $\mathcal{L}_{HF}$ and other sublanguages of $\mathcal{L}_{HC}$ of particular interest. The use of the $\mathcal{L}_A$ permits a uniform treatment of these special cases. §3 begins with the natural assignment of ordinals to infinite formulas and derivations. The main work in §3 is the proof of the cut-elimination theorem for $\mathcal{L}_{HC}$. In general, cuts are eliminated from a derivation only at the cost of a considerable increase in its length; explicit ordinal bounds are given for this increase.

§4 presents generalizations of interpolation theorems for $\mathcal{L}_{HF}$ to the languages $\mathcal{L}_A$. In addition to the usual requirements, particular attention is paid to restrictions which can be imposed on an interpolant in terms of the sorts of

---

4) If one is concerned only with model-theoretic applications of proof-theory, these combinatorial aspects are of no concern. Moreover, it is possible to take a short-cut for these purposes by an outright proof of the completeness of $\mathcal{F}''$; this is touched on at various points in the text.

variables and forms of quantifiers which appear in it. Model-theoretic applications of these theorems are then given in §5. First, diverse results characterizing the formulas of $\mathcal{L}_A$ which are persistent and invariant for extensions are recaptured by a uniform argument and are also generalized. Second, a general theorem is obtained giving simple sufficient conditions for eliminating quantifiers relative to a set S of axioms in $\mathcal{L}_A$.

§6 serves only as an introduction to the use of infinitary languages (contained in $\mathcal{L}_{HC}$) for extensions of Hilbert's program.[5] Formal systems Z of elementary number theory and R_0 of arithmetical analysis (ramified analysis of rank 0) are used as illustrations. The ordinal bounds of §3 are used to recover the known sharp bounds for the provable well-orderings of Z and R_0. The transformation of this work into proofs of consistency by constructive instances of transfinite induction is sketched.

These lectures were written up for a fairly wide audience. First, they should be accessible to readers with some background in mathematical logic (corresponding to the major portions of the texts [Kℓ], [Sh]) and model theory (corresponding to [Sh], Ch. 5 or the basic material of the texts [K,K], [R]). No knowledge of proof theory (or infinitary languages) is assumed. Thus the arguments in §§3,4 are presented in some detail, containing the various points that must be attended to. The reader who is familiar with the corresponding arguments in the finitary languages will readily see much that is repeated and that can be skipped. The new material which may be of interest to the expert is contained in the results stated in §§4,5 and the arguments given in §§5,6.

Some words are in order as to what is not done here. We do not go into Gentzen-type formalizations which have been obtained for a wide variety of fragments of the classical calculus and for non-classical (intuitionistic and modal) calculi; the interested reader is referred to [Pr] as a principal source. As implicitly suggested above, the arguments for the various known cut-elimination and interpolation theorems follow certain familiar patterns. This suggests the

---

[5] A further development of this approach will appear elsewhere.

possibility of more abstract formulations of these theorems which would not require complete specifications of language and rules. This is a matter of current research; no contribution is made to it here (except insofar as the statements of §§3,4 provide further specific cases to be accounted for).

No comparison is made of Gentzen's approach with the other principal approaches of proof theory. Nor is any attempt made to make explicit the notions and principles to be employed in constructive consistency proofs such as are aimed at in §6. Finally, we do not go into the basic question as to what one hopes to accomplish by proof-theory in this direction. In these cases we have an excellent excuse for the omissions. The survey paper [Kr3], to appear, contains a searching analysis of the aims of proof theory in the foundations of mathematics and a detailed assessment of its accomplishments. It is recommended reading as part of any serious study of the subject.

1(b) Background on infinitely long formulas. Detailed references will be given in the text for the sources of various results. This section is intended to trace some of the principal stages in the developments which led to the use of $\mathcal{L}_{HC}$ here. A reading of it is not presupposed in the remainder.[6)]

The different approaches in proof theory all produced much the same information when applied to the classical first-order predicate calculus (with finite formulas), and certain weak subsystems of Z . Moreover, combinatorial methods sufficed for the treatment of these systems. New difficulties arose in attempts to prove the consistency of axiomatic systems at least as strong as Z by these methods. A theoretical explanation for this block was subsequently provided by Gödel's second underivability theorem [Gd1], for the kinds of combinatorial arguments which had previously been carried out or were envisaged could be represented in Z .

The following was the immediate obstacle in the case of Gentzen's approach. Derivability in the predicate calculus from an arbitrary set of axioms S can be

6) It is addressed primarily to beginners in the subject with a taste for history.

transformed into derivability in a sequential calculus simply by adjoining as axioms all sequents $(0 \supset \phi)$ for $\phi \in S$ (0 = empty sequent). However, the cut-elimination theorem will not in general extend to such systems of derivation. Some modification of this approach is thus necessary if one is to analyze derivability from S, or even just prove the consistency of S.

Gentzen developed a special technique in [Gn2] to prove the consistency of Z. First he assigned ordinals less than ϵ_0 to derivations $\mathcal{D}$ from Z, as a certain measure of complexity of these derivations. ϵ_0 is the limit of the ordinals ω_n where $\omega_0 = \omega$, $\omega_{n+1} = \omega^{\omega_n}$. The well-ordering of ordinals $< \epsilon_0$ is isomorphic in a natural way (using Cantor's normal form) to a primitive recursive relation $\prec$ with field equal to the natural numbers. The assignment which Gentzen gave was actually a primitive recursive assignment $O(\mathcal{D})$ to derivations $\mathcal{D}$ of an element of ω. The main result of [Gn2] (by a combinatorial argument) was that with each derivation $\mathcal{D}$ of a contradiction, say $(\bar{0} \neq \bar{0})$, could be associated another derivation $\mathcal{D}'$ of the same, with $O(\mathcal{D}') \prec O(\mathcal{D})$. The consistency of Z follows by the principle of transfinite induction for $\prec$. Gentzen also showed that (in view of Gödel's result) this is optimal, in the sense that the principle of transfinite induction for each initial segment $\prec_1$ of $\prec$ can be derived in Z.

It was difficult for some time to assess the significance of Gentzen's work on the consistency of Z and to extend it. First of all, its use of transfinite induction up to ϵ_0 (though applied only to a primitive recursive predicate) went beyond what had hitherto been recognized as evidently finitist reasoning. Thus it did not clearly constitute a contribution to Hilbert's probram as originally understood. However, the theoretical limitations placed on that program by Gödel's theorem would lead to considering an extension of Hilbert's program in which the requirement that all reasoning of proof theory be finitist is relaxed to the requirement that all such reasoning be constructive. Gentzen's argument [Gn4] for the well-foundedness of $\prec$ is certainly of this character. But when the scope of proof-theory is enlarged in this way, it would be hoped to treat formal systems much stronger than Z. Unfortunately, the guiding idea of Gentzen's assignment of ordinals was difficult to understand and the details of his argument did not suggest the direction of further work.

One way ahead came from an at first sight quite distant quarter. In the early logical literature, one had spoken of universal (existential) quantification as being a kind of infinite conjunction (disjunction). This analogy was defective for the predicate calculus, since the conjunctions and disjunctions considered would have to change with each domain. However, the analogy is reasonable whenever (in a possibly many-sorted language) one of the sorts of quantifiers is interpreted as ranging over a fixed denumerable domain, each of whose elements is denoted by a term in the language. For example, in the case of Z this suggests consideration of the ω-rule:

$$\frac{\Gamma \supset \Delta, \phi(\overline{0})\ ;\ \ldots\ ;\ \Gamma \supset \Delta, \phi(\overline{n})\ ;\ \ldots}{\Gamma \supset \Delta, \bigwedge x\, \phi(x)}$$

Then it is no longer necessary to consider the instances of induction as axioms, since these are now derivable by means of the ω-rule. However, derivations must now be regarded as infinite trees. These do not provide a formal system in the usual sense of the word, but rather what is sometimes called a semi-formal or infinitary system $Z^{(\infty)}$. Nevertheless, such systems can be dealt with in a constructive framework as long as one restricts oneself to constructively described infinite derivations.

Certain parts of analysis, the so-called (formal) systems R_α of ramified analysis, also have definite denumerable interpretations. The quantifiers here range over collections of sets successively definable by quantification over previously introduced collections (dealt with in the systems R_β for $\beta < \alpha$). Then it is also possible to derive the axioms for these in corresponding infinitary systems $R_\alpha^{(\infty)}$.

The use of infinitary systems related to these to give consistence proofs of finitary formal systems was initiated independently by Novikov [N] for arithmetic and Lorenzen [Ln] for ramified systems R_n, $n < \omega$. In a series of papers in the 1950's Schütte systematically reworked these in Gentzen form, and showed how to extend these to other parts of analysis, principally ramified systems R_α of transfinite rank. He furthermore showed how ordinals entered in a natural way and how to use these methods to put bounds on the provable well-orderings of the formal systems treated. This work has been brought together in his book [Sch 1].[7]

The basic result for each of the infinitary systems $S^{(\infty)}$ considered is that <u>the cut-elimination theorem</u> holds for $S^{(\infty)}$, i.e. for each derivation $\mathcal{D}$ of $S^{(\infty)}$ we can find a cut-free $S^{(\infty)}$ derivation $\mathcal{D}'$ of the same conclusion. More precise information is given by assigning ordinals $od(\mathcal{D})$ as lengths to derivations. This is simply a special case of the natural assignment of ordinals $od(T)$ to arbitrary well-founded trees T, given in §3(a) below.

Now consider the special case of Z. Each derivation of a formula ϕ in Z has associated with it a simple derivation $\mathcal{D}$ of $(0 \supset \phi)$ in $Z^{(\infty)}$, with $od(\mathcal{D}) < \omega \cdot 2$. Then it is shown that in this case a cut-free derivation $\mathcal{D}'$ of $(0 \supset \phi)$ can be found with $od(\mathcal{D}') < \epsilon_0$. It follows by the subformula property and proof by transfinite induction up to ϵ_0 that $(\overline{0} \neq \overline{0})$ is not provable in Z. (We shall go through a version of this entire argument in §6.) Related arguments are used to obtain ϵ_0 as a bound for the provable well-orderings in Z.

Schütte likewise obtained consistency proofs and upper bounds for the provable well-orderings of systems R_α of ramified analysis by means of his cut-elimination results for the corresponding infinitary systems. These and further results will be discussed at the end of §6.

All these extensions of Gentzen's method depend on the use of special infinitary rules for quantification like the ω-rule. Returning to the original idea of infinite conjunctions $\prod$ and disjunctions $\sum$, Tait [Tt] showed how the various ordinal bounds could also be obtained by cut-elimination results in the <u>propositional calculus with (countably) infinitely long formulas</u>. This is accomplished by associating with each formula ϕ of the systems treated a certain infinitely long

---

7) Schütte's calculi in [Sch 1] do not, strictly speaking, deal with sequents. However, they provide rules for deriving individual formulas ϕ which are very closely related to rules for deriving sequents. These rules are formulated in terms of the notions of <u>positive</u> and <u>negative</u> parts of ϕ. In particular, the positive parts of $(\theta_1 \wedge \ldots \wedge \theta_n \rightarrow \Psi_1 \vee \ldots \vee \Psi_m)$ include $\Psi_1, \ldots, \Psi_m$, and the negative parts include $\theta_1, \ldots, \theta_n$. Then a Gentzen rule for deriving sequents $(\theta_1, \ldots, \theta_n \supset \Psi_1, \ldots, \Psi_m)$ has associated with it a rule for deriving formulas $\phi(\theta_1^-, \ldots, \theta_n^-, \Psi_1^+, \ldots, \Psi_m^+)$.

propositional formula $\emptyset^*$. For example, in the case of Z , take $(\bigwedge u \emptyset(u))^* = \prod_{n<\omega}(\emptyset(\overline{n}))^*$. Now the rules for $\prod$ and $\sum$ are natural extensions of the rules for finite conjunction $\wedge$ and disjunction $\vee$.

It is then an immediate step to include consideration of quantifiers as well. In this case, there is no need (at least at the outset) to consider propositional interpretations of formulas. We have a simple set of axioms in $\mathcal{L}_{HC}$ (also to be denoted by $Z^{(\infty)}$), including one infinitary axiom $\bigwedge u \sum_{n<\omega}(u=\overline{n})$, from which all the axioms of Z are derivable. This is the point of departure of §6.

We now turn to quite different work which has led to centering attention on $\mathcal{L}_{HC}$ and its sublanguages. There has been much interest in the last dozen years in treating the model theory of languages with greater expressive power than the finitary languages. This is done in one way, in effect, by restricting the class of models considered, e.g. to full higher type models, or to ω-models, etc. In another way it is accomplished by extending the language, permitting generalized quantifiers or infinitely long conjunctions and disjunctions, infinitely long quantifier sequences, etc. For the most part, only moderate success has been obtained in these various directions. We concentrate here on the languages $\mathcal{L}_{\kappa,\lambda}$ introduced by Tarski (κ,λ regular cardinals). These permit formation of $\prod_{\xi<\alpha} \emptyset_\xi$ when $\alpha < \kappa$ and $(\bigwedge u_0 \ldots \bigwedge u_\xi \ldots)\emptyset$ where $\xi < \beta < \lambda$ (as well as the dual operations). Thus $\mathcal{L}_{\omega,\omega}$ corresponds to $\mathcal{L}_{HF}$ here and $\mathcal{L}_{\omega_1,\omega}$ to $\mathcal{L}_{HC}$.

Karp [Kp] provides a fairly recent survey of the work on the $\mathcal{L}_{\kappa,\lambda}$. There does not seem to be a satisfactory generalization of the <u>completeness theorem</u> in most cases. Hanf showed that the mainstay of $\mathcal{L}_{\omega,\omega}$ model theory, the <u>compactness theorem</u>, fails (when generalized in the obvious way in terms of cardinality) except for unusually large inaccessible cardinals. Malitz [Mℓ] showed that Craig's interpolation theorem (expressed in terms of validity) fails for $\mathcal{L}_{\kappa,\omega}$ whenever $\kappa > \omega_1$ and for $\mathcal{L}_{\kappa,\lambda}$ whenever $\kappa \geq \lambda \geq \omega_1$.

Already before this, Kreisel [Kr1] and Scott [Sc1] suggested that one could not expect a good theory for $\mathcal{L}_{\kappa,\lambda}$ with $\lambda > \omega$. Scott's argument was that in permitting quantification $(\bigwedge u_0 \ldots \bigwedge u_n \ldots)\emptyset$ one is permitting quantification over arbitrary infinite sequences of a domain, which is a second-order concept. If one expects anything like first-order model theory, attention should be restricted to the languages $\mathcal{L}_{\kappa,\omega}$. In addition, Kreisel stressed that cardinality considerations in the choice of language and formulation of compactness theorems were too crude to give proper generalizations of the theory for $\mathcal{L}_{\omega,\omega}$. He suggested instead a treatment guided by considerations of definability of the basic syntactic and semantic notions for the language.

These ideas have been extremely fruitful, at least for $\mathcal{L}_{\omega_1,\omega}$ and a number of its sublanguages. The completeness of a deductive system of the standard kind for $\mathcal{L}_{\omega_1,\omega}$ was known from [Kp] . However, in order to extend the interpolation theorem to this language, Lopez-Escobar was led to establish in [LE1] the completeness of a cut-free sequential calculus in $\mathcal{L}_{\omega_1,\omega}$. We thus have here a confluence of the two streams of interest, the one from proof-theory as discussed above, the other from model theory. As discussed in §5(a) below, such a meeting already took place in $\mathcal{L}_{\omega,\omega}$. But it turned out possible in that case to give a strictly model-theoretic development of results first found by proof-theoretical arguments. There has so far been no corresponding extension of strictly model-theoretic methods to $\mathcal{L}_{\omega_1,\omega}$. In the meantime, refinements of the interpolation theorem obtained by proof-theoretical arguments (as in §4 below) have provided efficient machinery for the extension of some of the basic results of finitary model-theory. (Cf. §5(a) for a more detailed discussion of the background and specific references).

It is more difficult to bring out briefly the ideas stemming from [Kr1] but we shall try to indicate some of the considerations concerning the syntax of infinitary languages. These are intimately tied up with generalizations of ordinary recursion theory. Any non-trivial generalization of this theory must supply not only a notion to take the place of recursiveness, but also one or more notions to take the place of finiteness, for the latter figures in an essential way in standard arguments. One way to see the connection between the two is as follows. Let S be a consistent set of axioms in a number-theoretical language containing, say, R. M. Robinson's axioms Q ; to begin with S is assumed finite, but this requirement can then be loosened. Every model $\mathfrak{M}$ of S can be thought of as containing ω . Let $X \subseteq \omega$. X is said to be invariantly definable (relative to S) on ω if there is a formula $\phi(u)$ such that in each model $\mathfrak{M}$ of S , $X = \{n : \models_{\mathfrak{M}} \phi(\bar{n})\}$. X is said to be absolutely invariantly definable (relative to S) if there is a formula $\phi(u)$ such that in each model $\mathcal{M}$ of S , X = {x: x satisfies $\emptyset$(u) in $\mathcal{M}$} . Then standard arguments give: X is inv. def. on ω if and only if X is recursive, and X is abs. inv. def. if and only if X is finite.

The paper [GMR] had attempted to generalize recursiveness by considering a second-order language and restricting attention to ω-models. Note that the above distinction between two notions of definability disappears, since the first order objects consist of ω in all these models. Now one obtains, X is inv. def. (relative to sufficiently strong consistent axioms) if and only if X is hyperarithmetic. This led the authors of [GMR] to treat the latter notion as a generalization of the notion of being recursive. However, this analogy turned out to be defective in several respects. This was due to the following oversight. In the case of elementary number theory, the natural numbers form the hard core [Kr1] of the models considered, and one is considering definability of subsets of this hard core. In the latter case, the hard core contains much more than the natural numbers, in fact all hyperarithmetic subsets of ω . It can be shown that the abs. inv. def. subsets of this hard core are the hereditarily hyperarithmetic sets. It is also possible to characterize the sets inv. def. on this hard core; we omit this for simplicity. At any rate this suggests that the hyperarithmetic sets are to be compared with the finite rather than the recursive sets; this analogy has been successful.

A more elegant way of putting the matter is to deal with generalizations of recursion theory to subsets of a transitive set A , where A (or rather the structure (A, $\in\restriction$A)) is the hard core of certain axioms in abstract set theory. For the most elementary axioms this hard core is just the set HF of hereditarily finite sets. For certain stronger axioms, it is the set HH of hereditarily hyperarithmetic sets, and so on. In each of these particular cases, it turns out that the abs. inv. def. subsets of A are just the members of A .

The notion of hard core has not been easy to work with in general, but it has turned out possible to avoid it. By adapting the generalization of recursiveness relative to a structure given by Fraïssé [Fr] using implicit invariant definability, Kreisel [Kr2] was able to simplify somewhat further the treatment of the notions of recursiveness and finiteness in structures (A, $\in\restriction$A) . This was subsequently developed extensively by Kunen [Ku] and Barwise [Bw] . Besides giving the expected answers for $A = HF$, $A = HH$, this work yielded simple characterizations of the basic notions for a wide range of A , including $A = HC$ = the collection of hereditarily countable sets. The details will be explained in §2(d); we also indicate there connections with other generalizations of recursion theory.

We can now relate this work to questions concerning infinitary languages. Formulas can be identified in a natural way with sets, so that infinite conjunctions $\prod \emptyset[\emptyset \in K]$ and disjunctions $\sum \emptyset[\emptyset \in K]$ are the results of operations on sets K . Suppose a set A is closed under these operations and that the generalization of finiteness to subsets of A gives simply the members of A . In this case we consider a language $\mathcal{L}_A$ whose formulas are exactly those which, regarded as sets, belong to A . Then $\mathcal{L}_{HF}$ and $\mathcal{L}_{HC}$ are just versions of $\mathcal{L}_{\omega,\omega}$ and $\mathcal{L}_{\omega_1,\omega}$ resp., justifying the notation. But there are now many other languages to consider as well, including $\mathcal{L}_{HH}$. The question is whether one gets satisfactory generalizations of basic results about $\mathcal{L}_{HF}$ to these other languages. Barwise's work sketched in §2(d) and §3(c) below amply shows this to be the case for a wide class of $A \subseteq HC$, the so-called admissible sets. Among other things he obtains both completeness and suitable compactness theorems for these.

The extension of logic to these languages is thus seen to be in a very satisfactory state. It should not be concluded that this represents the limit of what one can hope to accomplish with infinitary languages. For example, it may be possible to obtain satisfactory generalizations of ordinary logic to $\mathcal{L}_A$ for sets A which do not coincide with their A-finite subsets. For uncountable languages, the failure to generalize the interpolation theorem may be due to a deficiency in the basic logical structure, since there are many conceivable logical operations besides the ones employed. These and a number of other interesting suggestions and questions are raised by Kreisel in [Kr3], §13. It is to be expected that much progress is yet to be made with infinitary languages and that proof theory will continue to contribute to and benefit from this.

§2. Preliminaries

2(a) Many-sorted structures. Consideration of structures $(\langle M_j\rangle_{j\in J},\ldots)$ involving several different basic domains or sorts of objects M_j is natural and commonplace in various branches of mathematics. For example, in geometry one deals with points, lines, planes, etc. In algebra, one deals with vector spaces, comprised of vectors and scalars. In analysis, one deals with real numbers, sets of real numbers, functions of real numbers, etc.; from the set-theoretical point of view this is part of a finite type structure over the natural numbers.

The first-order model theory of such many-sorted structures has been treated by a number of authors for the case of finite formulas (cf. e.g. [O], [Wg], [K,K] ch. 5). The basic results of this theory (particularly completeness and compactness) can be obtained by direct extension of usual arguments in the model theory of single-sorted structures. The standard alternative procedure is (for simplicity) to reduce the theory to the single-sorted case by the method of unification of domains. That is, one associates with any many-sorted structure $\mathfrak{M} = (\langle M_j\rangle_{j\in J},\ldots)$ the single-sorted structure $\mathfrak{M}^* = (\bigcup_{j\in J} M_j\ ;\ \langle M_j\rangle_{j\in J},\ldots)$ with additional unary relations M_j ; then corresponding to any sentence $\emptyset$ concerning $\mathfrak{M}$ is an equivalent sentence $\emptyset^*$ concerning $\mathfrak{M}^*$, got by relativizing quantifiers in $\emptyset$ to new unary predicate symbols $\underset{\sim}{M}_j(\ )$. However, we do not know of any way to reduce the interpolation theorem for many-sorted logic directly to Craig's interpolation theorem, even in the case of finitary languages. Since this is essential in our treatment of model-theoretic applications, we deal with the more general situation throughout.[8)]

The similarity class of a many-sorted structure

$$\mathfrak{M} = (\langle M_j\rangle_{j\in J}\ ,\ \langle R_i\rangle_{i\in I_0}\ ,\ \langle a_i\rangle_{i\in I_1}\ ,\ \langle F_i\rangle_{i\in I_2})$$

is specified by means of a signature

---

8) We do not give detailed references here to prior work on many-sorted logic. The essentially new work is that presented in §§4-5. The question as to a possible alternative approach to these results by reduction to known work is discussed in more detail there.

$$\sigma = (J,\ I_o,\ I_1,\ I_2,\ \langle k_i\rangle_{i\in I_o}\ ,\ \langle \ell_i\rangle_{i\in I_1}\ ,\ \langle m_i\rangle_{i\in I_2}\ \langle \gamma_i\rangle_{i\in I_2})$$

where

$(i)_\sigma$ $J \neq 0$;

$(ii)_\sigma$ for each $i \in I_o$, $0 < k_i < \omega$;

$(iii)_\sigma$ for each $i \in I_1$, $\ell_i \in J$; and

$(iv)_\sigma$ for each $i \in I_2$, $0 < m_i < \omega$ and $\gamma_i \in J^{m_i+1}$.[9)]

For $\mathfrak{M}$ to be of signature σ , the following conditions must be satisfied:

(i) $M_j \neq 0$ for each $j \in J$;

(ii) $R_i \subseteq M^{k_i}$ for each $i \in I_o$, where $M = \bigcup M_j[j\in J]$;

(iii) $a_i \in M_{\ell_i}$ for each $i \in I_1$; and

(iv) $F_i\colon (M_{j_1} \times \ldots \times M_{j_{m_i}}) \to M_{j_{m_i+1}}$ for each $i \in I_2$ when $\gamma_i = (j_1,\ldots,j_{m_i},j_{m_i+1})$.

J is called the collection of sorts and M_j the set of <u>elements of sort j in $\mathfrak{M}$</u>. We write $x \in M$ or $x \in \mathfrak{M}$ for $x \in \bigcup M_j$ $[j \in J]$. To simplify some details of syntax below it is convenient to assume in addition to $(i)_\sigma$ - $(iv)_\sigma$ throughout the following that $(v)_\sigma$ $J,\ I_o,\ I_1,\ I_2$ <u>are initial segments of ω and σ is recursively given.</u>

---

9) To be specific concerning some of the set-theoretical notation used here and throughout the following , we take $(a,b) = \{\{a\},\{a,b\}\}$, $(a_1,\ldots,a_{n+1}) = ((a_1,\ldots,a_n),\ a_{n+1})$, $\mathcal{D}(\varphi)$ is the domain of a function φ , $\mathcal{R}(\varphi)$ is its range, $\langle a_i\rangle_{i\in I} = \langle a_i : i\in I\rangle$ is a function φ with $\mathcal{D}(\varphi) = I$, $\varphi(i) = a_i$ for each $i \in I$, and $\langle a_{i_1},\ldots,a_{i_n}\rangle = \langle a_i\rangle_{i\in\{i_1,\ldots,i_n\}}$.

The following examples illustrate these notions. A vector space can be considered to be a structure $\mathcal{M}$ of signature σ where $J = \{0,1\}$, $I_o = 0$, $I_1 = \{0,1,2\}$, $\ell_o = 0$, $\ell_1 = \ell_2 = 1$, $I_2 = \{0,1,2,3\}$, $m_o = m_1 = m_2 = m_3 = 2$, $\gamma_o = (0,0,0)$, $\gamma_1 = \gamma_2 = (1,1,1)$ and $\gamma_3 = (1,0,0)$. There are two basic domains M_o (vectors), M_1 (the field of scalars), three basic individual $a_o \in M_o$ (the zero vector), $a_1, a_2 \in M_1$ (zero and unit of M_1, resp.) and four basic operations $F_o : M_o^2 \to M_o$ (vector addition) $F_1 : M_1^2 \to M_1$ (field addition), $F_2 : M_1^2 \to M_1$ (field multiplication) and $F_3 : M_1 \times M_0 \to M_0$ (multiplication of a vector by a scalar).

A signature σ suitable for simple type theory over the natural numbers is specified by $J = \omega$, $I_o = I_1 = I_2 = \{0\}$, $k_o = 2$, $\ell_o = 0$, $m_o = 1$, and $\gamma_o = (0,0)$. Interesting examples of structures $\mathcal{M}$ with this signature are provided by $M_o = \omega$, M_{j+1} any subset of $\mathcal{P}(M_j)$, $R_o = \{(x,y)$: for some j, $x \in M_j$, $y \in M_{j+1}$ and $x \in y\}$, $a_o = 0$, and $F_o(x) = x+1$ for $x \in M_o$. ($\mathcal{P}(M)$ = set of all subsets of M.) Alternatively, as is customary, one can introduce for a slightly modified signature a separate membership relation $R_j = \{(x,y): x \in M_j$, $y \in M_{j+1}$ and $x \in y\}$ for each j. To treat cumulative type theory, where one is interested in the cases $M_{j+1} \subseteq M_j \cup \mathcal{P}(M_j)$ one would take the signature first given and $R_o = \{(x,y) : x,y \in M$ and $x \in y\}$.

A signature σ and structure $\mathcal{M}$ of signature σ are said to be <u>relational</u> if $I_2 = 0$. In such a case we simply denote $\mathcal{M}$ by $(\langle M_j\rangle_{j\in J}, \langle R_i\rangle_{i\in I_o}, \langle a_i\rangle_{i\in I_1})$; we similarly drop off $\langle m_i\rangle_{i\in I_2}$ and $\langle \gamma_i\rangle_{i\in I_2}$ in σ. With each σ and $\mathcal{M}$ are associated corresponding relational σ_{Rel} and $\mathcal{M}_{Rel}$ in the obvious way, having the same domains and individuals and having relations $\langle R_i'\rangle_{i\in I_o \cup I_2}$ with $R_i' = R_i$ for $i \in I_o$ and $R_i' = F_i$ for $i \in I_2$. (first disjointing I_o and I_2).

Unless otherwise specified, σ is an arbitrary signature throughout the following.

2(b) <u>Basic syntactic and semantic notions for $\mathcal{L}_{HC}$</u>. The first-order countably infinitary language $\mathcal{L}_{HC}$ appropriate to structures of signature σ is described as follows. (i) For each $j \in J$, we have variables $v_{j,o},\ldots,v_{j,i},\ldots$ of sort j.

There are infinitely many variables of each sort, and variables of distinct sorts are distinct. (In the following we shall use $u, w, u_1, w_1, \ldots$ to range over variables of arbitrary sort.) There are (ii) a symbol for equality, = , (iii) for each $i \in I_o$, a k_i-ary relation symbol r_i , (iv) for each $i \in I_1$, a constant symbol c_i and (v) for each $i \in I_2$ an m_i-ary function symbol, f_i . The terms and their corresponding sorts are defined inductively by: (i) each variable of sort j is a term of sort j ; (ii) each constant c_i is a term of sort ℓ_i ; (iii) if $\gamma_i = (j_1, \ldots, j_{m_i}, j_{m_i+1})$ and $t_1, \ldots, t_{m_i}$ are terms of sorts $j_1, \ldots, j_{m_i}$ resp. then $f_i\ (t_1, \ldots, t_{m_i})$ is a term of sort j_{m_i+1} . The atomic formulas are the formulas $(t_1 = t_2)$ for any terms t_1, t_2 and $r_i(t_1, \ldots, t_{k_i})$ for any terms $t_1, \ldots, t_{k_i}$.[10)]

Arbitrary formulas are built up according to the following procedures: (i) every atomic formula is a formula; (ii) if $\emptyset$ is a formula, so is $\sim \emptyset$; (iii) if K is any set of formulas with $0 < \overline{\overline{K}} \leq \omega$ then $\sum K$ and $\prod K$ are formulas ; (iv) if $\emptyset$ is a formula and u is any variable then $\bigvee u\ \emptyset$ and $\bigwedge u\ \emptyset$ are formulas.

For most purposes, it is not important just what kinds of objects terms and formulas are taken to be or just how the operations are defined on them. The main thing to insure is that the unique readability conditions hold, so that these are determined up to isomorphism. For some special purposes below, we shall construe terms and formulas as certain sets and the syntactic operations as certain simple operations on sets. This is done in §2(d), which may be consulted at this point if desired.

We shall also write $\sum_{\emptyset \in K} \emptyset$ for $\sum K$ and $\prod_{\emptyset \in K} \emptyset$ for $\prod K$. Given an ordinal $\alpha \geq 1$ and $K = \{\emptyset_\xi : \xi \in C\}$, we shall write $\sum_{\xi \in C} \emptyset_\xi, \prod_{\xi \in C} \emptyset_\xi$ for $\sum_{\emptyset \in K} \emptyset, \prod_{\emptyset \in K} \emptyset$ resp. We also write: $(\emptyset_o \vee \emptyset_1)$ for $\sum_{\xi < 2} \emptyset_\xi$, $(\emptyset_o \wedge \emptyset_1)$ for $\prod_{\xi < 2} \emptyset_\xi$, $(\emptyset_o \rightarrow \emptyset_1)$ for

---

10) In practice, one is only concerned with identity formulas $(t_1 = t_2)$ in which t_1, t_2 are of the same sort and relations between terms of specified combinations of sorts. We shall refer to this as the restricted many-sorted language. The more general class of identity formulas is needed to cover such situations as cumulative type theory. Its use also greatly simplifies the model-theoretic applications below. We shall indicate within the text how various questions concerning the restricted language can be dealt with in this framework.

$(\sim \emptyset_0 \vee \emptyset_1)$, and $(\emptyset_0 \leftrightarrow \emptyset_1)$ for $(\emptyset_0 \rightarrow \emptyset_1) \wedge (\emptyset_1 \rightarrow \emptyset_0)$.

The notions: u <u>is a free (bound) variable of</u> $\emptyset$, are defined in the usual way. The notion: Ψ <u>is a subformula of</u> $\emptyset$ is also defined inductively in the usual way, taking it here in its ordinary sense. By a <u>sentence</u> we mean a formula without any free variables. Every subformula of a sentence has only a finite number of free variables; more generally, the class of formulas with finitely many free variables is closed under subformulas. <u>We shall assume throughout the following that each formula</u> $\emptyset$ <u>dealt with has only finitely many free variables.</u>

Let $\mathcal{M} = (\langle M_j \rangle_{j \in J}, \ldots)$, be any structure of signature σ . By an <u>assignment</u> x in $\mathcal{M}$ we mean a sequence $x = \langle x_u \rangle$, u ranging over all variables, such that when u is of sort j , $x_u \in M_j$. The notion: <u>the value of</u> t <u>in</u> $\mathcal{M}$ <u>under the assignment</u> x is defined exactly as for finitary languages; this is denoted by $Val_{\mathcal{M},x}(t)$. Then the relation: x satisfies $\emptyset$ in $\mathcal{M}$ is defined inductively by: (i) x sat. $(t_1 = t_2)$ iff $Val_{\mathcal{M},x}(t_1) = Val_{\mathcal{M},x}(t_2)$; (ii) x sat. $r_i(t_1, \ldots, t_{k_i})$ iff $(Val_{\mathcal{M},x}(t_1), \ldots, Val_{\mathcal{M},x}(t_{k_i})) \in R_i$; (iii) x sat. $\sim \emptyset$ iff x does not sat. $\emptyset$; (iv) x sat. $\sum K$ ($\prod K$) iff for some $\emptyset \in K$ (for all $\emptyset \in K$) , x sat. $\emptyset$; (v) x sat. $\bigvee u \emptyset$ ($\bigwedge u \emptyset$) iff for some assignment (for all assignments) y in $\mathcal{M}$ for which $y_w = x_w$ for each variable w distinct from u , we have y sat. $\emptyset$. We say that $\emptyset$ is <u>valid</u> in $\mathcal{M}$ and that $\mathcal{M}$ is a <u>model</u> of $\emptyset$, in symbols $\models_{\mathcal{M}} \emptyset$, if every assignment x in $\mathcal{M}$ satisfies $\emptyset$. $\mathcal{M}$ is a model of a set S of formulas if it is a model of each member of S . We write $S \Vdash \emptyset$ if every model of S is a model of $\emptyset$. In case $S = 0$ we simply write $\Vdash \emptyset$.

As usual, the question as to whether x satisfies $\emptyset$ in $\mathcal{M}$ depends only on the values x_u for u free in $\emptyset$. To any ordering $u_1, \ldots, u_n$ of the distinct free variables of $\emptyset$ corresponds an n-ary relation in $\mathcal{M}$ consisting of all $(z_1, \ldots, z_n)$ such that z_i and u_i are of the same sort and an assignment x with $x_{u_i} = z_i$ satisfies $\emptyset$ in $\mathcal{M}$. If $\emptyset'$ is obtained from $\emptyset$ by renaming the free and bound variables of $\emptyset$ in any one-to-one manner, the same n-ary relations correspond to $\emptyset'$ as to $\emptyset$ in each structure $\mathcal{M}$.

To avoid some technical annoyances, we now suppose that the set of variables is divided into two disjoint sets FV and BV , each containing infinitely many variables of each sort, e.g. $FV = \{v_{j,2i} : j \in J , i \in \omega\}$, $BV = \{v_{j,2i+1} ; j \in J , i \in \omega\}$. Let us say that a formula ϕ has the FV-BV <u>property</u> if every free variable of ϕ belongs to FV and every bound variable of ϕ belongs to BV . With each ϕ can be associated a ϕ' having the FV-BV property and which determines the same relations as ϕ in each structure. Without loss of expressive generality, we can thus <u>assume from now on that each formula dealt with has the FV-BV property</u>.

The notations $\phi(u_1,\ldots,u_n)$ and $\phi(t_1,\ldots,t_n)$ will be used with the following implicit assumptions: $u_1,\ldots,u_n$ are distinct and in FV and each variable occurring in each t_i belongs to FV . Then $\phi(t_1,\ldots,t_n)$ represents the result of simultaneously substituting $t_1,\ldots,t_n$ for $u_1,\ldots,u_n$ resp. throughout ϕ . Unless otherwise specified, it is not assumed that $u_1,\ldots,u_n$ are only or all the free variables of ϕ , nor that u_i , t_i are of the same sort. When $u_1,\ldots,u_n$ are just the free variables of ϕ and an assignment of $z_1,\ldots,z_n$ to $u_1,\ldots,u_n$ resp. in a structure $\mathfrak{M}$ satisfies ϕ , we write $\models_{\mathfrak{M}} \phi[z_1,\ldots,z_n]$.

We conclude this section with some simple examples of the expressive power of $\mathcal{L}_{HC}$ (cf. also [K,K],Ch. 7). Suppose σ is a signature for single-sorted structures. We can express that a structure $\mathfrak{M}$ is finitely generated, i.e. find a sentence ϕ whose models are exactly such $\mathfrak{M}$, as follows. Let Tm_n be the collection of all terms containing at most $v_0,\ldots,v_n$ free; take $\phi = \sum_{n<\omega} \bigvee v_0 \ldots \bigvee v_n \bigwedge v_{n+1} \sum_{t \in Tm_n} (v_{n+1} = t)$. It is easy to express of a structure $\mathfrak{M}$ (of suitable signature) that it is an Archimedean ordered field. The reader will readily construct further examples of this type.

Consider a language suitable in part for discussing number theory. That is, one of the sorts of variables, say sort 0 , is intended to range over the natural numbers. There is a constant symbol c of sort 0 and a unary function symbol f_0 from sort 0 to sort 0 . Write $\overline{0}$ for c , t' for $f_0(t)$ and define $\overline{n+1} = (\overline{n})'$. Then the models of the sentence $\bigwedge v_{0,0} \sum_{n<\omega} (v_{0,0} = \overline{n}) \wedge \prod_{n<m<\omega} \sim(\overline{n} = \overline{m})$ are just the ω-models. Further structure on ω is easily reflected by additional sentences. For example, if a binary function symbol f_1 is intended to correspond to addition, write $(t_1 + t_2)$ for $f_1(t_1,t_2)$ and adjoin the sentence $\prod_{n,m<\omega} (\overline{n}+\overline{m} = \overline{n+m})$.

Note that various restrictions on connections between domains or on sorts of objects admitted to relations are easily expressed by use of the unrestricted equality relation. For example, $\bigwedge v_{j_1,o} \bigwedge v_{j_2,o} (v_{j_1,o} \neq v_{j_2,o})$ expresses in each $\mathfrak{M}$ that $M_{j_1} \cap M_{j_2} = 0$ and $\bigwedge v_{j_1,o} \bigvee v_{j_2,o} \; (v_{j_1,o} = v_{j_2,o})$ expresses that $M_{j_1} \subseteq M_{j_2}$. The conjunction of all sentences $\bigwedge u_1 \ldots \bigwedge u_{k_1} [r_1(u_1,\ldots,u_{k_1}) \rightarrow \bigvee v_{j_1,o} \ldots \bigvee v_{j_{k_1},o} (u_1 = v_{j_1,o} \wedge \ldots \wedge u_1 = v_{j_{k_1},o})]$ where $u_1,\ldots,u_{k_1}$ are variables of any combination of sorts expresses that $R_1 \subseteq M_{j_1} \times \ldots \times M_{j_{k_1}}$.

The expressive power of this language would not be diminished if we eliminated either of $\sum, \prod$ in favor of the other and either of $\bigvee, \bigwedge$ in favor of the other. For example, with choice of basic connectives $\sim, \sum, \bigvee$ we would define $\prod_{\phi \in K} \phi$ to be $\sim \sum_{\phi \in K} \sim \phi$ and $\bigwedge u \phi$ to be $\sim \bigvee u \sim \phi$. Complete cut-free systems can be obtained for any such choices (e.g. [LE1]) . The utilization of all the operations $\sim, \sum, \prod, \bigvee$, and $\bigwedge$ simplifies some points in our proofs of interpolation theorems below.

The following gives some indication of the limits of expressibility in $\mathcal{L}_{HC}$. Consider single-sorted structures with one binary relation. For each countable ordinal α there is a sentence ϕ_α which expresses that a structure is well-ordered of order-type α . However, it can be shown that there is no such sentence for $\alpha = \omega_1$. Scott [Sc1] p. 331 explains why there is not even a <u>set</u> S of sentences of $\mathcal{L}_{HC}$ whose models are exactly the well-ordered structures.

2(c) <u>A Gentzen system for $\mathcal{L}_{HC}$; completeness</u>. By a <u>sequent</u> we mean a pair (Γ,Δ) where Γ and Δ are <u>finite</u> sequences (possibly empty) of formulas.[11] We shall write $(\Gamma \supset \Delta)$ for (Γ,Δ) ; Γ is called the <u>antecedent</u> and Δ the <u>consequent</u>. The empty sequence is denoted by 0 . A sequence $\langle \phi \rangle$ consisting of one formula is simply written ϕ ; concatenation of sequences $\Gamma^\frown\Gamma'$ is written Γ,Γ' . Thus notations such as: Γ,ϕ , etc. A variable u is said to be free in Γ $(\Gamma \supset \Delta)$ if it is free in at least one formula of Γ (Γ,Δ) . Suppose $\Gamma = \langle \phi_1,\ldots,\phi_n \rangle$, $\Delta = \langle \psi_1,\ldots,\psi_m \rangle$, and that x is an assignment $\mathfrak{M}$. We say x satisfies $(\Gamma \supset \Delta)$

---

11) The Gentzen calculi for $\mathcal{L}_{HC}$ dealt with in [Bx],[LE1] make use of infinite sequences or sets. This is necessary for their formulation of certain of the rules for $\sum, \prod$. However, suitable alternative rules using only finite sequences are just as well given (as below). This simplifies some questions of effectiveness later.

if either $n \neq 0$ and for some i, x does not satisfy ϕ_i or $m \neq 0$ and for some j, x satisfies Ψ_j. We write $\models_{\mathfrak{M}}(\Gamma \supset \Delta)$ and say $\mathfrak{M}$ is a model of $(\Gamma \supset \Delta)$ if every assignment satisfies $(\Gamma \supset \Delta)$ in $\mathfrak{M}$. If $\mathfrak{S}$ is a set of sequents, $\mathfrak{M}$ is a model of $\mathfrak{S}$ if it is a model of each member of $\mathfrak{S}$. $\mathfrak{S} \Vdash (\Gamma \supset \Delta)$ if each model of $\mathfrak{S}$ is a model of $(\Gamma \supset \Delta)$; in case $\mathfrak{S} = 0$, we simply write $\Vdash (\Gamma \supset \Delta)$. These extend the previous notions since $\models_{\mathfrak{M}} \phi$ iff $\models_{\mathfrak{M}} (0 \supset \phi)$ and $S \Vdash \phi$ iff $(0 \supset S) \Vdash (0 \supset \phi)$ where $(0 \supset S) = \{0 \supset \Psi : \Psi \in S\}$.

A <u>rule</u> is a relation (R) between a non-empty sequence of <u>hypotheses</u> $\langle \Gamma_h \supset \Delta_h : h \in H \rangle$ and a <u>conclusion</u> $(\Gamma \supset \Delta)$. When the relation holds, this is indicated by

$$(R) \quad \frac{\Gamma_h \supset \Delta_h \quad (h \in H)}{\Gamma \supset \Delta}$$

and we say that $(\Gamma \supset \Delta)$ is <u>inferred</u> by the rule (R) from the $(\Gamma_h \supset \Delta_h)$, for $h \in H$. When $H = \{0\}$, this is simply indicated by

$$(R) \quad \frac{\Gamma_o \supset \Delta_o}{\Gamma \supset \Delta}$$

and when $H = \{0,1\}$ by

$$(R) \quad \frac{\Gamma_o \supset \Delta_o \qquad \Gamma_1 \supset \Delta_1}{\Gamma \supset \Delta}$$

We call the rules for the Gentzen calculus in $\mathcal{L}_{HC}$ the G-<u>rules</u>. Except for the <u>structural rule</u> (S) and the <u>cut rule</u> (C) each of these is a rule for introducing a formula into an antecedent or consequent. In the case of <u>weakening</u> this is indicated by $(W\supset)$, $(\supset W)$, resp. Otherwise the notation depends on the principal connective, $\sim$, $\sum$, $\prod$, $\bigvee$, or $\bigwedge$ of the formula introduced. Thus, for example, $(\sim \supset)$ is the rule for introduction of a formula $\sim \phi$ in an antecedent, and $(\supset \sim)$ for introduction into a consequent, etc. We shall also give a numbering of the rules (R_1)-(R_{14}) for simplicity of coding later.

The usual formulation of the cut-rule, as we have given it in §1, is

$$(C_o) \quad \frac{\Gamma \supset \Delta, \phi \qquad \Gamma', \phi \supset \Delta'}{\Gamma, \Gamma' \supset \Delta, \Delta'}$$

It is convenient for the elimination argument in §3 to formulate it in a stronger way. Given $\Gamma = \langle \phi_1, \ldots, \phi_n \rangle$ we say that ϕ is in Γ if $\phi = \phi_i$ for some i. Then $\Gamma - \phi$ is the sequent (possibly empty) obtained from Γ by deleting all terms ϕ_i such that $\phi = \phi_i$. It is easily seen by the rules (W), (S) that the same formulas are derivable using the full rule (C) below as are derivable using the rule (C_o).

The G-rules now follow.

1. $(\sim \supset)\ \dfrac{\Gamma \supset \Delta, \phi}{\Gamma, \sim \phi \supset \Delta}$

2. $(\supset \sim)\ \dfrac{\Gamma, \phi \supset \Delta}{\Gamma \supset \Delta, \sim \phi}$

3. $(\sum \supset)\ \dfrac{\Gamma, \phi \supset \Delta \ (\phi \in K)}{\Gamma, \sum K \supset \Delta}$

4. $(\supset \sum)\ \dfrac{\Gamma \supset \Delta, \psi}{\Gamma \supset \Delta, \sum K}$ (when $\psi \in K$)

5. $(\prod \supset)\ \dfrac{\Gamma, \psi \supset \Delta}{\Gamma, \prod K \supset \Delta}$ (when $\psi \in K$)

6. $(\supset \prod)\ \dfrac{\Gamma \supset \Delta, \phi \ (\phi \in K)}{\Gamma \supset \Delta, \prod K}$

7*. $(\bigvee \supset)\ \dfrac{\Gamma, \phi(w) \supset \Delta}{\Gamma, \bigvee u \phi(u) \supset \Delta}$

8‡. $(\supset \bigvee)\ \dfrac{\Gamma \supset \Delta, \phi(t)}{\Gamma \supset \Delta, \bigvee u \phi(u)}$

9‡. $(\bigwedge \supset)\ \dfrac{\Gamma, \phi(t) \supset \Delta}{\Gamma, \bigwedge u \phi(u) \supset \Delta}$

10*. $(\supset \bigwedge)\ \dfrac{\Gamma \supset \Delta, \phi(w)}{\Gamma \supset \Delta, \bigwedge u \phi(u)}$

* where w is of the same sort as u, and is not free in the conclusion of the rule.

‡ where t is of the same sort as u.

11. $(W \supset)\ \dfrac{\Gamma \supset \Delta}{\Gamma, \phi \supset \Delta}$

12. $(\supset W)\ \dfrac{\Gamma \supset \Delta}{\Gamma \supset \Delta, \phi}$

13. $(S)\ \dfrac{\Gamma \supset \Delta}{\Gamma' \supset \Delta'}$ (when the set of formulas in Γ is the same as in Γ' and the set of formulas in Δ is the same as in Δ')

14. $(C)\ \dfrac{\Gamma \supset \Delta \quad \Gamma' \supset \Delta'}{\Gamma, \Gamma' - \phi \supset \Delta - \phi, \Delta'}$ (when ϕ is in Δ and in Γ').

Let us say that χ is a subformula of $\emptyset$ in the wider sense if for some subformula Ψ of $\emptyset$, χ is the result of substituting terms for one or more of the free variables of $\emptyset$. Then all the G-rules except (C) have the subformula property: any formula in any of the hypotheses is a subformula in the wider sense of some formula in the conclusion.

As axioms for the predicate calculus of sequents in $\mathcal{L}_{HC}$ we shall take all sequents $(\emptyset \supset \emptyset)$ for $\emptyset$ atomic. In a moment we shall take up axioms for equality. Let $\mathfrak{S}$ be any set of sequents. We say that $(\Gamma \supset \Delta)$ is G_{PC}-derivable from $\mathfrak{S}$ if it is either an axiom, a member of $\mathfrak{S}$, or is inferred by one of the G-rules from sequents $(\Gamma_h \supset \Delta_h)$, $h \in H$, each of which is G_{PC}-derivable from $\mathfrak{S}$. We write $\mathfrak{S} \vdash_{PC} (\Gamma \supset \Delta)$ if this holds, and $\vdash_{PC} (\Gamma \supset \Delta)$ when $\mathfrak{S} = 0$. If S is a set of formulas and $\emptyset$ a formula we write $S \vdash_{PC} (\Gamma \supset \Delta)$ if $(0 \supset S) \vdash_{PC} (\Gamma \supset \Delta)$ and $S \vdash_{PC} \emptyset$ if $S \vdash_{PC} (0 \supset \emptyset)$.

It is an easy exercise in the use of the G-rules to show that $\vdash_{PC} (\emptyset \supset \emptyset)$ for any formula $\emptyset$. Further exercises are provided by checking the facts about derivability stated in the proof of 2.4 below.

Our first result is a form of the Deduction Theorem.

2.1 Theorem. Suppose $\mathfrak{S}$ is any set of sequents and S is a countable non-empty set of sentences. Then $\mathfrak{S} \cup (0 \supset S) \vdash_{PC} (\Gamma \supset \Delta)$ if and only if $\mathfrak{S} \vdash_{PC} (\prod S, \Gamma \supset \Delta)$.

Proof. $\mathfrak{S} \cup (0 \supset S) \vdash_{PC} (0 \supset \prod S)$. Hence from $\mathfrak{S} \vdash_{PC} (\prod S, \Gamma \supset \Delta)$ we get $\mathfrak{S} \cup (0 \supset S) \vdash_{PC} (\Gamma \supset \Delta)$ by the cut-rule. For the converse we argue by induction on the sequents derivable from $\mathfrak{S} \cup (0 \supset S)$; details are left to the reader. The hypothesis that S consists only of sentences is used in applications of $(\bigvee \supset)$ or $(\supset \bigwedge)$.

For a complete system it would be sufficient to take as axioms for equality the set $(0 \supset Eq_1)$ where Eq_1 consists of all of the following sentences: $\bigwedge u(u = u)$; $\bigwedge u_1 \bigwedge u_2 [u_1 = u_2 \rightarrow u_2 = u_1]$; $\bigwedge u_1 \bigwedge u_2 \bigwedge u_3 [u_1 = u_2 \wedge u_2 = u_3 \rightarrow u_1 = u_3]$ (u, u_1, u_2, u_3 of any sorts); for each $i \in I_0$, $\bigwedge u_1 \ldots \bigwedge u_{k_i} \bigwedge w_1 \ldots \bigwedge w_{k_i}$

$$[u_1 = w_1 \wedge \ldots \wedge u_{k_i} = w_{k_i} \wedge r_i(u_1, \ldots, u_{k_i}) \rightarrow r_i(w_1, \ldots, w_{k_i})],$$

where $u_1, \ldots, u_{k_i}$ and $w_1, \ldots, w_{k_i}$ are of any sorts; and for each $i \in I_2$,

$\bigwedge u_1 \ldots \bigwedge u_{m_i} \bigwedge w_1 \ldots \bigwedge w_{m_i} [u_1 = w_1 \wedge \ldots \wedge u_{m_i} = w_{m_i} \rightarrow f_i(u_1,\ldots,u_{m_i}) = f_i(w_1,\ldots,w_{m_i})]$,

where $u_1,\ldots,u_{m_i}$ and $w_1,\ldots,w_{m_i}$ are of the sorts for which this is an admitted formula.

However, in order to obtain a complete system for which the cut-elimination theorem holds, it is necessary to adjoin a number of immediate consequences of $(0 \supset Eq_1)$ obtained by successive applications of the cut-rule. This is the approach taken in [M,T] . As an example with $k_i = w$, and any terms $t_1, t_2, t_1', t_2', t_1''$, the following is such a consequence: $t_1 = t_1''$, $t_1' = t_1''$, $t_2' = t_2$, $r_i(t_1,t_2) \supset r_i(t_1',t_2')$. Alternatively, one can extend the rules used by some special substitution rules involving equality; this is the approach taken in [LEi]. We follow the former approach here.

Let t, t' be any terms. By an <u>equality chain from</u> t <u>to</u> t' , $C(t,t')$ we mean any sequence of equations of the form $s_i = s_{i+1}$, where $s_1,\ldots,s_p$ $(p \geq 0)$ is a sequence of terms with first term t and last term t' . By an <u>equality chain from</u> $t_1,\ldots,t_k$ <u>to</u> $t_1',\ldots,t_k'$ we mean a sequence $C(t_1,t_1'),\ldots,C(t_k,t_k')$ where for each i, $C(t_i,t_i')$ is some equality chain from t_i to t_i' and t_i is the same as t_i' when $C(t_i,t_i')$ is empty. We now define Eq to consist of all sequents $(0 \supset t = t)$ for any term t , and all sequents

$$E(t_1,\ldots,t_k ; t_1',\ldots,t_k') , \emptyset(t_1,\ldots,t_k) \supset \emptyset(t_1',\ldots,t_2') ,$$

where $\emptyset(u_1,\ldots,u_k)$ is any <u>atomic</u> formula containing free variables $u_1,\ldots,u_k$ and $E(t_1,\ldots,t_k ; t_1',\ldots,t_k')$ is any equality chain from $t_1,\ldots,t_k$ to $t_1',\ldots,t_k'$. Observe that Eq contains all the axioms for the predicate calculus.

The following is an immediate consequence of the fact that each member of Eq is G_{PC}-derivable from $(0 \supset Eq_1)$ and conversely.

2.2 <u>Lemma</u> $\mathfrak{S} \cup Eq \vdash_{PC} (\Gamma \supset \Delta)$ <u>if and only if</u> $\mathfrak{S} \cup (0 \supset Eq_1) \vdash_{PC} (\Gamma \supset \Delta)$

We now say that $(\Gamma \supset \Delta)$ is <u>G-derivable from</u> $\mathfrak{S}$, in symbols $\mathfrak{S} \vdash (\Gamma \supset \Delta)$, if $\mathfrak{S} \cup Eq \vdash_{PC} (\Gamma \supset \Delta)$. By the Deduction Theorem and 2.2 this is equivalent to $\mathfrak{S} \vdash_{PC} (\prod Eq_1 , \Gamma \supset \Delta)$. We write $\mathfrak{S} \vdash \emptyset$ if $\mathfrak{S} \vdash (0 \supset \emptyset)$, and $S \vdash (\Gamma \supset \Delta)$, resp.

$S \vdash \emptyset$ if $(0 \supset S) \vdash (\Gamma \supset \Delta)$, resp $(0 \supset S) \vdash \emptyset$.

We now take up the completeness of the G-system. This can be obtained from some completeness results for certain deductive systems in $\mathcal{L}_{HC}$ already available in the literature, due to Karp [Kp] , and then Lopez-Escobar [LE1] . However, we shall sketch an argument which may be found more accessible, assuming knowledge of the now quite familiar Henkin-style proof of completeness of the usual finitary calculi (cf., e.g. [Mn] , [Sh]). In comparison, it should be remarked that Karp's proof makes use of a great deal of Boolean-algebraic mechinery which is mainly developed to deal with the more general languages $\mathcal{L}_{\kappa,\lambda}$. On the other hand Lopez-Escobar's result is much stronger than the one to be dealt with in a moment, since it establishes the completeness of a Gentzen type system withuut the cut-rule. Since the cut-rule preserves validity (2.3 below), this shows that the cut-rule is eliminable as a rule of inference in deriving valid sequents. Unfortunately, this (non-constructive) proof of the cut-elimination theorem does not provide the additional information about ordinal lengths of derivations, etc. which is used in consistency proofs. The syntactic proof of cut-elimination in §3 meets this purpose.

The easy half of completeness holds in extended form, relative to any set of sequents, and is proved by a straightforward induction on the sequents derivable from $\mathfrak{S}$.

2.3 Theorem If $\mathfrak{S} \vdash (\Gamma \supset \Delta)$ then $\mathfrak{S} \Vdash (\Gamma \supset \Delta)$.

As a corollary, we have that if $\mathfrak{S}$ has a model then $(0 \supset 0)$ is not G-derivable from $\mathfrak{S}$.

The harder part holds only for countable $\mathfrak{S}$; we shall reduce this to the case $\mathfrak{S} = 0$.

2.4 Theorem If $\Vdash (\Gamma \supset \Delta)$ then $\vdash (\Gamma \supset \Delta)$.

Proof If $\Vdash (\Gamma \supset \Delta)$ then $\Gamma \neq 0$ or $\Delta \neq 0$. Let $\Gamma = \langle \emptyset_1, \ldots, \emptyset_n \rangle$, $\Delta = \langle \Psi_1 \ldots \Psi_m \rangle$. Then one first verifies that

(1) $\Vdash (\Gamma \supset \Delta)$ iff $\Vdash (\sim \emptyset_1 \vee \ldots \vee \sim \emptyset_n \vee \Psi_1 \vee \ldots \vee \Psi_m)$ and

$\vdash (\Gamma \supset \Delta)$ iff $\vdash (\sim \emptyset_1 \vee \ldots \vee \sim \emptyset_n \vee \Psi_1 \vee \ldots \vee \Psi_m)$.

Next observe that for any formula θ and free variable w and bound variable u of the same sort,

(2) $\Vdash \theta(w)$ iff $\Vdash \bigwedge u\, \theta(u)$ and $\vdash \theta(w)$ iff $\vdash \bigwedge u\, \theta(u)$.

By successive universal quantification, it is seen to be sufficient to show, for any <u>sentence</u> ϕ ,

(3) if $\Vdash \phi$ then $\vdash \phi$.

This, in turn, reduces to showing that if $\Vdash \phi$ then $Eq_1 \vdash_{PC} \phi$. Call a set S of formulas PC-<u>consistent</u> if $\neg(S \vdash_{PC} 0 \supset 0)$; when S contains just one formula θ , we say that θ is PC-consistent. A countable set S is PC-consistent iff $\prod S$ is PC-consistent.

If $\neg(Eq_1 \vdash_{PC} \phi)$ then $Eq_1 \cup \{\sim \phi\}$ is PC-consistent, and $\phi_1 = \prod_{\chi \in Eq_1 \cup \{\sim \phi\}} \chi$ is PC-consistent. We shall show that if any sentence θ is PC-consistent then it has a model in the sense of the predicate calculus, i.e. there is a structure $\mathfrak{M}$ and relation $(x \equiv y)$ between arbitrary elements of $\mathfrak{M}$ such that θ is satisfied in $\mathfrak{M}$ (by any assignment) when $=$ is interpreted as $\equiv$. In particular, for the sentence ϕ_1 above, $\equiv$ will be a congruence relation in $\mathfrak{M}$ since all members of Eq_1 will be satisfied in $(\mathfrak{M}, \equiv)$. Then the corresponding structure $\mathfrak{M}/\equiv$ of equivalence classes will also be a model of ϕ_1 and hence of $\sim \phi$. In that case $\neg \Vdash \phi$.

In the remainder of the proof we assume that θ is any PC-consistent sentence; the preceding designations of formulas are disregarded. The following facts will be needed and are left to the reader.

(4) Suppose c is a constant symbol not occurring in any formula of S or in $\chi(w)$ and of the same sort as w ; if $S \vdash_{PC} \chi(c)$ then $S \vdash_{PC} \chi(w)$.

(5) Suppose c is a constant symbol not occurring in any formula of S or in $\Psi(w)$ and of the same sort as w and suppose that $\Psi(c)$ is a sentence; if S is PC-consistent then $S \cup \{\bigvee u\, \Psi(u) \rightarrow \Psi(c)\}$ and $S \cup \{\Psi(c) \rightarrow \bigwedge u\, \Psi(u)\}$ are PC-consistent.

(6) If S is PC-consistent and K is a non-empty countable set of sentences then for some $\chi_0 \in K$, $S \cup \{\sum K \to \chi_0\}$ is PC-consistent and for some $\chi_1 \in K$, $S \cup \{\chi_1 \to \prod K\}$ is PC-consistent.

(7) Suppose S is PC-consistent. If $S \vdash \Psi$ then $S \cup \{\Psi\}$ is PC-consistent; if Ψ is a sentence and $\neg(S \vdash \Psi)$ then $S \cup \{\sim \Psi\}$ is PC-consistent.

Now extend the language (for the given signature σ) by adjunction of a collection of new distinct constant symbols $c_{j,i}$ of sort j for each $j \in J$ and $i < \omega$. Formulas of the original language are referred to as σ-formulas, of the extended language as σ^+-formulas. By induction on arbitrary formulas we obtain

(8) the collection of all σ^+-subformulas, in the wider sense, of θ is countable.

Let $\theta_0, \ldots, \theta_n, \ldots$ be an enumeration without repetition of all such σ^+-subformulas of θ satisfying the following two additional conditions: (i) each θ_n is a sentence, and (ii) there are at most a finite number of constants $c_{j,i}$ which occur in θ_n. We take $\theta_0 = \theta$. We shall now define a sequence of formulas θ'_m where each $\theta'_{2n} = \theta_n$ or $\theta'_{2n} = \sim \theta_n$ and each initial segment $S_m = \{\theta'_0, \ldots, \theta'_m\}$ is PC-consistent. Suppose given $\theta'_0, \ldots, \theta'_{2n-1}$. In case $n = 0$, we take $\theta'_0 = \theta_0$. For $n > 0$, either $S_{2n-1} \cup \{\theta_n\}$ is PC-consistent or $S_{2n-1} \cup \{\sim \theta_n\}$ is PC-consistent. by (7); if the former holds take $\theta'_{2n} = \theta_n$, otherwise $\theta'_{2n} = \sim \theta_n$. For any n, if θ_n is atomic or $\theta_n = \sim \Psi_n$, take $\theta'_{2n-1} = \theta'_{2n}$. If $\theta_n = \bigvee u\, \Psi_n(u)$ take $\theta'_{2n+1} = (\bigvee u\, \Psi_n(u) \to \Psi_n(c_{j,i}))$, where j is the sort of u and i is the least index such that $c_{j,i}$ does not occur in any sentence of S_{2n}. If $\theta_n = \bigwedge u\, \Psi_n(u)$ take $\theta'_{2n+1} = (\Psi_n(c_{j,i}) \to \bigwedge u\, \Psi_n(u))$, where $c_{j,i}$ is obtained in the same way. If $\theta_n = \sum K_n$, take $\Psi'_{2n+1} = (\sum K_n \to \theta_t)$ for the first θ_t in K_n such that $S_{2n} \cup \{\sum K_n \to \theta_t\}$ is PC-consistent. If $\theta_n = \prod K_n$, take $\theta_{2n+1} = (\theta_t \to \prod K_n)$ for the first θ_t in K_n such that $S_{2n} \cup \{\theta_t \to \prod K_n\}$ is PC-consistent. Such θ_t exist by (6). The choice of each θ'_m is now uniquely determined.

Let $S = \bigcup_{m<\omega} S_m$. It has the following properties.

(9) (i) Either $\theta_n \in S$ or $(\sim \theta_n) \in S$, but not both.

(ii) For $\theta_n = \bigvee u\, \Psi_n(u)$, $\bigvee u\, \Psi_n(u) \in S$ iff for some closed term t of the same sort as u , $\Psi_n(t) \in S$.

(iii) For $\theta_n = \bigwedge u\, \Psi_n(u)$, $\bigwedge u\, \Psi_n(u) \in S$ iff for all closed terms t of the same sort as u , $\Psi_n(t) \in S$.

(iv) For $\theta_n = \sum K_n$, $\sum K_n \in S$ iff $K_n \cap S \neq 0$.

(v) For $\theta_n = \prod K_n$, $\prod K_n \in S$ iff $K_n \subseteq S$.

This leads directly to the definition of the required structure $(\mathfrak{M}, \equiv)$. M_j is taken to be the set of closed terms of sort j . $(t_1,\ldots,t_{k_i}) \in R_i$ iff $r_i(t_1,\ldots,t_{k_i}) \in S$, and $(t_1 \equiv t_2)$ iff $(t_1 = t_2) \in S$. The individuals are taken to be the constants. Finally , $F_i(t_1,\ldots,t_{m_i})$ is taken to be the term $f_i(t_1,\ldots,t_{m_i})$ (when it is a term). Then it is proved by induction on the structure of the sentences θ_n that for any n

(10) $\theta_n \in S$ iff θ_n is satisfied in $(\mathfrak{M}, \equiv)$. Hence $(\mathfrak{M}, \equiv)$ provides the required model of θ.[12)]

2.5 <u>Corollary</u> <u>If $\mathfrak{S}$ is countable and $\mathfrak{S} \Vdash (\Gamma \supset \Delta)$ then $\mathfrak{S} \vdash (\Gamma \supset \Delta)$</u> .

<u>Proof.</u> As at the beginning of the proof of (2.4) this is reduced to the case of a countable set of sentences S . Suppose that $S \Vdash (\Gamma \supset \Delta)$. Then $\Vdash (\prod S, \Gamma \supset \Delta)$, hence $\vdash (\prod S, \Gamma \supset \Delta)$ and then $S \vdash (\Gamma \supset \Delta)$ by (2.1), (2.2).

2(d) <u>Set-theoretical representation of syntax: the languages $\mathcal{L}_A$</u> . Sets are taken to be built up from the empty set. A set <u>A</u> is <u>transitive</u> if $y \in x$, $x \in A$

12) A proof of this kind was mentioned by Scott in [Sci]. It is essentially the same sort of proof as given for the completeness of ω-logic in [GMK] with $\sum$, $\prod$ corresponding to existential and universal quantification over the natural numbers. The success of the final part of the proof hinges directly on the fact (8) and this in turn makes essential use of the countability of all disjunctions and conjunctions in the language.

implies $y \in A$. By the transitive closure of A, $TC(A)$, is meant the smallest set which contains A and which is transitive. A is hereditarily finite if $TC(A)$ is finite; HF is the collection of all hered. finite sets. $A \in HF$ if A is finite and $A \subseteq HF$. A is hereditarily countable if $TC(A)$ is countable; HC is the collection of all hered. countable sets. $A \in HC$ if A is countable and $A \subseteq HC$. Ordinals are taken so that each ordinal is the set of its predecessors.

As suggested in §2(b), we now represent terms and formulas of $\mathcal{L}_{HC}$ as certain sets and the operations used to build these as certain operations on sets. (i) $v_{j,i}$ is taken to be $(0,j,i)$, (ii) c_i is taken to be $(1,i)$; (iii) for any sets $t_1,\ldots,t_{n_i}$, $f_i(t_1,\ldots,t_{m_i})$ is taken to be $(2,i,t_1,\ldots,t_{m_i})$. Then the set of terms and their sorts is defined inductively as before. Since J, I_0, I_1, I_2 are taken to be ordinals $\leq \omega$, it follows that each term is a member of HF.

Next, (iv) for any sets t_1, t_2, take $(t_1 = t_2)$ to be $(3,t_1,t_2)$ and (v) for any sets $t_1,\ldots,t_{k_i}$, take $r_i(t_1,\ldots,t_{k_i})$ to be $(4,i,t_1,\ldots,t_{k_i})$. Then the set of atomic formulas is defined as before; hence also each atomic formula is in HF. Finally, for any sets ϕ and K, (vi) $\sim\phi$ is taken to be $(5,\phi)$, (vii) $\sum K$ is taken to be $(6,K)$, (viii) $\prod K$ is taken to be $(7,K)$, (ix) $\bigvee u\,\phi$ is taken to be $(8,u,\phi)$, and (x) $\bigwedge u\,\phi$ is taken to be $(9,u,\phi)$. Then the set of formulas is defined inductively as before. Note that under this set-theoretical representation, the immediate subformulas Ψ of a formula ϕ all belong to $TC(\phi)$. This inductive definition can be presented as follows:

ϕ is a formula iff

- (i) ϕ is atomic, or
- (ii) for some $\Psi \in TC(\phi)$, Ψ is a formula and $\phi = \sim\Psi$, or
- (iii) for some $K \in TC(\phi)$, $K \neq 0$, K is countable, each member of K is a formula and $\phi = \sum K$ or $\phi = \prod K$, or
- (iv) for some $u, \Psi \in TC(\phi)$, u is a variable and Ψ is a formula and $\phi = \bigvee u\Psi$ or $\Psi = \bigwedge u\Psi$.

It follows by induction that each formula ϕ is in HC.

Following [Bw], given any set A we shall say that ϕ is a formula of $\mathcal{L}_A$

or is an A-formula if $\phi \in A$. The following are suitable conditions for a coherent development of syntax in $\mathcal{L}_A$, in which membership in A plays the same role as (hereditary) finiteness in usual syntax; these conditions are assumed from now on. $(i)_A$ $0 \in A$, $(ii)_A$ $x, y \in A$ implies $\{x,y\} \in A$, $(iii)_A$ A is transitive, $(iv)_A$ $A \subseteq HC$. By $(i)_A$, $(ii)_A$ we have $HF \subseteq A$ and hence each atomic formula is an A-formula. Then by $(ii)_A$ and $(iii)_A$, for $\phi = \sim \Psi$ or $\phi = \bigvee u\, \Psi$ or $\phi = \bigwedge u\, \Psi$ (u a variable), ϕ is an A-formula if and only if Ψ is an A-formula. By $(ii)_A$ $(iv)_A$, for $\phi = \sum K$ or $\phi = \prod K$, ϕ is an A-formula if and only if $K \neq 0$, $K \in A$ and each $\Psi \in K$ is an A-formula. If each formula of a sequent $(\Gamma \supset \Delta)$ is an A-formula then also the finite sequences $\Gamma, \Delta \in A$; in such a case we shall say that $(\Gamma \supset \Delta)$ is an A-sequent.

The conditions on A are fulfilled by $A = HF$ and $A = HC$ and by many intermediate sets. The formulas of the usual finitary predicate calculus built up using $\sim, \vee, \wedge, \bigvee$ and $\bigwedge$ are represented by HF-formulas, and each HF-formula expresses the same as one of these. In $\mathcal{L}_{HF}$ derivability can also be treated in terms of derivations, i.e. certain finite sequences or trees of formulas. This suggests consideration more generally of a notion of A-derivation; this will in turn lead us to consider further closure conditions on A.

The (set-theoretical representation of) the notion: $\mathcal{D}$ is a derivation of $(\Gamma \supset \Delta)$ (from a set $\mathfrak{S}$ of sequents) is defined inductively as follows. (i) If $(\Gamma \supset \Delta) \in Eq$ (or $(\Gamma \supset \Delta) \in \mathfrak{S}$) then $(0, \Gamma \supset \Delta)$ is a derivation of $(\Gamma \supset \Delta)$(from $\mathfrak{S}$). (ii) If for each $h \in H$, $\mathcal{D}_h$ is a derivation of $(\Gamma_h \supset \Delta_h)$ (from $\mathfrak{S}$) and $(\Gamma \supset \Delta)$ is inferred from $(\Gamma_h \supset \Delta_h)$, $h \in H$ by the kth G-rule, $(k = 1,\ldots,14)$, then $(k, \Gamma \supset \Delta, \langle \mathcal{D}_h : h \in H \rangle)$ is a derivation of $(\Gamma \supset \Delta)$(from $\mathfrak{S}$). $\mathcal{D}$ is a derivation (from $\mathfrak{S}$) if it is a derivation of its second term (from $\mathfrak{S}$). For example, if $\mathcal{D}_\phi$ is a derivation of $\Gamma, \phi \supset \Delta$ for each $\phi \in K$ then $(3, (\Gamma, \sum K \supset \Delta), \langle \mathcal{D}_\phi : \phi \in K \rangle)$ is a derivation. If ϕ is in Δ and in Γ' and $\mathcal{D}_0$ is a derivation of $(\Gamma \supset \Delta)$ and $\mathcal{D}_1$ is a derivation of $(\Gamma' \supset \Delta')$ then $(14, (\Gamma, \Gamma' - \phi \supset \Delta - \phi, \Delta'), \langle \mathcal{D}_0, \mathcal{D}_1 \rangle)$ is a derivation. It is easy to see that if $\mathcal{D}$ is a derivation of this form then ϕ is uniquely determined by $\mathcal{D}$ (if it were not, this would have to be supplied as additional information).

If $\mathcal{D}_0$ is a derivation of $(\Gamma, \phi \supset \Delta)$ then $(2, (\Gamma \supset \Delta, \sim \phi), \langle \mathcal{D}_0 \rangle)$ is a derivation. It would be more natural in this case to identify the derivation with $(2, (\Gamma \supset \Delta, \sim \emptyset), \mathcal{D}_o)$, and similarly for the other one-hypothesis rules. However, we shall not make any special exception for these.

We shall indicate that $\mathcal{D}$ is a derivation of $(\Gamma \supset \Delta)$(from $\mathfrak{S}$) by the diagram:

$$\begin{array}{c} \vdots\, \mathcal{D} \\ \downarrow \\ \Gamma \supset \Delta \end{array}$$

If $\mathcal{D}$ has the form $(k, \Gamma \supset \Delta, \langle \mathcal{D}_h : h \in H \rangle)$ we shall indicate this by the diagram:

$$\left.\begin{array}{c} \vdots\, \mathcal{D}_h (h \in H) \\ \downarrow \\ (R_k)\ \dfrac{\Gamma_h \supset \Delta_h}{\Gamma \supset \Delta} \end{array}\right\} \mathcal{D}$$

$\Gamma \supset \Delta$ is called the <u>conclusion of</u> $\mathcal{D}$ and (R_k) the <u>last rule of</u> $\mathcal{D}$. In the cases that H consists of one or two elements, these are indicated, explicitly, for example as in the following two cases.

$$\left.\begin{array}{c} \vdots\, \mathcal{D}_o \\ \downarrow \\ (\supset \sim)\ \dfrac{\Gamma, \emptyset \supset \Delta}{\Gamma \supset \Delta, \sim \emptyset} \end{array}\right\} \mathcal{D} \qquad \left.\begin{array}{c} \vdots\, \mathcal{D}_o \qquad\qquad \vdots\, \mathcal{D}_1 \\ \downarrow \qquad\qquad\quad \downarrow \\ (C)\ \dfrac{\Gamma \supset \Delta \qquad \Gamma' \supset \Delta'}{\Gamma, \Gamma' - \emptyset \supset \Delta - \emptyset, \Delta'} \end{array}\right\} \mathcal{D}$$

The notion of <u>subderivation</u> is defined inductively: For $\mathcal{D} = (0, \Gamma \supset \Delta)$, $\mathcal{D}$ has itself as unique subderivation: for $\mathcal{D} = (k, \Gamma \supset \Delta, \langle \mathcal{D}_h : h \in H \rangle)$, $\mathcal{D}'$ is a subderivation of $\mathcal{D}$ if $\mathcal{D}' = \mathcal{D}$ or $\mathcal{D}'$ is a subderivation of at least one $\mathcal{D}_h$, $h \in H$. Then a derivation $\mathcal{D}$ (from $\mathfrak{S}$) is said to be <u>cut-free</u> if there is no subderivation $\mathcal{D}'$ of $\mathcal{D}$ with (C) as its last rule. Equivalently, the notion of cut-free derivation (from $\mathfrak{S}$) can be defined inductively, permitting only $k = 1,\ldots,13$ in (ii).

We now begin consideration of questions of completeness for the $\mathcal{L}_A$; however, the conclusions will not be obtained until §3(c). $\mathcal{D}$ is said to be an A-<u>derivation</u> (from $\mathfrak{S}$) , if $\mathcal{D}$ is a derivation (from $\mathfrak{S}$) such that $\mathcal{D} \in A$. When $\mathfrak{S} = (0 \supset S)$ we shall say that $\mathcal{D}$ is an A-derivation from S . The extended completeness theorem for $\mathcal{L}_{HF}$ can be put in the sharp form: if $\mathfrak{S} \Vdash (\Gamma \supset \Delta)$ where all members of $\mathfrak{S} \cup \{\Gamma \supset \Delta\}$ are HF-sequents then there is HF-derivation $\mathfrak{D}$ of $(\Gamma \supset \Delta)$ from $\mathfrak{S}$. Taking membership in A as a generalization of finiteness this suggests the question: for which A and $\mathfrak{S}$ is it the case that whenever all members of $\mathfrak{S} \cup \{\Gamma \supset \Delta\}$ are A-sequents and $\mathfrak{S} \Vdash (\Gamma \supset \Delta)$ then there is an A-derivation $\mathfrak{D}$ of $(\Gamma \supset \Delta)$ from $\mathfrak{S}$? For A = HC and $\mathfrak{S}$ countable the answer is provided by the completeness theorem (2.5) using the following fact: if a sequent is derivable from an arbitrary set $\mathfrak{S}$ then it has an HC-derivation from $\mathfrak{S}$. It also follows from this fact that if a sequent is derivable from $\mathfrak{S}$ then it is derivable from some countable subset $\mathfrak{S}_1$ of $\mathfrak{S}$. Because of the failure of the compactness theorem for $\mathcal{L}_{HC}$ [Sci], this shows the extended completeness theorem does not hold in $\mathcal{L}_{HC}$ for arbitrary $\mathfrak{S}$.

It can be shown that the completeness theorem does not hold for $\mathcal{L}_A$ without additional conditions on A. Barwise [Bw] has shown that suitable sufficient conditions are found in the work of Kripke [Kk] and Platek [Pl] generalizing recursion theory to sets. These conditions are formulated in the language of abstract set theory, i.e. a single-sorted language of finite formulas with = and one binary relation symbol ε. Given a formula $\emptyset$ of this language, a quantifier occurrence in this formula is said to be <u>restricted</u> if it has the form $\bigvee u\ [u \varepsilon w \wedge \ldots]$ or $\bigwedge u\ [u \varepsilon w \rightarrow \ldots]$ where w is distinct from u; such occurrences are abbreviated as $\bigvee u \varepsilon w\ (\ldots)$ and $\bigwedge u \varepsilon w\ (\ldots)$ resp. A formula is said to be <u>completely restricted</u>, or Δ_o in the notation of Lévy [Lv], if every quantifier occurrence in it is restricted. As an example, $\bigwedge u_2 \varepsilon u_o \bigvee u_3 \varepsilon u_2 \bigwedge u_4 \varepsilon u_1\ \Psi(u_o, u_1, u_2, u_3, u_4)$ is completely restricted when Ψ is quantifier-free. By a <u>$\sum_1$-formula</u> we mean (as in [Lv]) one of the form $\bigvee u\ \Psi$ where Ψ is Δ_o.

Given a set A, a formula is said to be <u>valid in</u> A if it is valid in the structure $(A, \in \upharpoonright A)$. <u>We now assume the following additional hypothesis on</u> A:

$(v)_A$ <u>for each</u> $x \in A$, $TC(x) \in A$.

$(vi)_A$ (Δ_o-<u>separation axiom</u>)

$$\bigwedge u \bigvee w \bigwedge v\ [v \varepsilon w \leftrightarrow v \varepsilon u \wedge \emptyset\ (u, v, \ldots)]$$

is valid in A for each Δ_o-formula $\emptyset$, (where w is not free in $\emptyset$).

(vii)$_A$ ($\sum_1$-axiom of choice)

$$\bigwedge v_o \in u \bigvee v_1 \emptyset (v_o, v_1, \ldots) \rightarrow \bigvee w \{ \bigwedge v_o \in u \bigvee! v_1 (\langle v_o, v_1 \rangle \in w) \wedge$$

$$\wedge \bigwedge v_o \in u \bigwedge v_1 [(v_o, v_1) \in w \rightarrow \emptyset(v_o, v_1, \ldots)]\}$$

is valid in A for each $\sum_1$-formula $\emptyset$ (where w is not free in $\emptyset$).

(The abbreviation $(v_o, v_1) \in w$ is introduced in the customary way, as is the quantifier $\bigvee! v_1$ of unique existence.) The conclusion in (vii)$_A$ expresses that there is a choice function w such that for each $v_o \in u$, $\emptyset(v_o, w(v_o))$.

These hypotheses are obviously satisfied for A = HF and A = HC. However, there are also a number of interesting intermediate cases for which they hold. It is shown in the developments of Kripke and Platek that these are satisfied for $A = L_\alpha$ = the collection of sets constructible (in Gödel's sense) in less than α steps, whenever α is recursively regular or admissible in their sense. The collection of ordinals α for which this holds is unbounded and thus can be enumerated in a transfinite increasing sequence τ_ξ with $\tau_o = \omega$; moreover, the τ_ξ for $\xi < \omega_1$ are all countable. τ_1 turns out to be the least ordinal not isomorphic to a recursive well-ordering in the natural numbers, the so-called Church-Kleene ω_1, or $\omega_1^{(R)}$ as we shall denote it here. Then L_{τ_1} coincides with the collection HH of hereditarily hyperarithmetic sets. The reader may refer to [Pl] for more details containing these notions and results. It should be observed that the sentences expressing algebraic and number-theoretic properties discussed at the end of §2(b) (as well as most sentences met in applications to algebra) all belong to HH, since even more they have simple recursive structure.

Let $R \subseteq A^{n+1}$, $n \geq 0$; R is said to be $\sum_1^{(A)}$-definable if there is a $\sum_1$-formula $\emptyset$ with just the free variables $v_o, \ldots, v_n$ such that for all $x_o, \ldots, x_n \in A$,

$$(x_o, \ldots, x_n) \in R \text{ if and only if } (x_o, \ldots, x_n) \text{ satisfies } \emptyset \text{ in } A.$$

In the generalizations of recursion theory mentioned, the $\sum_1^{(A)}$-definable relations play the same role as the recursively enumerable relations in the natural numbers; thus the $\sum_1^{(A)}$-definable relations are also said to be A-recursively enumerable. Then a relation R is said to be A-recursive if both R and $A^{n+1}-R$ are A-rec. enum. There are a number of good reasons for this choice of generalization to $A \subseteq HC$ of the notions of usual recursion theory, but which we cannot go into here. (cf. also the end of §1(b).) It is not difficult to see by standard definability arguments that a subset of ω is HF-recursive (HF- rec. enum.) if and only if it is recursive (rec. enum.) in the usual sense. Moreover, it can be shown using results of Kleene and Spector that a subset of ω is HH-rec. enum. if and only if it is $\prod_1^1$, in accordance with the metarecursion theory of [K,S]. Following results of Fraissé [Fr] and Kreisel [Kr2] a general theoretical justification in terms of invariant and semi-invariant definability for the choice of the above notions (for $A \subseteq HC$ satisfying $(i)_A$-$(vi)_A$ and a weaker $(vii)_A$) is provided by a basic result of Kunen [Ku]. Continuing along the same lines, a result of Barwise [Bw] justifies the generalization of (hereditary) finiteness to membership in A (for the same sets A).

Just as the basic syntactic relations for the language $\mathcal{L}_{HF}$ are all recursive when the signature of the language is recursively given, so also the basic syntactic relations for $\mathcal{L}_A$ are all A-recursive. Specifically, let Fm_A be the set of all A-formulas, Sqt_A the set of all A-sequents, Der_A the binary relation consisting of all (x,y) for which x is an A-derivation of (the A-sequent) y and $Der_A^{(-C)}$ of all (x,y) for which x is a cut-free A-derivation of y. Further, for each set $\mathfrak{S}$ of A-sequents, let $Der_A(\mathfrak{S})$ consist of all (x,y) such that x is an A-derivation of y from $\mathfrak{S}$.

2.6 Lemma. The relations Fm_A, Sqt_A, Der_A, and $Der_A^{(-C)}$ are all A-recursive. If $\mathfrak{S}$ is an A-rec. enum. set of A-sequents then $Der_A(\mathfrak{S})$ is A-rec. enum.

Proof. We shall only sketch the argument for Fm_A. The proofs for the other notions involve similar considerations, and the details of these will be left to the reader (or cf. [Bw]). Let F be the characteristic function of Fm_A. It is sufficient to show that the relation $F(x) = z$ for $x \in A$, $z = 0,1$ is $\sum_1^{(A)}$-definable. The inductive definition of Fm_A gives (for $x \in A$)

$$F(x) = \begin{cases} 1 & \text{if } x \text{ is an atomic formula, or for some } y \in TC(x), \\ & F(y) = 1 \text{ and } x = (5,y), \text{ or for some } K \in TC(x), \\ & K \neq 0 \text{ and } F(y) = 1 \text{ for each } y \in K \text{ and } x = (6,K) \\ & \text{or } x = (7,K), \text{ or for some } u,y \in TC(x), u \text{ is a} \\ & \text{variable and } F(y) = 1 \text{ and } x = (8,u,y) \text{ or } x = (9,u,y); \\ 0 & \text{otherwise}. \end{cases}$$

This can be put in the form

$$F(x) = G(x, F \upharpoonright TC(x))$$

where the relation $G(x,f) = z$ is $\Sigma_1^{(A)}$-definable (in fact, $\Delta_0^{(A)}$-definable in this case). Now one proves by induction on the set-theoretical rank of x_1 that for each $x_1 \in A$, there is unique $f \in A$ which satisfies the recursion conditions on $TC(x_1)$, i.e. such that $\mathcal{D}(f) = TC(x_1)$ and

$$f(x) = G(x, f \upharpoonright TC(x)) \text{ for all } x \in TC(x_1).$$

At the inductive step, one uses the conditions $(vi)_A$, $(vii)_A$. From this it follows that F itself is $\Sigma_1^{(A)}$-definable: $F(x) = z$ if and only if there is $f \in A$ which satisfies the recursion conditions on $TC(\{x\})$ such that $f(x) = z$.

Now to get a completeness theorem for $\mathcal{L}_A$, one might first attempt to proceed by induction on derivations $\mathcal{D}$ of A-sequents $(\Gamma \supset \Delta)$, to show that for each such there is an A-derivation $\mathcal{D}'$ of the same conclusion. When $\mathcal{D} = (k, \Gamma \supset \Delta, \langle \mathcal{D}_h : h \in H \rangle)$ and $k \neq 14$ there is no problem, since the conclusion of each $\mathcal{D}_h$ is an A-sequent by the subformula property. However if the last rule of $\mathcal{D}$ is cut, $\mathcal{D} = (14, \Gamma \supset \Delta, \langle \mathcal{D}_0, \mathcal{D}_1 \rangle)$, it is not necessary that the conclusions of $\mathcal{D}_0, \mathcal{D}_1$ be A-sequents. We must instead begin by making use of the fact [LEI] that each derivable sequent has a <u>cut-free</u> derivation $\mathcal{D}$. This will be established here in the next section (Theorem 3.4). The discussion of completeness for the $\mathcal{L}_A$ will then be concluded in §3(c).

Correction to §2(c). The axioms Eq given above are adequate only if the signature is relational. In general, one must also adjoin as axioms substitution instances of $[u_1 = w_1 \wedge \ldots \wedge u_{m_i} = w_{m_i} \rightarrow f_i(u_1,\ldots,u_{m_i}) = f_i(w_1,\ldots,w_{m_i})]$ and consequences by (C) of these and the other axioms shown. This is done as follows.

Let Eq_2 consist of all sequents of the form: (i) $[0 \supset t = t]$,

(ii) $[t_1 = t_2 \supset t_2 = t_1]$, (iii) $[t_1 = t_2 , t_2 = t_3 \supset t_1 = t_3]$,

(iv) $[t_1 = t_1' ,\ldots, t_{k_i} = t_{k_i}' , r_i(t_1,\ldots,t_{k_i}) \supset r_i(t_1',\ldots,t_{k_i}')]$,

(v) $[t_1 = t_1',\ldots,t_{m_i} = t_{m_i}' \supset f_i(t_1,\ldots,t_{m_i}) = f_i(t_1',\ldots,t_{m_i}')]$,

where throughout $t, t_1, \ldots, f_i(t_1,\ldots)$, $f_i(t_1',\ldots)$ are terms of the language. Then let Eq consist of all sequents which can be obtained from Eq_2 simply by successive applications of (C) . It is easily seen that every sequent in Eq is of one of the forms $[\Gamma \supset t = t']$ or $[\Gamma, r_i(t_1,\ldots,t_{k_i}) \supset r_i(t_1',\ldots,t_{k_i}')]$ where Γ is a sequence (possibly empty) of equations; in fact Eq consists just of the valid sequents of these forms. Moreover, this definition of Eq reduces to the one in the text when σ is relational, up to applications of the structural rule (S) . With this change in the definition of Eq , no further corrections are necessary in the text, either in §2 or the further sections. (One must consider more general subcases than indicated for Case I in the proof of (3.3) below, but now the argument can even be simplified. Nothing is changed in (4.3) since it is necessary to assume there, in any case, that the signature is relational.)

§3. The cut-elimination theorem with ordinal bounds.

3(a) Ordinals of formulas and derivations. The main reason for considering an assignment of ordinals to formulas and derivations is that it provides basic information for consistency proofs. However, its use also simplifies the inductive arguments in the next section.

Given a formula $\emptyset$ of $\mathcal{L}_{HC}$, the subformulas of $\emptyset$ form a tree, under the subformula relation, which is inverted (downward branching) in our usual intuitive picture and well-founded (no infinite descending paths).

For any such tree, there is a natural assignment of ordinals $od(\nu)$ to the nodes ν of the tree as follows. If ν is a minimal node, $od(\nu) = 0$; if ν has as immediate predecessors ν_p , $p \in P$, then $od(\nu) = \sup(od(\nu_p) + 1 : p \in P)$. This is an order-preserving map of the tree-ordering into the ordinals. Specifically, for formulas, this leads to the following assignment.

(i) If $\emptyset$ is atomic then $od(\emptyset) = 0$.

(ii) $od(\sim \emptyset) = od(\bigvee u\ \emptyset) = od(\bigwedge u\ \emptyset) = od(\emptyset) + 1$.

(iii) $od(\sum K) = od(\prod K) = \sup(od(\emptyset) + 1 : \emptyset \in K)$.

This generalizes the usual assigment of ranks to HF-formulas.

As an example, for the sentence $\emptyset$ of §2(b) expressing that a structure is finitely generated, we have $od(\emptyset) = \omega$ (note that $od(\sum_{t \in Tm_n}(v_{n+1} = t)) = 1$) . Then the sentence $\sim \emptyset$ has $od(\sim \emptyset) = \omega + 1$, etc. The following is easily proved by induction on formulas.

3.1 Lemma. (i) If Ψ is a proper subformula of $\emptyset$ then $od(\Psi) < od(\emptyset)$.

(ii) $od(\emptyset(t)) = od(\emptyset(u))$.

Hence if Ψ is a subformula in the wider sense of $\emptyset$ then $od(\Psi) \leq od(\emptyset)$.

Given an HC-derivation $\mathcal{D}$, the subderivations of $\mathcal{D}$ form a tree, under the subderivation relation, which is upward branching in our usual intuitive picture and well-founded (when inverted). The principle of assignment of ordinals is the same, except that we take 1 instead of 0 to begin with. Specifically, using the set-theoretical representation of derivations of §2(d), we define:

(i) If $\mathcal{D}$ is a derivation of the form $(0, \Gamma \supset \Delta)$ then $od(\mathcal{D}) = 1$.

(ii) If $\mathcal{D} = (k, \Gamma \supset \Delta, \langle \mathcal{D}_h : h \in H \rangle)$ <u>then</u> $od(\mathcal{D}) = \sup(od(\mathcal{D}_h) + 1 : h \in H)$.

Again, it is obvious that if $\mathcal{D}'$ is a proper subderivation of $\mathcal{D}$ then $od(\mathcal{D}') < od(\mathcal{D})$.

Consider a language which contains number theory as a part, as described at the end of §2(b). Let u be the variable $v_{0,0}$. It is an instructive exercise to find for each $\phi(u)$ a derivation $\mathcal{D}_\phi$ of

$$\bigwedge u \sum_{n<\omega} (u = \bar{n}), \phi(\bar{0}), \bigwedge u(\phi(u) \rightarrow \phi(u')) \supset \bigwedge u\, \phi(u),$$

and to compute $od(\mathcal{D}_\phi)$.

With respect to the question of cut-elimination, the complexity of a derivation $\mathcal{D}$ is measured by the complexities of all the formulas ϕ which are <u>cut-formulas</u> (i.e. removed) in some application of (C) in $\mathcal{D}$. Let

$$\rho(\mathcal{D}) = \sup(od(\phi) + 1 : \phi \text{ is a cut-formula of } \mathcal{D})$$

We call this the <u>cut-rank</u> of $\mathcal{D}$; if $\mathcal{D}$ is cut-free, we take $\rho(\mathcal{D}) = 0$. The following is an alternative inductive definition of $\rho(\mathcal{D})$.

(i) If $\mathcal{D} = (0, \Gamma \supset \Delta)$, $\rho(\mathcal{D}) = 0$.

(ii) If $\mathcal{D} = (k, \Gamma \supset \Delta, \langle \mathcal{D}_h : h \in H \rangle)$ and $k = 1, \ldots, 13$, then $\rho(\mathcal{D}) = \sup(\rho(\mathcal{D}_h) : h \in H)$.

(iii) If $\mathcal{D} = (14, (\Gamma, \Gamma' - \phi \supset \Delta - \phi, \Delta'), \langle \mathcal{D}_0, \mathcal{D}_1 \rangle)$ where ϕ is in Γ' and in Δ and $\mathcal{D}_0$ is a derivation of $(\Gamma \supset \Delta)$ and $\mathcal{D}_1$ is a derivation of

$(\Gamma' \supset \Delta')$ then

$$\rho(\mathcal{D}) = \max\,(\rho(\mathcal{D}_0)\ ,\ \rho(\mathcal{D}_1)\ ,\ \mathrm{od}(\emptyset) + 1)\ .$$

(Recall that $\emptyset$ is uniquely determined by $\mathcal{D}$ in (iii).)

The following lemma will be very useful. Given a variable w (in FV) which may occur free in one or more of the formulas of a sequent $(\Gamma \supset \Delta)$, and a term t (all of whose variables are in FV) , we shall also write $[\Gamma(w) \supset \Delta(w)]$ for $(\Gamma \supset \Delta)$ and $[\Gamma(t) \supset \Delta(t)]$ for the result of substituting t for all occurrences of w in all formulas of $(\Gamma \supset \Delta)$.

3.2 Lemma. Suppose $\mathcal{D}$ is a derivation of $[\Gamma(w) \supset \Delta(w)]$, and that t is of the same sort as w . Then we can find a derivation $\mathcal{D}'$ of $[\Gamma(t) \supset \Delta(t)]$ with $\mathrm{od}(\mathcal{D}') = \mathrm{od}(\mathcal{D})$ and $\rho(\mathcal{D}') = \rho(\mathcal{D})$.

Proof. This is by induction on $\mathrm{od}(\mathcal{D})$. The only cases which require a little more care are the quantification rules. Consider the rules for $\bigvee$; those for $\bigwedge$ are treated dually. If the last rule of $\mathcal{D}$ is $(\bigvee\supset)$, $\mathcal{D}$ has the form

$$\begin{array}{c} \vdots\ \mathcal{D}_0 \\ \downarrow \\ \dfrac{\Gamma_0(w),\ \emptyset(w_0,w) \supset \Delta(w)}{\Gamma_0(w), \bigvee u\emptyset(u,w) \supset \Delta(w)} \end{array}$$

w must be distinct from w_0 in this case, since w_0 may not appear free in the conclusion. By induction, we can get a derivation $\mathcal{D}_0'$ of $[\Gamma_0(w),\emptyset(\bar{w}_0,w) \supset \Delta(w)]$ where $\bar{w}_0$ is a free variable of the same sort as w_0 not occurring in $\Gamma,\emptyset,\Delta$ or t, and $\mathrm{od}(\mathcal{D}_0') = \mathrm{od}(\mathcal{D}_0)$, $\rho(\mathcal{D}_0') = \rho(\mathcal{D}_0)$. Then we can apply induction again to get a derivation $\mathcal{D}_0''$ of $[\Gamma_0(t)\ ,\ \emptyset(\bar{w}_0,t) \supset \Delta(t)]$ with the same values of od,ρ. Now the rule $(\bigvee\supset)$ may be applied to give $[\Gamma_0(t), \bigvee u\ \emptyset(u,t) \supset \Delta(t)]$.

If the last rule of $\mathcal{D}$ is $(\supset\bigvee)$, $\mathcal{D}$ has the form

$$\begin{array}{c} \vdots\ \mathcal{D}_0 \\ \downarrow \\ \dfrac{\Gamma(w) \supset \Delta_0(w),\ \phi(t_0,w)}{\Gamma(w) \supset \Delta_0(w),\ \bigvee u\phi(u,w)} \end{array}$$

It is possible that w occurs in t_0 ; we write $t_0(w)$ to indicate this. Then $t_0(t)$ is the result of substituting t for w in t_0 . By induction we can form a derivation $\mathcal{D}'_0$ of $[\Gamma(t) \supset \Delta_0(t), \phi(t_0(t),t)]$ with $od(\mathcal{D}'_0) = od(\mathcal{D}_0)$, $\rho(\mathcal{D}'_0) = \rho(\mathcal{D}_0)$. Then $[\Gamma(t) \supset \Delta_0(t), \bigvee u\, \phi(u,t)]$ follows by one application of $(\supset \bigvee)$.[13]

We shall need to make use of some special functions of ordinals. For more information on these the reader should consult [Ba] or [Ff2] . In this discussion, all ordinals can be taken to be countable. We assume knowledge of the functions $\alpha + \beta$, $\alpha \cdot \beta$, ω^α and of the Cantor normal form. The function $\alpha + \beta$ is only strictly increasing in β . It is useful to have a function $\alpha \# \beta$ of the same order of magnitude which is strictly increasing in both arguments. This can be defined as follows. Write

$$\alpha = \omega^{\gamma_0}\cdot k_0 + \ldots + \omega^{\gamma_n}\cdot k_n$$

$$\beta = \omega^{\gamma_0}\cdot \ell_0 + \ldots + \omega^{\gamma_n}\cdot \ell_n$$

---

13) The separation of free from bound variables made in §2(b) becomes essential at this point. If no such restriction were made we would not be able to obtain a suitable version of this lemma (substituting terms t free for w in all its free occurrences). The inductive argument for such would break down as follows. Suppose we have inferred $[\Gamma(w) \supset \Delta_0(w), \bigvee w \bigvee u\, \Psi(w,u)]$ from $[\Gamma(w) \supset \Delta_0(w), \bigvee u\, \Psi(w,u)]$ by $(\supset \bigvee)$. If we take t to be the term u , we would want to be able to obtain $[\Gamma(u) \supset \Delta_0(u), \bigvee w \bigvee u\, \Psi(w,u)]$. But we cannot apply the inductive hypothesis here, since u is not free for w in $\bigvee u\, \Psi(w,u)$. A lemma of this sort is essential to the cut-elimination theorem. In fact, it is not difficult to see that there is no cut-free derivation of $[r(v_0,v_1) \supset \bigvee v_1 \bigvee v_0\, r(v_1,v_0)]$ (for r a binary relation symbol) in a system with no restriction on free and bound variables.

where $\gamma_o > \ldots > \gamma_n$, $0 \leq k_i < \omega$, $0 \leq \ell_i < \omega$ and for each i, $k_i \neq 0$ or $\ell_i \neq 0$; the γ_i, k_i, ℓ_i are uniquely determined by these requirements. Then take

$$\alpha \# \beta = \omega^{\gamma_o} \cdot (k_o + \ell_o) + \ldots + \omega^{\gamma_n} \cdot (k_n + \ell_n)$$

Thus $(\alpha \# \beta) = (\beta \# \alpha)$ and $\beta < \beta'$ implies $\alpha \# \beta < \alpha \# \beta'$. Note that $(\alpha + 1) \# \beta = (\alpha \# \beta) + 1$ and that $\alpha \# \beta < \max(\alpha, \beta) \cdot 3$.

A function φ from ordinals to ordinals is called normal if it is increasing, i.e. $\alpha < \beta$ implies $\varphi(\alpha) < \varphi(\beta)$, and continuous, i.e. $\varphi(\alpha) = \sup(\varphi(\xi): \xi < \alpha)$ for α a limit number. A set X of ordinals is closed if for each bounded non-empty set $Y \subseteq X$, we have $(\sup Y) \in X$. Then X is closed and unbounded if and only if it is the range of a normal function φ. φ is uniquely determined by X; we say that $\varphi(\alpha)$ is the αth member of X (in order of magnitude). If φ is a normal function, the set $X = \{\xi: \varphi(\xi) = \xi\}$ of its fixed points is closed and unbounded; the normal function which enumerates it is denoted by φ', and is called the critical function of φ. Thus

$\varphi'(\alpha) = \alpha$th fixed point of $\varphi = \alpha$th solution ξ of $\varphi(\xi) = \xi$. For $\varphi(\alpha) = \omega^\alpha$ we write $\varphi'(\alpha) = \epsilon_\alpha = \alpha$th epsilon number. We can continue this procedure. Let $\kappa^{(0)}(\alpha) = \omega^\alpha$, $\kappa^{(1)}(\alpha) = \epsilon_\alpha$, and in general $\kappa^{(n+1)} = (\kappa^{(n)})'$. Each $\kappa^{(n)}$ is normal and $\kappa^{(n+1)}$ enumerates the fixed points of $\kappa^{(n)}$, hence $\mathcal{R}(\kappa^{(n+1)}) \subseteq \mathcal{R}(\kappa^{(n)})$. Let $X = \bigcap_{n<\omega} \mathcal{R}(\kappa^{(n)})$. It can also be shown that X is closed and unbounded; let the normal function $\kappa^{(\omega)}$ enumerate this set. Thus

$$\kappa^{(\omega)}(\alpha) = \alpha\text{th common solution } \xi \text{ of the equations } \kappa^{(n)}(\xi) = \xi$$
$$\text{for all } n < \omega .$$

In general we can define for each (countable) ordinal ν a normal function

$$\kappa^{(\nu)}(\alpha) = \alpha\text{th common solution } \xi \text{ of the equations}$$
$$\kappa^{(\mu)}(\xi) = \xi \text{ for all } \mu<\nu .$$

The sequence of $\kappa^{(\nu)}$ is called the hierarchy of critical functions associated with $\kappa^{(0)}$. From the definition we see that if $\mu < \nu$ then $\kappa^{(\mu)}(\kappa^{(\nu)}(\alpha)) = \kappa^{(\nu)}(\alpha)$.

Also $\alpha \leq \kappa^{(\nu)}(\alpha)$, so $\kappa^{(\mu)}(\alpha) \leq \kappa^{(\mu)}(\kappa^{(\nu)}(\alpha)) = \kappa^{(\mu)}(\alpha)$. For further details on the functions $\kappa^{(\nu)}$, consult [Sch2] or [Ff2] . In particular, it is shown there that also the function of ν , $\kappa^{(\nu)}(0)$, is normal. Its least fixed point is denoted by Γ_0 , $\kappa^{(\Gamma_0)}(0) = \Gamma_0$. Equivalently, Γ_0 is the least ordinal $\gamma > 0$ satisfying: if ν , $\alpha < \gamma$ then $\kappa^{(\nu)}(\alpha) < \gamma$. A specific primitive recursive well-ordering of order type Γ_0 can be given ([Sch2], [Ff2]) . Thus $\Gamma_0 < \omega_1^{(R)}$.

3(b) <u>Eliminating cuts</u>. The proof of the main result will be by induction. If all cuts above a given one can be eliminated in a derivation we are then in a situation as considered in the following theorem. This cut is not eliminated outright, but it is shown how to lower the cut-rank. The main theorem will then follow by an iteration of this procedure.

3.3 <u>Theorem</u>. <u>Suppose $\mathcal{D}^+$ is a derivation of the following form,</u>

$$\left.(c)\ \dfrac{\overset{\vdots\ \mathcal{D}}{\Gamma \supset \Delta} \qquad \overset{\vdots\ \mathcal{D}'}{\Gamma' \supset \Delta'}}{\Gamma, \Gamma' - \emptyset \supset \Delta - \emptyset, \Delta'}\ \right\}\ \mathcal{D}^+$$

<u>where $\emptyset$ is in Γ' and in Δ and $\rho(\mathcal{D}) \leq od(\emptyset)$, $\rho(\mathcal{D}') \leq od(\emptyset)$. Then we can find a derivation $\mathcal{D}^*$ of the same conclusion and lower cut-rank, i.e.</u> $\rho(\mathcal{D}^*) < \rho(\mathcal{D}^+) = od(\emptyset) + 1$, <u>and with</u> $od(\mathcal{D}^*) \leq \omega \cdot (od(\mathcal{D}) \# od(\mathcal{D}'))$.

<u>Proof</u>. In the following we shall denote $\mathcal{D}^+$ by $C_\emptyset(\mathcal{D}, \mathcal{D}')$ and the new derivation $\mathcal{D}^*$ by $L_\emptyset(\mathcal{D}, \mathcal{D}')$. The proof is by induction on $od(\mathcal{D}) \# od(\mathcal{D}')$; the inductive hypothesis will thus apply in particular to all derivations $C_\emptyset(\mathcal{D}_0, \mathcal{D}_0')$ for which $od(\mathcal{D}_0) < od(\mathcal{D})$ and $od(\mathcal{D}_0') \leq od(\mathcal{D}')$, or $od(\mathcal{D}_0') < od(\mathcal{D}')$ and $od(\mathcal{D}_0) \leq od(\mathcal{D})$. The factor of ω in the bound on $od(\mathcal{D}^*)$ will allow us to insert finitely many steps as needed.

If $od(\mathcal{D}) = 1$, $\Gamma \supset \Delta$ is an axiom ; if $od(\mathcal{D}) > 1$, then $\mathcal{D} = (k, (\Gamma \supset \Delta), \langle \mathcal{D}_h : h \in H \rangle)$ where $k = 1, \ldots, 14$. Similarly for $\mathcal{D}'$. We consider possible combinations of these cases. We shall in each case show how to form $\mathcal{D}^* = L_\phi(\mathcal{D},\mathcal{D}')$ with $\rho(\mathcal{D}^*) < \rho(\mathcal{D}^+)$; it will be left to the reader in most cases to verify the bound for $od(\mathcal{D}^*)$.

Case I. $od(\mathcal{D}) = 1$, $(\Gamma \supset \Delta)$ is an axiom.

Subcase I(a). $(\Gamma \supset \Delta)$ is $(0 \supset \phi)$ where ϕ is of the form $(t=t)$. The form of $C_\phi(\mathcal{D},\mathcal{D}')$ is then

$$\text{(C)}\ \frac{0 \supset \phi \qquad \overset{\mathcal{D}'}{\Gamma' \supset \Delta'}}{\Gamma'-\phi \supset \Delta'}$$

The following lemma is easily proved by induction: if $\mathcal{D}'$ is a derivation of $(\Gamma' \supset \Delta')$ and $(t=t)$ is in Γ' then we can find a derivation $\mathcal{D}^*$ of $(\Gamma'-(t=t)\supset\Delta')$ with $od(\mathcal{D}^*) \leq \omega \cdot od(\mathcal{D}')$ and $\rho(\mathcal{D}^*) \leq \rho(\mathcal{D})$.

Subcase I(b). $(\Gamma \supset \Delta)$ is of the form $(E(t_1,\ldots,t_k;t_1',\ldots,t_k'),\Psi(t_1,\ldots t_k)\supset\Psi(t_1',\ldots t_k')$ where $\Psi(u_1,\ldots,u_k)$ is atomic. We take up only the special case where the equality chain E is empty; the more general case is left to the reader. In this case, $(\Gamma \supset \Delta)$ is $(\phi \supset \phi)$ and ϕ is atomic. $C_\phi(\mathcal{D},\mathcal{D}')$ has the form

$$\text{(C)}\ \frac{\phi \supset \phi \qquad \overset{\mathcal{D}'}{\Gamma' \supset \Delta'}}{\phi,\ \Gamma'-\phi \supset \Delta'}$$

But $(\phi,\ \Gamma'-\phi \supset \Delta')$ follows directly from $(\Gamma' \supset \Delta')$ by one application of the structural rule (S) .

Case I'. $od(\mathcal{D}') = 1$. This is entirely symmetric to case I . /In all the further cases it is assumed that $od(\mathcal{D}) > 1$ and $od(\mathcal{D}') > 1$.

Case II. The last rule of $\mathcal{D}$ is (S) . For certain Γ_0 with the same set of formulas as Γ and Δ_0 with the same set of formulas as Δ , $\mathcal{D}$ has the form

$$\begin{array}{c} \vdots\ \mathcal{D}_o \\ \downarrow \\ \text{(S)}\ \dfrac{\Gamma_o \supset \Delta_o}{\Gamma \supset \Delta} \end{array}$$

and $od(\mathcal{D}) = od(\mathcal{D}_o) + 1$. By inductive hypothesis, we can find a derivation $\mathcal{D}_o^* = L_\phi(\mathcal{D}_o, \mathcal{D}')$ of

$$\Gamma_o\ ,\ \Gamma'-\phi \supset \Delta_o-\phi\ ,\ \Delta'$$

with $od(\mathcal{D}_o^*) \leq \omega \cdot (od(\mathcal{D}_o) \# od(\mathcal{D}'))$. Then $\Gamma, \Gamma'-\phi \supset \Delta-\phi, \Delta'$ follows by one additional application of (S) , giving the desired $\mathcal{D}^*$.

<u>Case</u> II'. The last rule of $\mathcal{D}'$ is (S) . Symmetric.

<u>Case</u> III. The last rule of $\mathcal{D}$ is (C) . Then $\mathcal{D}$ has the form

$$\text{(C)}\ \dfrac{\begin{array}{c}\vdots\ \mathcal{D}_o \\ \downarrow \\ \Gamma_o \supset \Delta_o\end{array} \qquad \begin{array}{c}\vdots\ \mathcal{D}_1 \\ \downarrow \\ \Gamma_1 \supset \Delta_1\end{array}}{\Gamma_o,\ \Gamma_1-\Psi \supset \Delta_o-\Psi,\ \Delta_1}$$

where Ψ is in Γ_1 and Δ_o and $\Gamma = \Gamma_o, \Gamma_1-\Psi$ and $\Delta = \Delta_o-\Psi, \Delta_1$. By hypothesis $\rho(\mathcal{D}) \leq od(\phi)$ so $od(\Psi) < od(\phi)$. Then ϕ is in at least one of Δ_o, Δ_1 and $\Delta-\phi = (\Delta_o-\Psi)-\phi\ ,\ \Delta_1-\phi$. Suppose, for example, that ϕ is in Δ_1, but not in Δ_o . By inductive hypothesis form the derivation $L_\phi(\mathcal{D}_1, \mathcal{D}')$ of

$$\Gamma_1,\ \Gamma'-\phi \supset \Delta_1-\phi,\ \Delta'\ .$$

Then we apply cut, removing Ψ, to $\mathcal{D}_o$ and $L_\phi(\mathcal{D}_1, \mathcal{D}')$ to give

$$\Gamma_o,\ \Gamma_1-\Psi,\ (\Gamma'-\phi)-\Psi \supset \Delta_o-\Psi,\ \Delta_1-\phi,\ \Delta'\ .$$

Now by weakening we can restore any occurrences of Ψ that may have been removed from $(\Gamma'-\emptyset)$, giving $[\Gamma, \Gamma'-\emptyset \supset \Delta-\emptyset, \Delta']$. Similarly if $\emptyset$ is in Δ_0 but not in Δ_1, we first form $L_\emptyset(\mathcal{D}_0, \mathcal{D}')$ and then cut Ψ from this with $\mathcal{D}_1$. If $\emptyset$ is in both Δ_0 and Δ_1, we cut Ψ from $L_\emptyset(\mathcal{D}_0, \mathcal{D}')$ and $L_\emptyset(\mathcal{D}_1, \mathcal{D}')$.

Case III'. The last rule of $\mathcal{D}'$ is (C). Symmetric.

We can thus assume in all the remaining cases that the last rule of $\mathcal{D}$ is one of the G-rules (R_k) for $k = 1,\ldots,12$, and the same for $\mathcal{D}'$. The active formula θ of an application of one of these rules is defined to be the rightmost formula of the antecedent (consequent) of the conclusion of the rule when k is odd (even). When this application is the last rule in $\mathcal{D}$ and k is odd (even) we shall say that θ is the active formula on the left (the right) of $(\Gamma \supset \Delta)$ in $\mathcal{D}$; similarly for $\mathcal{D}'$ and $\Gamma' \supset \Delta'$.

Case IV. The active formula θ of $(\Gamma \supset \Delta)$ in $\mathcal{D}$ is distinct from $\emptyset$, or $\emptyset$ is the active formula on the left of $(\Gamma \supset \Delta)$ in $\mathcal{D}$. There are a number of possibilities, all treated in the same way, namely by permuting the applications of the rules. For example, suppose the last rule of $\mathcal{D}$ is $(\supset \sim)$ with $\theta = \sim \Psi$, distinct from $\emptyset$; we have $\Delta = \Delta_0, \sim \Psi$ and $\mathcal{D}$ has the form

$$\begin{array}{c} \vdots \; \mathcal{D}_0 \\ \downarrow \\ \Gamma, \Psi \supset \Delta_0 \\ \hline \Gamma \supset \Delta_0, \sim\Psi \end{array}$$

Since $\emptyset \neq \sim \Psi$, we have $\emptyset$ in Δ_0. By inductive hypothesis, we can get a derivation $L_\emptyset(\mathcal{D}_0, \mathcal{D}')$ of

$$\Gamma, \Psi, \Gamma'-\emptyset \supset \Delta_0-\emptyset, \Delta' .$$

Using (S) this is followed by

$$(\supset \sim) \frac{\Gamma,\ \Gamma'\text{-}\phi,\ \Psi \supset \Delta_0\text{-}\phi,\ \Delta'}{\Gamma,\ \Gamma'\text{-}\phi \supset \Delta_0\text{-}\phi,\ \Delta',\ \sim\Psi}$$
$$(S) \frac{\Gamma,\ \Gamma'\text{-}\phi \supset \Delta_0\text{-}\phi,\ \Delta',\ \sim\Psi}{\Gamma,\ \Gamma'\text{-}\phi \supset \Delta\text{-}\phi,\ \Delta'}$$

As an example with ϕ active on the left of $(\Gamma \supset \Delta)$ in $\mathcal{D}$, suppose $\phi = \bigwedge u\ \Psi(u)$ and the last rule of $\mathcal{D}$ is $(\bigwedge\supset)$; with $\Gamma = \Gamma_0$, $\phi = \Gamma_0, \bigwedge u\ \Psi(u)$, $\mathcal{D}$ has the form

$$\begin{array}{c} \vdots\ \mathcal{D}_0 \\ \downarrow \end{array}$$
$$(\bigwedge\supset) \frac{\Gamma_0,\ \Psi(t) \supset \Delta}{\Gamma_0, \bigwedge u\ \Psi(u) \supset \Delta}$$

By inductive hypothesis we are able to construct a derivation:

$$\begin{array}{c} \vdots\ L_\phi(\mathcal{D}_0, \mathcal{D}') \\ \downarrow \end{array}$$
$$\frac{\Gamma_0,\ \Psi(t),\ \Gamma'\text{-}\phi \supset \Delta\text{-}\phi,\ \Delta'}{\overline{\Gamma_0,\ \bigwedge u\ \Psi(u),\ \Gamma'\text{-}\phi \supset \Delta\text{-}\phi,\ \Delta'}}$$

where the double lines indicate successive finite number of applications of rules other than (C) , in this case (S) , $(\bigwedge\supset)$, and (S) again.
The case of ϕ active on the left of $(\Gamma \supset \Delta)$ in $\mathcal{D}$ with $\phi = \bigvee u\ \Psi(u)$ and $(\bigvee\supset)$ the last rule of $\mathcal{D}$ is slightly more delicate. Here $\Gamma = \Gamma_0$, $\phi = \Gamma_0, \bigvee u\ \Psi(u)$ and we have derivations

$$\begin{array}{c} \vdots\ \mathcal{D}_0 \\ \downarrow \end{array} \qquad\qquad \begin{array}{c} \vdots\ \mathcal{D}' \\ \downarrow \end{array}$$
$$(\bigvee\supset) \frac{\Gamma_0,\ \Psi(w) \supset \Delta}{\Gamma_0,\ \bigvee u\ \Psi(u) \supset \Delta} \qquad\qquad \Gamma' \supset \Delta'$$

where w is not free in $\Gamma_0, \bigvee u\ \Psi(u)$ or Δ . However w may be free in Γ', Δ'. Let $\overline{w}$ be a variable which does not occur free in any of these sequents. By (3.2) we can find a derivation $\overline{\mathcal{D}}_0$ of $[\Gamma_0,\ \Psi(\overline{w}) \supset \Delta]$ with $od(\overline{\mathcal{D}}_0) = od(\mathcal{D}_0)$, and $\rho(\overline{\mathcal{D}}_0) = \rho(\mathcal{D}_0)$. Then inductive hypothesis gives us a derivation satisfying the required conditions and of the following form:

$$\Big\downarrow L_\phi(\bar{\mathcal{D}}_o, \mathcal{D}')$$

$$\frac{\Gamma_o,\ \Psi(\bar{w}),\ \Gamma'-\phi \supset \Delta-\phi,\ \Delta'}{\Gamma_o,\ \forall u\ \Psi(u),\ \Gamma'-\phi \supset \Delta-\phi,\ \Delta'}$$

The remaining possibilities in this case are treated similarly.

<u>Case</u> IV'. The active formula θ of $(\Gamma' \supset \Delta')$ in $\mathcal{D}'$ is distinct from ϕ, or ϕ is the active formula on the right of $(\Gamma' \supset \Delta')$ in $\mathcal{D}'$. This is symmetric to case IV .

Thus from now on we can assume that ϕ is the active formula on the right of $(\Gamma \supset \Delta)$ in $\mathcal{D}$ and ϕ is the active formula on the left of $(\Gamma' \supset \Delta')$ in $\mathcal{D}'$.

<u>Case</u> V . The last rule of $\mathcal{D}$ is $(\supset W)$. Then $\Delta = \Delta_o, \phi$ and $\mathcal{D}$ has the form

$$\Big\downarrow$$

$$\frac{\Gamma \supset \Delta_o}{\Gamma \supset \Delta_o,\ \phi}$$

If ϕ occurs in Δ_o, inductive hypothesis gives a derivation of $[\Gamma,\Gamma'-\phi \supset \Delta_o-\phi,\Delta']$. But $\Delta_o-\phi = \Delta-\phi$. If ϕ does not occur in Δ_o then $\Delta-\phi = \Delta_o$ and we wish to infer $[\Gamma,\ \Gamma'-\phi \supset \Delta_o,\ \Delta']$ in this case. This is obtained by a finite number of applications of $(W \supset)$ and $(\supset W)$ from $(\Gamma \supset \Delta_o)$.

<u>Case</u> V'. The last rule of $\mathcal{D}'$ is $(W \supset)$. Symmetric.

Thus the only remaining cases are those in which ϕ is active on the right of $(\Gamma \supset \Delta)$ by one of the rules (R_k), $k = 2,4,6,8,10$ and on the left of $(\Gamma' \supset \Delta')$ by one of the rules (R_k), $k = 1,3,5,7,9$. Which of these rules is applied, is completely determined by the syntactic form of ϕ. In the following let $\alpha = od(\mathcal{D})$, $\alpha' = od(\mathcal{D}')$.

<u>Case</u> VI. $\phi = \sim \Psi$. Then $\mathcal{D}$, $\mathcal{D}'$ have the form

$$\left.\begin{array}{c} \vdots\ \mathcal{D}_0 \\ \downarrow \\ \dfrac{\Gamma,\ \Psi \supset \Delta_0}{\Gamma \supset \Delta_0,\ \emptyset} \end{array}\right\}\mathcal{D} \qquad\qquad \left.\begin{array}{c} \vdots\ \mathcal{D}_0' \\ \downarrow \\ \dfrac{\Gamma_0' \supset \Delta',\ \Psi}{\Gamma_0',\ \emptyset \supset \Delta'} \end{array}\right\}\mathcal{D}'$$

where $\Delta = \Delta_0, \sim \Psi$ and $\Gamma' = \Gamma_0', \sim \Psi$. We must consider four subcases here, depending on whether or not $\emptyset$ is in Δ_0 and whether or not $\emptyset$ is in Γ_0'. Let $\alpha_0 = od(\mathcal{D}_0)$, $\alpha_0' = od(\mathcal{D}_0')$, so $\alpha = \alpha_0 + 1$ and $\alpha' = \alpha_0' + 1$.

Subcase VI(a). $\emptyset$ is not in Δ_0 and $\emptyset$ is not in Γ_0'. Then $\Delta\text{-}\emptyset = \Delta_0$, $\Gamma'\text{-}\emptyset = \Gamma_0'$. The following derivation $\mathcal{D}^*$ is then built:

$$\begin{array}{c} \begin{array}{cc} \vdots\ \mathcal{D}_0' & \vdots\ \mathcal{D}_0 \\ \downarrow & \downarrow \end{array} \\ (C,W)\ \dfrac{\Gamma_0' \supset \Delta',\ \Psi \qquad \Gamma,\ \Psi \supset \Delta_0}{(S)\ \dfrac{\Gamma_0',\ \Gamma \supset \Delta',\ \Delta_0}{\Gamma,\ \Gamma_0' \supset \Delta_0,\ \Delta'}} \end{array}$$

That is, the simple removal of Ψ is effected by full cut followed by a finite number of weakenings; then (S) is applied to conclude $[\Gamma, \Gamma'\text{-}\emptyset \supset \Delta\text{-}\emptyset, \Delta']$. $\rho(\mathcal{D}^*) = \max(\rho(\mathcal{D}_0'), \rho(\mathcal{D}_0), od(\Psi) + 1) = od(\emptyset)$ and $od(\mathcal{D}^*) < \max(\alpha_0, \alpha_0') + \omega \leq \omega(\alpha \# \alpha')$.

Subcase VI(b). $\emptyset$ is in Δ_0 but not in Γ_0'. Then $\Delta\text{-}\emptyset = \Delta_0\text{-}\emptyset$ and $\Gamma'\text{-}\emptyset = \Gamma_0'$. Using inductive hypothesis we first form the "cross-cut" $L_\emptyset(\mathcal{D}_0, \mathcal{D}')$ and then remove Ψ:

$$\left.\begin{array}{c} \begin{array}{cc} \vdots\ \mathcal{D}_0' & \vdots\ L_\emptyset(\mathcal{D}_0, \mathcal{D}') \\ \downarrow & \downarrow \end{array} \\ (C,W)\ \dfrac{\Gamma_0' \supset \Delta',\ \Psi \qquad \Gamma,\ \Psi,\ \Gamma_0' \supset \Delta_0\text{-}\emptyset,\ \Delta'}{(S)\ \dfrac{\Gamma_0',\ \Gamma,\ \Gamma_0' \supset \Delta',\ \Delta_0\text{-}\emptyset,\ \Delta'}{\Gamma,\ \Gamma_0' \supset \Delta_0\text{-}\emptyset,\ \Delta'}} \end{array}\right\}\mathcal{D}^*$$

Again the cut-rank is lowered. Now $od(L_\emptyset(\mathcal{D}_0, \mathcal{D}')) \leq \omega(\alpha_0 \# \alpha')$ so

$od(\mathcal{D}^*) < \max(\alpha'_o, \omega \cdot (\alpha_o \# \alpha')) + \omega = \omega \cdot (\alpha_o \# \alpha') + \omega = \omega(\alpha \# \alpha')$.

Subcase VI(c). $\emptyset$ is in Γ'_o but not in Δ_o . Symmetric to V(b) .

Subcase VI(d). $\emptyset$ is both in Δ_o and in Γ'_o . Then $\Delta\text{-}\emptyset = \Delta_o\text{-}\emptyset$ and $\Gamma'\text{-}\emptyset = \Gamma'_o\text{-}\emptyset$. We now perform two cross-cuts and then remove Ψ :

$$\left.\begin{array}{c} \begin{array}{cc} \vdots L_\emptyset(\mathcal{D}, \mathcal{D}'_o) & \vdots L_\emptyset(\mathcal{D}_o, \mathcal{D}') \\ \Gamma, \Gamma'_o\text{-}\emptyset \supset \Delta_o\text{-}\emptyset, \Delta', \Psi & \Gamma, \Psi, \Gamma'_o\text{-}\emptyset \supset \Delta_o\text{-}\emptyset, \Delta' \end{array} \\ \hline\hline (C,W)\quad \Gamma, \Gamma'_o\text{-}\emptyset, \Gamma, \Gamma'_o\text{-}\emptyset \supset \Delta_o\text{-}\emptyset, \Delta', \Delta_o\text{-}\emptyset, \Delta' \\ \hline (S)\quad \Gamma, \Gamma'_o\text{-}\emptyset \supset \Delta_o\text{-}\emptyset, \Delta' \end{array}\right\} \mathcal{D}^*$$

The cut-rank is again lowered and inductive hypothesis gives

$od(\mathcal{D}^*) < \max(\omega(\alpha \# \alpha'_o), \omega(\alpha_o \# \alpha')) + \omega = \omega(\alpha \# \alpha')$.

The arguments for this case are typical for the remaining cases.

Case VII. $\emptyset = \sum K$. Then $\mathcal{D}, \mathcal{D}'$ have the form

$$\left.\begin{array}{c} \vdots \mathcal{D}_{\Psi_o} \\ \Gamma \supset \Delta_o, \Psi_o \\ \hline \Gamma \supset \Delta_o, \emptyset \end{array}\right\}\mathcal{D} \qquad\qquad \left.\begin{array}{c} \vdots \mathcal{D}'_\Psi \ (\Psi \text{ in } K) \\ \Gamma'_o, \Psi \supset \Delta' \\ \hline \Gamma'_o, \emptyset \supset \Delta' \end{array}\right\}\mathcal{D}'$$

Here Ψ_o is some specific member of K . In the simplest subcase we need only deal with $C_{\Psi_o}(\mathcal{D}_{\Psi_o}, \mathcal{D}'_{\Psi_o})$. In the remaining subcases we first form cross-cuts $L_\emptyset(\mathcal{D}_{\Psi_o}, \mathcal{D}')$, $L_\emptyset(\mathcal{D}, \mathcal{D}'_{\Psi_o})$, as needed and then remove Ψ_o by cut. The details are left to the reader.

Case VIII. $\emptyset = \prod K$. This is dual to case VI .

Case IX. $\emptyset = \bigvee u\, \Psi(u)$. Then $\mathcal{D}, \mathcal{D}'$ have the form

$$\left.\begin{array}{c} \vdots \mathcal{D}_o \\ \Gamma \supset \Delta_o, \Psi(t) \\ \hline \Gamma \supset \Delta_o, \emptyset \end{array}\right\}\mathcal{D} \qquad\qquad \left.\begin{array}{c} \vdots \mathcal{D}'_o \\ \Gamma'_o, \Psi(w) \supset \Delta' \\ \hline \Gamma'_o, \emptyset \supset \Delta' \end{array}\right\}\mathcal{D}'$$

Here t and w are of the same sort as u and w is not free in Γ'_o, $\emptyset$ or Δ'. By (3.2) we can form a derivation $\mathcal{D}_o^{(t)}$ of $(\Gamma'_o, \Psi(t) \supset \Delta')$ with $od(\mathcal{D}_o^{(t)}) = od(\mathcal{D}'_o)$, and $\rho(\mathcal{D}_o^{(t)}) = \rho(\mathcal{D}_o)$. Then we can proceed as before. The cut-rank is lowered because $od(\Psi(t)) = od(\Psi(w))$ by (3.1).

<u>Case</u> X. $\emptyset = \bigwedge u\, \Psi(u)$. Dual to case IX.

This completes the proof.

The cut-elimination theorem (with ordinal bounds) for $\mathcal{L}_{HC}$ can now be established without much additional argument.

3.4 <u>Theorem</u>. <u>Suppose</u> $\mathcal{D}$ <u>is a derivation with</u> $od(\mathcal{D}) = \alpha$ <u>and</u> $\rho(\mathcal{D}) = \nu$. <u>Then we can find a cut-free derivation</u> $\mathcal{D}^*$ <u>of the same conclusion with</u> $od(\mathcal{D}^*) \leq \kappa^{(\nu)}(\alpha)$.

<u>Proof</u>. This is proved by induction on ν, and for fixed ν by induction on α. We may assume $od(\mathcal{D}) > 1$, $\mathcal{D} = (k, (\Gamma \supset \Delta), \langle \mathcal{D}_h : h \in H \rangle)$. Let $\alpha_h = od(\mathcal{D}_h)$ and $\nu_h = \rho(\mathcal{D}_h)$; then $\alpha = \sup(\alpha_h + 1 : h \in H)$. Suppose $k \neq 14$, i.e. the last rule of $\mathcal{D}$ is not (C). In this case $\nu = \sup(\nu_h : h \in H)$. By inductive hypothesis we can find for each $h \in H$ a derivation $\mathcal{D}_h^*$ of the same conclusion as $\mathcal{D}_h$ with $od(\mathcal{D}_h^*) \leq \kappa^{(\nu_h)}(\alpha_h) \leq \kappa^{(\nu)}(\alpha_h)$. Let $\mathcal{D}^* = (k, (\Gamma \supset \Delta), \langle \mathcal{D}_h^* : h \in H \rangle)$. Then $\mathcal{D}^*$ is cut-free and $od(\mathcal{D}^*) = \sup(od(\mathcal{D}_h^*) + 1 : h \in H) \leq \sup(\kappa^{(\nu)}(\alpha_h) + 1 : h \in H)$

$$\leq \sup(\kappa^{(\nu)}(\alpha_h + 1) : h \in H) = \kappa^{(\nu)}(\alpha$$

Suppose now that the last rule of $\mathcal{D}$ is (C). $\mathcal{D}$ has the form

$$\left.\begin{array}{c} \begin{array}{cc} \vdots\, \mathcal{D}_o & \vdots\, \mathcal{D}_1 \\ \downarrow & \downarrow \end{array} \\ (C)\ \dfrac{\Gamma_o \supset \Delta_o \qquad \Gamma_1 \supset \Delta_1}{\Gamma_o,\ \Gamma_1 - \emptyset \supset \Delta_o - \emptyset,\ \Delta_1} \end{array}\right\} \mathcal{D}$$

where $\emptyset$ is in Δ_o and Γ_1. We use the same notation as above, with $H = \{0,1\}$. Thus $\nu_h \leq \nu$ for $h = 0,1$ and $od(\emptyset) + 1 \leq \nu$. Let $\beta_h = \kappa^{(\nu)}(\alpha_h)$. Now form the derivation the derivation

$$\left.\begin{array}{c} \vdots\mathcal{D}_0^* \qquad\qquad \vdots\mathcal{D}_1^* \\ (C)\ \dfrac{\Gamma_0 \supset \Delta_0 \qquad \Gamma_1 \supset \Delta_1}{\Gamma_0,\ \Gamma_1\text{-}\emptyset \supset \Delta_0\text{-}\emptyset,\ \Delta_1} \end{array}\right\} \mathcal{D}^+$$

Then $\rho(\mathcal{D}^+) = \mathrm{od}\,(\emptyset) + 1$. By theorem 3.3 we can find a derivation $\bar{\mathcal{D}}$ of $[\Gamma_0,\ \Gamma_1\text{-}\emptyset \supset \Delta_0\text{-}\emptyset,\ \Delta_1]$ with $\rho(\bar{\mathcal{D}}) < \nu$ and $\mathrm{od}(\bar{\mathcal{D}}) \leq \omega(\beta_0 \# \beta_1)$. Let $\mu = \rho(\bar{\mathcal{D}})$, $\gamma = \omega(\beta_0 \# \beta_1)$. Since $\mu < \nu$, inductive hypothesis can be applied to find a cut-free derivation $\mathcal{D}^*$ of the same conclusion as $\bar{\mathcal{D}}$ with $\mathrm{od}(\mathcal{D}^*) \leq \kappa^{(\mu)}(\gamma)$. Now if, say, $\alpha_0 \leq \alpha_1$ we have $\beta_0 \# \beta_1 \leq \kappa^{(\nu)}(\alpha_1) \# \kappa^{(\nu)}(\alpha_1) \leq \kappa^{(\nu)}(\alpha_1) \cdot 3 \leq$ $\leq \kappa^{(\nu)}(\alpha_1 + 1) = \kappa^{(\nu)}(\alpha)$ and $\omega \cdot (\beta_0 \# \beta_1) \leq \omega \cdot \kappa^{(\nu)}(\alpha) = \kappa^{(\nu)}(\alpha)$. Hence $\mathrm{od}(\mathcal{D}^*) \leq \kappa^{(\mu)}(\kappa^{(\nu)}(\alpha)) = \kappa^{(\nu)}(\alpha)$.

By the subformula property, this includes Tait's cut-elimination result [Tt] for quantifier-free sequents in $\mathcal{L}_{HC}$ as a special case. The ordinal bounds found here are of roughly the same order of magnitude as those given in [Tt] but, for simplicity, not quite as sharp ; cf. exercise (iii) at the end of this section for the latter.

3.5 Corollary. If $\mathrm{od}(\mathcal{D}) < \Gamma_0$ and $\rho(\mathcal{D}) < \Gamma_0$ then we can find a cut-free derivation $\mathcal{D}^*$ of the same conclusion as $\mathcal{D}$ with $\mathrm{od}(\mathcal{D}^*) < \Gamma_0$.

For, $\nu,\ \alpha < \Gamma_0$ implies $\kappa^{(\nu)}(\alpha) < \Gamma_0$.

Special assumptions on $\rho(\mathcal{D})$ can lead to improved bounds on $(\mathcal{D}^*)$. The following is a useful lemma for such.

3.6 Lemma. Suppose $\mathrm{od}(\mathcal{D}) = \alpha$, $\rho(\mathcal{D}) \leq \nu + 1$. Then we can find a derivation $\mathcal{D}^*$ of the same conclusion as $\mathcal{D}$ with $\mathrm{od}(\mathcal{D}^*) \leq \omega^{\omega+\alpha}$ and $\rho(\mathcal{D}^*) \leq \nu$.

Proof. The argument for this is similar to that in (3.4). Note that in the latter, we did not make full use of (3.3), applying it only to the case $C_\emptyset(\mathcal{D}_0^*, \mathcal{D}_1^*)$ where both $\mathcal{D}_0^*, \mathcal{D}_1^*$ are cut-free. This proof makes use of all the information in (3.3).

3.7 Corollary. If $\mathcal{D}$ has finite cut-rank and $od(\mathcal{D}) < \epsilon_0$ then we can find a cut-free derivation $\mathcal{D}^*$ of the same conclusion as $\mathcal{D}$ with $od(\mathcal{D}^*) < \epsilon_0$.

The following can be verified as exercises: (i) If $od(\mathcal{D}) = \alpha$, $\rho(\mathcal{D}) \leq \omega$ then we can find a cut-free derivation $\mathcal{D}^*$ of the same conclusion as $\mathcal{D}$ with $od(\mathcal{D}^*) \leq \epsilon_\alpha$. (ii) If $od(\mathcal{D}) < \epsilon_0$, $\rho(\mathcal{D}) < \omega \cdot 2$ then we can find a cut-free derivation $\mathcal{D}^*$ of the same conclusion as $\mathcal{D}$ with $od(\mathcal{D}^*) < \epsilon_{\epsilon_0}$. (iii) [Tt] If $od(\mathcal{D}) = \alpha$, $\rho(\mathcal{D}) \leq \nu_0 + \omega^\gamma$ where $\gamma > 0$, then we can find a derivation $\mathcal{D}^*$ of the same conclusion as $\mathcal{D}$ with $\rho(\mathcal{D}^*) \leq \nu_0$ and $od(\mathcal{D}^*) \leq \kappa^{(\gamma)}(\alpha)$. (Note that (iii) generalizes (i) and, by (3.6), it also holds for $\gamma = 0$ when $\omega + \alpha = \alpha$.) The results (i) and (ii) will be useful in the consistency proofs of §6.

§3(c). Completeness of $\mathcal{L}_A$. We can now bring the discussion of §2(d) to a conclusion with the results of Barwise [Bw].

3.8 Theorem. If $(\Gamma \supset \Delta)$ is an A-sequent and $\Vdash(\Gamma \supset \Delta)$ then there is a cut-free A-derivation of $(\Gamma \supset \Delta)$.

Proof. By (2.4) there is a derivation of $(\Gamma \supset \Delta)$. Then by (3.4) there is a cut-free derivation $\mathcal{D}$ of $(\Gamma \supset \Delta)$. We prove by induction on such $\mathcal{D}$ that there is a cut-free A-derivation $\mathcal{D}'$ of the same conclusion. Obviously $\mathcal{D} \in A$ if it is of the form $(0, \Gamma \supset \Delta)$. Otherwise $\mathcal{D} = (k, (\Gamma \supset \Delta), \langle \mathcal{D}_h : h \in H \rangle)$ where $1 \leq k \leq 13$ and for each $h \in H$, $\mathcal{D}_h$ is a cut-free derivation of $(\Gamma_h \supset \Delta_h)$ and $(\Gamma \supset \Delta)$ is inferred from the $(\Gamma_h \supset \Delta_h)$ for $h \in H$ by the kth G-rule. Inspection of these rules then shows that each $(\Gamma_h \supset \Delta_h)$ belongs to A and $H \in A$.[14] By inductive hypothesis, for each $h \in H$ there is at least one cut-free A-derivation $\mathcal{D}'$ of $(\Gamma_h \supset \Delta_h)$; in the notation of §2(d), for each $h \in H$ there exists $\mathcal{D}' \in A$ with $(\mathcal{D}', (\Gamma_h \supset \Delta_h)) \in Der_A^{(-C)}$. Since $Der_A^{(-C)}$ is $\sum_1^{(A)}$-definable by (2.6), we can apply the $\sum_1$-axiom of choice in A to conclude that there is a function $f \in A$ with $\mathcal{D}(f) = H$ such that for each $h \in H$, $\mathcal{D}'_h = f(h)$ is a cut-free A-derivation

[14] In most of the cases, $(\Gamma_h \supset \Delta_h)$ and H belong to $TC(\Gamma \supset \Delta)$, namely when the $(\Gamma_h \supset \Delta_h)$ consist of subformulas of formulas of $(\Gamma \supset \Delta)$. In the case of the quantification rules, however, the hypotheses involve subformulas only in the extended sense. This gives no trouble since substitution of a term in an A-formula yields an A-formula.

of $(\Gamma_h \supset \Delta_h)$. Hence $\mathcal{D}' = (k, (\Gamma \supset \Delta), \langle \mathcal{D}'_h : h \in H \rangle)$ is a cut-free A-derivation of $(\Gamma \supset \Delta)$.[15]

Note that this includes the completeness theorem for $\mathcal{L}_{HF}$ as a special case.

The proof of the following result from [Bw] is left to the reader; there is one essentially new step to be provided.

3.9 Theorem. Suppose $\mathfrak{S}$ is a countable A-rec. enum. set of A-sequents, and that $\mathfrak{S} \Vdash (\Gamma \supset \Delta)$ where $(\Gamma \supset \Delta)$ is an A-sequent. Then there is an A-derivation $\mathcal{D}$ of $(\Gamma \supset \Delta)$ from $\mathfrak{S}$. Moreover, for some $\mathfrak{S}_1 \subseteq \mathfrak{S}$, $\mathfrak{S}_1 \in A$ and $\mathcal{D}$ is an A-derivation of $(\Gamma \supset \Delta)$ from $\mathfrak{S}_1$.

This includes the completeness theorem (2.5) for $\mathcal{L}_{HC}$ as a special case. The latter part of this theorem immediately provides a compactness theorem for A-rec. enum. sets $\mathfrak{S}$. By suitable relativization of this result, Barwise [Bw] obtained a complete generalization of the extended completeness and compactness theorems for countable sets of axioms in $\mathcal{L}_{HF}$. Actually, he obtained such theorems for an even wider class of A than those satisfying the given conditions $(i)_A$-$(vii)_A$. These are the admissible A in the sense of Platek [Pl], which satisfy $(i)_A$-$(vi)_A$ and for which $(vii)_A$ is replaced by $(vii)'_A$, the so-called $\sum_1$-reflection principle. In fact, it is shown in [Bw] that the compactness theorem holds for all A-rec. enum. sets if and only if A is admissible.[16] The completeness theorem also generalizes to these A, but only if the notion of derivation is taken in a certain multi-valued sense. For this, clause (ii) of the definition of derivations is modified so that $\mathcal{D} = (k, \Gamma \supset \Delta, \langle B_h : h \in H \rangle)$ where for each $h \in H$, B_h is a non-empty set of derivations $\mathcal{D}_h$ of $(\Gamma_h \supset \Delta_h)$. This modification is necessary to get along without some form of the axiom of choice. We have treated the less general case here for simplicity and since it covers all the special cases of interest.

---

[15] It is seen that the proof will work with any reasonable set-theoretical representation of the set-theoretic notions such that a subformula (subderivation) of a formula (derivation) is in its transitive closure and for which we have the substitution property of ftn. 14.

[16] Given the analogy of membership in A to (hereditary) finiteness, this provides a partially satisfactory explanation for the choice of the admissibility conditions in dealing with logic on the $\mathcal{L}_A$.

The results (3.8) and (3.9) are the basis of a uniform application of proof theory to the model theory of the languages $\mathcal{L}_A$, including $\mathcal{L}_{HF}$, $\mathcal{L}_{HH}$ and $\mathcal{L}_{HC}$ as special cases.

The cut-elimination theorem for $\mathcal{L}_A$ is the following consequence of (3.8) : if $\mathcal{D}$ is an A-derivation of $(\Gamma \supset \Delta)$ then there is a cut-free A-derivation $\mathcal{D}^*$ of $(\Gamma \supset \Delta)$. However, it is not necessary to use the full force of the Δ_0-separation axiom and $\sum_1$-axiom of choice for A to derive this. By analyzing the set-theoretical facts concerning HC which are used implicitly in the arguments for (3.3) and (3.4) , it is seen that one need only know that A is closed under functions defined by certain double transfinite recursions. For example, it can be shown that the collection A of sets hereditarily hyperarithmetic before Γ_0 satisfies this closure condition (this is related to (3.5)). The cut-elimination theorem for $\mathcal{L}_{HF}$ is a form of Gentzen's Hauptsatz [Gn1] . The arguments of (3.3) and (3.4) for this special case can be made completely finitistic with little change (cf. also [Kl] Ch. XV). Suitable number-theoretical bounds are also easily given in this case.

§4. Some interpolation theorems. The theorems under this heading are given both syntactic and semantic formulations in the literature. However, the syntactic versions imply the semantic ones for complete systems. Moreover, there are also interpolation theorems for incomplete deductive systems, including non-classical ones (cf. e.g. [Sch4]) . This is one reason that the interpolation theorems considered here are presented in syntactic form. The other reason is that we only know proof-theoretical arguments for these particular results. Nevertheless, a strictly model-theoretic development of the corresponding semantic versions is to be hoped for, in view of their model-theoretic consequences such as presented in §5.

The theorems involve certain functions Sn from formulas to sets consisting of various kinds of syntactic objects. Given such functions $Sn_1,\dots,Sn_k$ and formulas $\emptyset,\Psi$, a formula θ is said to be an A-interpolant for $\emptyset,\Psi$ with respect to $Sn_1,\dots,Sn_k$ if there exist cut-free A-derivations of $(\emptyset \to \theta)$ and $(\theta \to \Psi)$ such that $Sn_i(\theta) \subseteq Sn_i(\emptyset) \cap Sn_i(\Psi)$ for $i = 1,\dots,k$. If this holds then $\emptyset,\Psi,\theta$ are necessarily A-formulas and there is a (cut-free) A-derivation of $(\emptyset \to \Psi)$; the latter is the basic hypothesis in each interpolation theorem. The theorems also involve some additional special hypotheses and some additional conclusions besides the existence of an A-interpolant with respect to given functions, which are stated as appropriate.

In the proof-theoretical approach via Gentzen calculi, the interpolation theorems are obtained as special cases of corresponding theorems for sequents. Each of the functions Sn_i considered has a certain extension to sequents. We say that θ is an A-interpolant for $(\Gamma \supset \Delta)$, $(\Gamma' \supset \Delta')$ with respect to $Sn_1,\dots,Sn_k$ if there exist cut-free A-derivations of $(\Gamma \supset \Delta, \theta)$ and $(\Gamma', \theta \supset \Delta')$ such that $Sn_i(\theta) \subseteq Sn_i(\Gamma \supset \Delta) \cap Sn_i(\Gamma' \supset \Delta')$ for $i = 1,\dots,k$. Again this implies that θ is an A-formula and $(\Gamma \supset \Delta)$, $(\Gamma' \supset \Delta')$ are A-sequents and that there is a (cut-free) A-derivation of $(\Gamma, \Gamma' \supset \Delta, \Delta')$.

We shall be particularly concerned with the following syntactic functions on formulas. $Fr(\emptyset)$ is the set of variables which occur free in $\emptyset$, $Cn(\emptyset)$ is the set of constant symbols which occur in $\emptyset$, $Rel(\emptyset)$ is the set of relation symbols, other than the equality symbol $=$, which occur in $\emptyset$, $Fn(\emptyset)$ is the set of function symbols which occur in $\emptyset$, $Sort(\emptyset)$ is the set of $j \in J$ such that a variable of sort j occurs (free or bound) in ϕ, and Tm-Sort(ϕ) is the set of $j \in J$ such that a term of sort j occurs in ϕ. Each of these functions Sn is extended to sequents by taking $Sn(\Gamma \supset \Delta) = \bigcup Sn(\phi)$ [ϕ is in Γ or in Δ].

We also make use of two special syntactic functions, Un(ϕ) and Ex(ϕ), which single out the sorts of bound variables which appear in some essentially universal, resp. existential quantification in ϕ. There are two ways to define these functions precisely. In the first way, one makes use of a formula $\phi^{(N)}$ logically equivalent to ϕ and in which $\sim$ is applied only to atomic formulas, called the <u>negation normal form</u> of ϕ; $\phi^{(N)}$ is defined inductively, using generalized DeMorgan's laws. Then put $j \in Un(\phi)$, resp. $j \in Ex(\phi)$, if $\bigwedge u(\ldots)$, resp. $\bigvee u(\ldots)$ occurs in ϕ for some variable u of sort j. The second way is directly by the following simultaneous inductive definition:

(i) $Un(\phi) = Ex(\phi) = 0$ for ϕ atomic;

(ii) $Un(\sim\phi) = Ex(\phi)$, $Ex(\sim\phi) = Un(\phi)$;

(iii) $Un(\sum K) = Un(\prod K) = \bigcup Un(\phi)[\phi \in K]$ and $Ex(\sum K) = Ex(\prod K) = \bigcup Ex(\phi)[\phi \in K]$;

(iv) $Un(\bigvee u\, \phi) = Un(\phi)$ and $Ex(\bigvee u\, \phi) = Ex(\phi) \cup \{j\}$ if u is of sort j;

(v) $Ex(\bigwedge u\, \phi) = Ex(\phi)$ and $Un(\bigwedge u\, \phi) = Un(\phi) \cup \{j\}$ if u is of sort j.

These functions are defined for sequents by taking

$Un(\Gamma \supset \Delta) = \bigcup Ex(\phi)[\phi \text{ is in } \Gamma] \cup \bigcup Un(\phi)[\phi \text{ is in } \Delta]$ and

$Ex(\Gamma \supset \Delta) = \bigcup Un(\phi)[\phi \text{ is in } \Gamma] \cup \bigcup Ex(\phi)[\phi \text{ is in } \Delta]$.

The proof of the following is along the same lines as 2.6.

4.1 <u>Lemma</u>. <u>Each of the functions</u> Fr, Cn, Rel, Fn, Sort, Tm-Sort, Un <u>and</u> Ex <u>is A-recursive when restricted to A-formulas and A-sequents</u>.

We can now express our first interpolation theorem here. A discussion of the sources for this result and of related results will be given following the proof of Theorem 4.3.

4.2 <u>Theorem</u>. <u>Assume that</u> σ <u>is a relational signature</u>. <u>Suppose</u> $(\phi \to \Psi)$ <u>has a cut-free</u> A-<u>derivation that</u> $Sort(\phi) \cap Ex(\Psi) \neq 0$ <u>or</u> $Sort(\Psi) \cap Un(\phi)$. <u>Then we can find an</u> A-<u>interpolant</u> θ, Ψ <u>with respect to</u> Fr, Cn, Rel <u>and</u> Sort <u>such that</u> $Un(\theta) \subseteq Un(\phi)$ <u>and</u> $Ex(\theta) \subseteq Ex(\Psi)$.

This is obtained directly from the next theorem for sequents. We make use in its formulation of the notion of a <u>mesh</u> of two sequences Γ and Γ' ; by this we mean any sequence having the terms of Γ and Γ' arbitrarily interspersed, but having the terms of Γ and Γ' arbitrarily interspersed, but otherwise maintaining their original order. In the following, $\Gamma \circ \Gamma'$ is taken to be any mesh of Γ and $\Delta \circ \Delta'$ is any mesh of Δ and Δ' .

4.3 <u>Theorem</u>. <u>Assume that</u> σ <u>is a relational signature</u>. <u>Suppose</u> $(\Gamma \circ \Gamma' \supset \Delta \circ \Delta')$ <u>has a cut-free</u> A-<u>derivation</u>. <u>Let</u> $w_o \in FV-[Fr(\Gamma \supset \Delta) \cup Fr(\Gamma' \supset \Delta')]$, <u>and let</u> j_o <u>be the sort of</u> w_o . <u>Then we can find an</u> A-<u>interpolant</u> θ <u>for</u> $(\Gamma \supset \Delta), (\Gamma' \supset \Delta')$ <u>with respect to</u> Cn, Rel <u>satisfying the following additional conditions</u>:

(i) $Fr(\theta) \subseteq Fr(\Gamma \supset \Delta) \cap Fr(\Gamma' \supset \Delta') \cup \{w_o\}$,

(ii) $Sort(\theta) \subseteq Sort(\Gamma \supset \Delta) \cap Sort(\Gamma' \supset \Delta') \cup \{j_o\}$, and

(iii) $Un(\theta) \subseteq Ex(\Gamma \supset \Delta)$ <u>and</u> $Ex(\theta) \subseteq Ex(\Gamma' \supset \Delta')$.

<u>Proof</u> of (4.2) from (4.3). Assume the hypotheses of (4.2). Suppose for example that $Sort(\phi) \cap Ex(\Psi) \neq 0$; in this case we take $j_o \in Sort(\phi) \cap Ex(\Psi)$, and w_o(of sort j_o) not free in ϕ or Ψ . Let $(\Gamma \supset \Delta) = (\phi \supset 0)$ and $(\Gamma' \supset \Delta') = (0 \supset \Psi)$. We see that $(\Gamma \circ \Gamma' \supset \Delta \circ \Delta')$ has a cut-free A-derivation. Let $\theta = \theta(w_o)$ satisfy the conclusion of this theorem for the given pair of sequents. Thus we have cut-free A-derivations of $(\phi \supset \theta(w_o))$ and $(\theta(w_o) \supset \Psi)$. Let $u_o \in BV$ be of sort j_o . Then it is easily seen that $\bigvee u_o\ \theta(u_o)$ is an A-interpolant for ϕ, Ψ with respect to Fr, Cn, Rel and Sort. Moreover,

$Un(\bigvee u_o\ \theta(u_o)) = Un(\theta) \subseteq Ex(\phi \supset 0) = Un(\phi)$, and

$Ex(\bigvee u_o\ \theta(u_o)) = Ex(\theta) \cup \{j_o\} \subseteq Ex(0 \supset \Psi) \cup \{j_o\} = Ex(\Psi) \cup \{j_o\} = Ex(\Psi)$.

when $Sort(\Psi) \cap Un(\phi) \neq 0$ we take $j_o \in Sort(\Psi) \cap Un(\phi)$ and $\bigwedge u_o\ \theta(u_o)$ instead.

<u>Proof</u> of (4.3). We proceed by induction on a cut-free A-derivation $\mathcal{D}$ with conclusion $\Gamma \circ \Gamma' \supset \Delta \circ \Delta'$. With no loss of generality we can assume that w_o is not free in any sequent of $\mathcal{D}$.

<u>Case</u> I. $\Gamma \circ \Gamma' \supset \Delta \circ \Delta'$ is an axiom .

<u>Subcase</u> I(a). The axiom has the form $(0 \supset t = t)$. Then $\Gamma = \Gamma' = 0$ and one of Δ, Δ' is $(t=t)$, the other is empty. In either case $Fr(\Gamma \supset \Delta) \cap Fr(\Gamma' \supset \Delta') = 0$. In the first case we want derivations of $(0 \supset t=t,\ \theta)$ and $(\theta \supset 0)$; here we take θ to be $\sim (w_o = w_o)$. In the second case we take θ to be $(w_o = w_o)$.

<u>Subcase</u> I(b). The axiom has the form

$$E(t_1,\ldots,t_k\ ;\ t_1',\ldots,t_k')\ ,\ \emptyset(t_1,\ldots,t_k) \supset \emptyset(t_1',\ldots,t_k')$$

where $\emptyset$ is atomic and each term is a variable or a constant. There are a number of possibilities to consider. We shall just treat one example, leaving the details of the full subcase to the reader. Suppose the axiom has the form

$$[t_1 = t_{11}\ ,\ t_{11} = t_{12}\ ,\ t_{13} = t_{12}\ ,\ t_{13} = t_1'\ ,\ t_2 = t_2'\ ,\ r(t_1,t_2) \supset r(t_1',t_2')]$$

and that $(\Gamma \supset \Delta)$ is $[t_1 = t_{11}\ ,\ t_{13} = t_1', r\ (t_1,t_2) \supset 0]$

and that $(\Gamma' \supset \Delta')$ is $[t_{11} = t_{12}\ ,\ t_{13} = t_{12}, t_2 = t_2' \supset r\ (t_1',t_2')]$

The common terms are t_{11} , t_{13} , t_1' and t_2 . We want θ which just contains these (as free variables or constants) for which we can derive

$[t_1 = t_{11},\ t_{13} = t_1',\ r(t_1,t_2) \supset \theta]$ and $[t_{11} = t_{12},\ t_{13} = t_{12},\ t_2 = t_2', \theta \supset r(t_1',t_2')]$.

We take θ to be $[t_{11} = t_{13} \rightarrow r(t_1',t_2)]$.

We can now assume $od(\mathcal{D}) > 1$; the last rule of $\mathcal{D}$ is one of (R_1) - (R_{13}) .

<u>Case</u> II. The last rule of $\mathcal{D}$ is (S) . Then $(\Gamma \circ \Gamma' \supset \Delta \circ \Delta')$ is deduced from $(\Gamma^* \supset \Delta^*)$ with the same sets of formulas on each side. Hence Γ^* is a mesh of Γ_1 , Γ_1' and Δ^* is a mesh of Δ_1, Δ_1' , where we can apply

$$(S)\ \frac{\Gamma_1 \supset \Delta_1}{\Gamma \supset \Delta} \qquad\qquad (S)\ \frac{\Gamma_1' \supset \Delta_1'}{\Gamma' \supset \Delta'}$$

By inductive hypothesis, choose θ to satisfy the conditions of the theorem for $(\Gamma_1 \circ \Gamma_1' \supset \Delta_1 \circ \Delta_1')$; then the same θ works for $(\Gamma \circ \Gamma' \supset \Delta \circ \Delta')$.

Case III. The last rule of $\mathfrak{D}$ is weakening. Suppose it is $(\supset W)$ with active formula ϕ . Since ϕ is the rightmost formula of $\Delta \circ \Delta'$, it is either the rightmost formula of Δ or of Δ' ; we shall say in similar cases below that it is counted with Δ or with Δ' , resp. Suppose it is counted with Δ' . Then $\Delta' = \Delta_1' , \phi$ and $\Delta \circ \Delta' = \Delta \circ \Delta_1' , \phi$. By inductive hypothesis we have an A-interpolant θ for $(\Gamma \supset \Delta)$ and $(\Gamma' \supset \Delta_1')$ satisfying the additional conditions. Since $(\Gamma', \theta \supset \Delta_1' , \phi)$ follows from $(\Gamma', \theta \supset \Delta_1')$ by $(\supset W)$, the same θ continues to work. The other possibilities are treated similarly.

Case IV. The last rule of $\mathfrak{D}$ is $(\sim \supset)$ or $(\supset \sim)$ with active formula $\sim \phi$. Suppose it is the latter and that $\sim \phi$ is counted with Δ' , so $\Delta' = \Delta_1' , \sim \phi$. $\mathfrak{D}$ has the form

$$\vdots$$
$$(\supset \sim)\ \frac{\Gamma \circ \Gamma' , \phi \supset \Delta \circ \Delta_1'}{\Gamma \circ \Gamma' \supset \Delta \circ \Delta_1' , \sim \phi}$$

We are free to consider $\Gamma \circ \Gamma' , \phi$ as $\Gamma \circ (\Gamma', \phi)$ or $(\Gamma, \phi) \circ \Gamma'$. In this case we take the former. By inductive hypothesis we can find an A-interpolant θ for $(\Gamma \supset \Delta)$ and $(\Gamma', \phi \supset \Delta_1')$ satisfying the additional conditions. Since we can infer $(\Gamma', \theta \supset \Delta_1', \sim \phi)$ from $(\Gamma', \phi, \theta \supset \Delta_1')$, the same θ works. The other possibilities here are treated in the same way.

Case V. The last rule of $\mathfrak{D}$ is $(\supset \sum)$, with active formula $\sum K$, which we suppose counted with Δ' . $\mathfrak{D}$ has the form

$$\vdots$$
$$(\supset \textstyle\sum)\ \frac{\Gamma \circ \Gamma' \supset \Delta \circ \Delta_1' , \Psi}{\Gamma \circ \Gamma' \supset \Delta \circ \Delta_1' , \sum K}$$

where Ψ is some member of K . We apply inductive hypothesis to find an A-interpolant for $(\Gamma \supset \Delta)$ and $(\Gamma' \supset \Delta_1', \Psi)$ satisfying the additional conditions. The same θ works for $(\Gamma \supset \Delta)$ and $(\Gamma' \supset \Delta_1', \sum K)$. Moreover, the same argument works when $\sum K$ is counted with Δ.

Case VI. The last rule of $\mathcal{D}$ is $(\sum \supset)$, with active formula $\sum K$.

Subcase VIa. $\sum K$ is counted with $\Gamma = \Gamma_1, \sum K$. $\mathcal{D}$ has the form

$$\begin{array}{c} \vdots\ (\phi \in K) \\ \downarrow \end{array}$$

$$(\textstyle\sum \supset)\ \frac{\Gamma_1 \circ \Gamma',\ \phi \supset \Delta \circ \Delta'}{\Gamma_1 \circ \Gamma',\ \sum K \supset \Delta \circ \Delta'}$$

By inductive hypothesis, for each $\phi \in K$ there exists a formula θ which is an A-interpolant for $(\Gamma_1, \phi \supset \Delta)$ and $(\Gamma' \supset \Delta')$ satisfying the additional conditions of the theorem. Considering $\Gamma_1, \Delta, \Gamma', \Delta'$ as fixed we indicate this by writing $Int_A(\theta,\phi)$. It is seen from (2.6) and (4.1) that the relation Int_A is $\sum_1^{(A)}$-definable. Since $\mathcal{D}$ is an A-derivation, each formula in $\mathcal{D}$ belongs to A; in particular, $\sum K$ is an A-formula and $K \in A$. Hence by the $\sum_1$-axiom of choice for A there is a function F with domain K such that for each $\phi \in K$, $Int_A(F(\phi),\phi)$. Let $K^* = \mathcal{R}(F)$; then also $K^* \in A$. Now we have for each $\phi \in K$ cut-free A-derivations of

$$[\Gamma_1, \phi \supset \Delta, F(\phi)] \text{ and } [\Gamma', F(\phi) \supset \Delta']$$

Then we can infer $[\Gamma_1, \phi \supset \Delta, \sum K^*]$ for each $\phi \in K$ and hence $[\Gamma_1, \sum K \supset \Delta, \sum K^*]$. On the other hand, since we have a cut-free A-derivation of $[\Gamma', \theta \supset \Delta']$ for each $\theta \in K^*$, we also get one of $[\Gamma', \sum K^* \supset \Delta']$. It is then seen that $\sum K^*$ is a suitable A-interpolant for $(\Gamma_1, \sum K \supset \Delta)$ and $(\Gamma' \supset \Delta')$.

Subcase VIb. $\sum K$ is counted with $\Gamma' = \Gamma_1', \sum K$. $\mathcal{D}$ has the form

$$\begin{array}{c} \vdots\ (\phi \in K) \\ \downarrow \end{array}$$

$$(\textstyle\sum \supset)\ \frac{\Gamma \circ \Gamma_1',\ \phi \supset \Delta \circ \Delta'}{\Gamma \circ \Gamma_1',\ \sum K \supset \Delta \circ \Delta'}$$

As in VIa we can obtain a function F with $\mathcal{D}(F) = K$, $\mathcal{R}(F) = K^* \in A$ and $F(\emptyset)$ an A-interpolant for $(\Gamma \supset \Delta)$ and $(\Gamma_1', \emptyset \supset \Delta)$ for each $\emptyset \in K$. This leads to cut-free A-derivations

$$(\supset\prod)\ \frac{\Gamma \supset \Delta,\ \theta \quad (\theta \in K^*)}{\Gamma \supset \Delta, \prod K^*} \qquad (S,\sum\supset)\ \frac{\dfrac{\Gamma_1',\ \emptyset, F(\emptyset) \supset \Delta'}{\Gamma_1',\ \emptyset, \prod K^* \supset \Delta'}\ (\text{each } \emptyset \in K)}{\Gamma_1',\ \sum K, \prod K^* \supset \Delta'}$$

Hence $\prod K^*$ is a suitable interpolant in this case.

<u>Case</u> VII. The last rule of $\mathcal{D}$ is $(\prod\supset)$. Dual to Case V.

<u>Case</u> VIII. The last rule of $\mathcal{D}$ is $(\supset\prod)$. Dual to Case VI.

<u>Case</u> IX. The last rule of $\mathcal{D}$ is $(\supset\bigvee)$ with active formula $\bigvee u\ \emptyset(u)$.

<u>Subcase</u> IXa. $\bigvee u\ \emptyset(u)$ is counted with $\Delta' = \Delta_1', \bigvee u\ \emptyset(u)$. $\mathcal{D}$ has the form

$$(\supset\bigvee)\ \frac{\Gamma \circ \Gamma' \supset \Delta \circ \Delta_1',\ \emptyset(t)}{\Gamma \circ \Gamma' \supset \Delta \circ \Delta_1', \bigvee u\ \emptyset(u)}$$

where t is a variable or constant of the same sort as u. Inductive hypothesis gives an A-interpolant θ for $(\Gamma \supset \Delta)$, $(\Gamma' \supset \Delta_1', \emptyset(t))$ satisfying the additional conditions. Then we have cut-free A-derivations of

$$(\Gamma \supset \Delta,\ \theta) \text{ and } (\Gamma',\ \theta \supset \Delta_1',\ \emptyset(t)).$$

From the latter we infer $(\Gamma',\ \theta \supset \Delta_1', \bigvee u\ \emptyset(u))$. If t is a constant, it is seen that the conditions on Fr and Cn as well as the other syntactic functions continue to apply, and we can take the same θ. However, if t is a variable w, the condition on Fr need not continue to apply. We know that

$$Fr(\theta) \subseteq Fr(\Gamma \supset \Delta) \cap Fr(\Gamma' \supset \Delta_1',\ \emptyset(w)) \cup \{w_0\}.$$

If $Fr(\theta) \not\subseteq Fr(\Gamma \supset \Delta) \cap Fr(\Gamma' \supset \Delta_1', \bigvee u\, \emptyset(u)) \cup \{w_o\}$, this can only mean, that $w \in Fr(\theta)$ and hence $w \in Fr(\Gamma \supset \Delta)$, but $w \notin Fr(\Gamma' \supset \Delta_1', \bigvee u\, \emptyset(u))$. Write $\theta = \theta(w)$. Then from $(\Gamma \supset \Delta, \theta(w))$ we infer $(\Gamma \supset \Delta, \bigvee u\, \theta(u))$ and from $(\Gamma', \theta(w) \supset \Delta_1', \bigvee u \emptyset(u))$ we can infer $(\Gamma', \bigvee u\, \theta(u) \supset \Delta_1', \bigvee u\, \emptyset(u))$ by $(\bigvee \supset)$, because the free variable condition is satisfied. Now one must check the functions Sort, Un and Ex. $Sort(\bigvee u\, \theta(u)) = Sort(\theta)$ since $w \in Fr(\theta)$, and $Un(\bigvee u\, \theta(u)) = Un(\theta)$. On the other hand, $Ex(\bigvee u\, \theta(u)) = Ex(\theta) \cup \{j\}$ where j is the sort of u; it is the case that $j \in Ex(\Gamma' \supset \Delta_1', \bigvee u\, \theta(u)) = Ex(\Gamma' \supset \Delta')$. Hence $\bigvee u\, \theta(u)$ is a suitable A-interpolant in this case.

Subcase IXb. $\bigvee u\, \emptyset(u)$ is counted with $\Delta = \Delta_1, \bigvee u\, \emptyset(u)$. In this case it can be seen that either the same θ continues to work or, dually to the preceding argument, $\bigwedge u\, \theta(u)$ works. Then $Un(\bigwedge u\, \theta(u)) = Un(\theta) \cup \{j\}$ and $j \in Ex(\Gamma \supset \Delta_1, \bigvee u\, \theta(u))$ as required.

Case X. The last rule of $\mathcal{D}$ is $(\bigvee \supset)$ with active formula $\bigvee u\, \emptyset(u)$. Suppose $\bigvee u\, \emptyset(u)$ is counted with $\Gamma' = \Gamma_1', \bigvee u\, \emptyset(u)$. $\mathcal{D}$ has the form

$$\vdots$$

$$(\bigvee \supset)\ \frac{\Gamma \circ \Gamma_1', \emptyset(w) \supset \Delta \circ \Delta'}{\Gamma \circ \Gamma_1', \bigvee u\, \emptyset(u) \supset \Delta \circ \Delta'}$$

where w is of the same sort as u and $w \notin Fr(\Gamma \circ \Gamma_1', \bigvee u\, \emptyset(u) \supset \Delta \circ \Delta')$. By inductive hypothesis, choose an A-interpolant θ for $(\Gamma \supset \Delta)$, $(\Gamma_1', \emptyset(w) \supset \Delta')$ satisfying the additional conditions. Then we have a cut-free A-derivation of $(\Gamma_1', \emptyset(w), \theta \supset \Delta')$ and $Fr(\theta) \subseteq Fr(\Gamma \supset \Delta) \cap Fr(\Gamma_1', \emptyset(w) \supset \Delta') \cup \{w_o\}$. $w \notin Fr(\theta)$ for otherwise $w \in Fr(\Gamma \supset \Delta)$ (recall that w_o was chosen so that w_o is distinct from w). Hence we can still make the inference $(\bigvee \supset)$ to $(\Gamma_1', \bigvee u\, \emptyset(u), \theta \supset \Delta')$. Thus the same θ continues to work here.

The argument is similar if $\bigvee u\, \emptyset(u)$ is counted with Γ.

Case XI. The last rule of $\mathcal{D}$ is $(\bigwedge \supset)$. Dual to Case IX.

Case XII. The last rule of $\mathcal{D}$ is $(\supset \bigwedge)$. Dual to Case X.

This completes the proof of the theorem.

In the following remarks we examine special cases, variants and (partial) generalizations of (4.2), in particular as these relate to work on interpolation theorems in the literature.

First observe that if $Rel(\phi) \cap Rel(\psi)$ contains at least one relation symbol r_{i_0} and the equality symbol is neither in ϕ nor in ψ then we can choose an A-interpolant θ satisfying the conditions of (4.2) and which also does not contain the equality symbol. In other words, we get an interpolation theorem for the predicate calculus in $\mathcal{L}_A$ without equality. To see this, note first that by the subformula property, if $\mathcal{D}$ is a cut-free derivation of a sequent which does not contain the symbol $(=)$ then no sequent in $\mathcal{D}$ contains $(=)$; in this case the only axioms in $\mathcal{D}$ are of the form $(\phi \supset \phi)$, where ϕ is an atomic formula $r_i(t_1,\ldots,t_{m_i})$. The theorem here is then obtained from a corresponding modification of (4.3), with θ not containing the $(=)$ symbol but $Rel(\theta) \subseteq Rel(\Gamma \supset \Delta) \cap Rel(\Gamma' \supset \Delta') \cup \{r_{i_0}\}$. However, if the equality symbol is in ϕ or ψ it may have to occur in an interpolant even if $Rel(\phi) \cap Rel(\psi) \neq 0$. (As an example in the single sorted case with unary relation symbols r and s , take ϕ to be $\bigwedge u_1 \bigwedge u_2 (u_1 = u_2) \wedge \bigwedge u[s(u) \vee \sim s(u)]$ and ψ to be $[\bigvee u\; r(u) \rightarrow \bigwedge u\; r(u)] \wedge \bigwedge u[s(u) \vee \sim s(u)]$.)

In the single-sorted case, where the function Sort is trivial, we can reduce the interpolation theorem for the predicate calculus with equality PCE to the one for the PC just described, if we also omit the conditions on Un and Ex. For let $Eq_1^{(\phi)}$ be the conjunction of the members χ of Eq_1 with $Rel(\chi) \subseteq Rel(\phi)$. Then from a derivation of $(\phi \rightarrow \psi)$ in the PCE we get a derivation of $[(Eq_1^{(\phi)} \wedge \phi) \rightarrow (Eq_1^{(\psi)} \rightarrow \psi)]$ in the PC . We now treat $(=)$ simply as a binary relation symbol and apply the result for PC .

A similar argument works for the <u>restricted</u> many-sorted case. Here one assigns each relation symbol r_i a specified sequence $\rho_i = (j_1,\ldots,j_{k_i})$ of sorts. The restricted atomic formulas are of the form $(t_1 = t_2)$ where t_1, t_2 are of the same sort and $r_i(t_1,\ldots,t_{k_i})$ where the sequence of sorts of $t_1,\ldots,t_{k_i}$ is ρ_i . The restricted formulas are then built up in the usual way from these atomic formulas. Then one can reduce an interpolation theorem with respect to Fr, Cn, Rel and Sort for the restricted many-sorted case in PCE to that in PC . However, it does not seem possible to make such a simple reduction of PCE to PC for the unrestricted many-sorted language which we use here.

It would be natural to try to reduce (4.2) (with or without mention of Un and Ex) to the corresponding result for the single-sorted calculus. One would begin with the standard association of a single-sorted formula $\emptyset^*$ with each $\emptyset$, got by the adjunction of new unary predicate symbols m_j $(j \in J)$ to the language and relativization of quantifiers. That is, each occurrence of a quantifier $\bigvee u(\ldots)$ in $\emptyset$ with u of sort j is replaced by $\bigvee u(m_j(u) \wedge \ldots)$, while each occurrence of $\bigwedge u(\ldots)$ is replaced by $\bigwedge u(m_j(u) \to \ldots)$. Then from a derivation of $(\emptyset \to \Psi)$ one gets a derivation of

$$[\prod_{j \in \mathrm{Sort}(\emptyset)} \bigvee u\, m_j(u) \wedge \emptyset^*] \to [\prod_{j \in \mathrm{Sort}(\Psi)} \bigvee u\, m_j(u) \to \Psi^*]$$

in the single-sorted calculus. An interpolant θ_1 here (at least with respect to Rel) would have $\mathrm{Rel}(\theta_1) \subseteq \mathrm{Rel}(\emptyset) \cap \mathrm{Rel}(\Psi) \cup \{m_j\colon j \in \mathrm{Sort}(\emptyset) \cap \mathrm{Sort}(\Psi)\}$. We could then recapture (4.2) if we knew that $\theta_1 = \theta^*$ for some θ . However, I don't know any argument in the single-sorted case which would insure this.

The hypothesis that σ is relational can be eliminated in (4.2) if we ignore Un and Ex . To be precise, one can obtain the following result. <u>Suppose</u> $(\emptyset \to \Psi)$ <u>has a cut-free A-derivation and that</u> $\mathrm{Tm\text{-}Sort}(\emptyset) \cap \mathrm{Tm\text{-}Sort}(\Psi) \neq 0$; <u>then we can find an A-interpolant for</u> $\emptyset, \Psi$ <u>with respect to</u> Fr, Cn, Fn, Rel <u>and</u> Tm-Sort. This is seen by the standard elimination of function symbols in favor of relation symbols. First one reduces all occurrences of terms to occurrences in atomic formulas of the simple form $f_i(t_1,\ldots,t_{m_i}) = t$ where $t_1,\ldots,t_{m_i}$, t are variables or constants; these are then replaced by atomic formulas $\bar{f}_i(t_1,\ldots,t_{m_i},t)$ where $\bar{f}_i$ is a new (m_i+1)-ary relation symbol associated with f_i . For each formula $\emptyset$, let $\bar{\emptyset}$ be the associated formula without function symbols. Then one applies (4.2) to

$$[\prod_{f_i \in \mathrm{Fn}(\emptyset)} \bigwedge u_1 \ldots \bigwedge u_{m_i} \bigvee! \, w\, \bar{f}_i(u_1,\ldots,u_{m_i},w) \wedge \bar{\emptyset}] \to$$

$$\to [\prod_{f_i \in \mathrm{Fn}(\Psi)} \bigwedge u_1 \ldots \bigwedge u_{m_i} \bigvee! \, w\, \bar{f}_i(u_1,\ldots,u_{m_i},w) \to \bar{\Psi}] \, .$$

(The sorts of the variables in the unicity hypotheses are restricted in accordance with the sequences γ_i in the signature σ.) Note that $Un(\emptyset_1)$ for the hypothesis $\emptyset_1$ of this implication may be larger than $Un(\emptyset)$; similarly for Ex. I don't know whether (4.2) including the conditions on Un and Ex holds for arbitrary σ when we replace Sort by Tm-Sort (or if there is any simple generalization of (4.2) without restriction on σ) .

For comparison with the literature, consider (4.2) for the single-sorted case, again ignoring Un and Ex . For A = HF this is simply Craig's interpolation theorem[17] [Cr1] , which was the first interpolation theorem stated. Craig's own demonstration of this was proof-theoretic, using a special system of natural deduction which he had developed. Craig's theorem turned out to be equivalent to a joint consistency theorem found independently by A. Robinson. A variety of model-theoretic proofs have been given for Craig's and Robinson's theorems; we mention only the different ones to be found in the texts of Robinson [R], Ch. 5, Kreisel-Krivine [K,K], Chs. 2-3 and in the paper of Keisler [Ka2]. The theorem for the case A = HC was obtained by Lopez-Escobar [LE1]. The case A = HH was announced in Feferman-Kreisel [F,K]. The general case was then treated by Barwise [Bw]. His proof uses an idea of Malitz [Ml] to restrict attention to formulas in negation normal form, for which a complete Gentzen calculus can be given. This avoids the partition problems of our argument for (4.3), and is thus simpler. (The proof here is along the lines given by Maehara [Mh].)

Next, consider the statement (4.2) including Un and Ex but still for the single-sorted case. This was obtained by Barwise [Bw]. This also generalized an interpolation theorem for universal formulas (defined below) in $\mathcal{L}_{HC}$ due to Malitz [Ml].

Now consider (4.2) for the full many-sorted case, but ignoring Un and Ex . This was obtained by the author for A = HF, HH and HC (announced in [F,K]), by a proof-theoretical argument similar to that given here for (4.3). A model-theoretic argument for the case A = HF can be found in [K,K], Ch. 5. Recently, Makkai [Mk]

[17] In this and further citations, we assume the statements of theorems are recast in terms of validity (if not already in this form) using completeness.

has found new proofs of this and other interpolation results for $A = HC$ based on methods of [Hn] and [Sm] for $\mathcal{L}_{HF}$.[18] Of course, (4.2) as given here involved a simple combination of the Un-Ex theorem from [Bw] with the many-sorted theorem.

There are various interpolation theorems for $\mathcal{L}_{HF}$ which are not directly generalized by (4.2) or its variants discussed above. Scott [S-2] gives a recent survey of these. We mention particularly the one due to Lyndon [Ly1]. It is noteworthy because of its model-theoretic applications (cf. §5(a) below). In place of Rel, the theorem treats the functions $Rel^{+}(\emptyset)$ and $Rel^{-}(\emptyset)$ which give the sets of relation symbols with at least one positive, resp. negative, occurrence in $\emptyset$. This has been shown to generalize to $\mathcal{L}_{HC}$ in [LE1] ; it is easily seen that it can be extended just as well to the $\mathcal{L}_A$ considered here.

As has already been indicated, sharper results on the form of an interpolant can be obtained when the syntactic form of the initial formulas is suitably restricted. We consider now formulas of least complexity, from the point of view of quantification. Let $J_o \subseteq J$. A formula $\emptyset$ is said to be existential/J_o , resp. universal/J_o , if $Un(\emptyset) \subseteq J_o$, resp $Ex(\emptyset) \subseteq J_o$. $\emptyset$ is said to be quantifier-free/J_o if it has both these properties. (When $J_o = 0$ we simply refer to existential, universal, and quantifier-free formulas.) Note that if $\emptyset(u)$ has any one of these three properties the same property holds of $\emptyset(t)$. Furthermore, if $\emptyset$ is of one of the forms $\bigvee u\, \Psi$, $\bigwedge u\, \Psi$, $\sum K$, $\prod K$ and $\emptyset$ has one of these three properties the same property holds of Ψ in the first two cases, and of each $\Psi \in K$ in the latter two. However, if $\emptyset = \sim \Psi$ and $\emptyset$ is existential/J_o (universal/J_o) then Ψ is universal/J_o (existential/J_o) .

In $\mathcal{L}_{HF}$ every existential formula Ψ is equivalent to a formula with the same symbols and free variables in prenex form $\bigvee u_1 \dots \bigvee u_n \Psi'$ where Ψ' is quantifier-free; similarly for universal formulas. By Herbrand's work [Hk], there is a derivation of $\bigvee u_1 \dots \bigvee u_n\, \chi(u_1,\dots,u_n)$ where χ is quantifier-free, if and only if there are terms $t_1^{(i)},\dots,t_n^{(i)}$ $(1 \leq i \leq m)$ and a derivation of $\sum_{1 \leq i \leq m} \chi(t_1^{(i)},\dots,t_n^{(i)})$.
(A model-theoretic proof of this result can be found in [K,K], where it is called the uniformity theorem.) Now if Ψ is existential, equivalent as above to

18) Makkai has informed me that his arguments also extend to the $\mathcal{L}_A$ considered here, making use only of the fact that the set of A-formulas which are valid is A-recursively enumerable.

$\bigvee u_1 \ldots \bigvee u_n\ \Psi'(u_1,\ldots,u_n)$ and $\emptyset$ is universal, equivalent to $\bigwedge u_{n+1} \ldots \bigwedge u_{n+k}\ \emptyset'(u_{n+1},\ldots,u_{n+k})$, then from a derivation of $(\emptyset \rightarrow \Psi)$ we can get a derivation of

$$\bigvee u_1 \ldots \bigvee u_n \bigvee u_{n+1} \ldots \bigvee u_{n+k}\ [\sim \emptyset'(u_{n+1},\ldots,u_{n+k}) \vee \Psi'(u_1,\ldots,u_n)] .$$

Then by Herbrand's uniformity theorem we can obtain a derivation of

$$\sum_{1 \leq i \leq m} [\sim \emptyset'(t_{n+1}^{(i)},\ldots,t_{n+k}^{(i)}) \vee \Psi'(t_1^{(i)},\ldots,t_n^{(i)})]$$

for suitable terms $t_j^{(i)}$. Let θ be $\sum_{1 \leq i \leq m} \Psi'(t_1^{(i)},\ldots,t_n^{(i)})$. θ is quantifier-free and we can derive $(\emptyset \rightarrow \theta)$ and $(\theta \rightarrow \Psi)$. This result was generalized to $\mathcal{L}_A$ by Barwise [Bw] for $J_o = 0$. With hardly any more trouble one obtains the following for any J_o .

4.4 <u>Theorem:</u> <u>Suppose that</u> $(\emptyset \rightarrow \Psi)$ <u>has a cut-free</u> A-<u>derivation, where</u> $\emptyset$ <u>is universal/</u>J_o <u>and</u> Ψ <u>is existential/</u>J_o . <u>Then we can find an</u> A-<u>interpolant</u> θ <u>for</u> $\emptyset, \Psi$ <u>with respect to</u> Rel <u>which is quantifier-free/</u>J_o .

Note that it is not necessary to assume that σ is relational here. Note also that we cannot include the function Fr (since the language may not include quantifier-free sentences).[19] As before, this is deduced from a corresponding result for sequents. We call a sequent $(\Gamma \supset \Delta)$ <u>existential/</u>J_o if each formula in Γ is universal/J_o and each formula in Δ is existential/J_o . Then (4.4) is an immediate corollary of the following.

4.5 <u>Theorem.</u> <u>Suppose that</u> $(\Gamma \circ \Gamma \supset \Delta \circ \Delta')$ <u>has a cut-free</u> A-<u>derivation where both</u> $(\Gamma \supset \Delta)$ <u>and</u> $(\Gamma' \supset \Delta')$ <u>are existential/</u>J_o . <u>Then we can find an</u> A-<u>interpolant</u> θ <u>for</u> $(\Gamma \supset \Delta)$, $(\Gamma' \supset \Delta')$ <u>with respect to</u> Rel <u>which is quantifier-free/</u>J_o .

---

[19] Also, we cannot expect to include Cn and Fn . As the argument above for $\mathcal{L}_{HF}$ indicates, there may be terms in θ which do not occur in $\emptyset$ or Ψ . We could include Tm-Sort by adding the hypothesis Tm-Sort($\emptyset$) $\cap$ Tm-Sort(Ψ) $\neq 0$; the only applications of (4.4) we know (cf. §5) do not use different sorts in any essential way.

Proof. By induction on a cut-free derivation $\mathcal{D}$ of $(\Gamma \circ \Gamma' \supset \Delta \circ \Delta')$. By the subformula properties mentioned above, each sequent in $\mathcal{D}$ is also existential/J_o. Consider, particularly, an inference step for $\sim$, say

$$\frac{\Gamma_1, \Psi \supset \Delta_1}{\Gamma_1 \supset \Delta_1, \sim \Psi}$$

If the conclusion is existential/J_o then Ψ is universal/J_o, so the hypothesis is also existential/J_o.

Now the inductive argument proceeds just as the proof of (4.3) for the cases that $(\Gamma \circ \Gamma' \supset \Delta \circ \Delta')$ is an axiom or the last rule of $\mathcal{D}$ is one of the rules (W), (S), or for $\sim$, $\sum$, $\prod$. The only cases where a slightly different strategy is required are those for $\bigvee$ and $\bigwedge$. We consider only those for $\bigvee$, the others being dual.

Suppose the last rule in $\mathcal{D}$ is $(\supset \bigvee)$ with active formula $\bigvee u\, \phi(u)$, counted with $\Delta' = \Delta_1', \bigvee u\, \phi(u)$. $\mathcal{D}$ has the form

$$\begin{array}{c}\vdots\\ (\supset \bigvee)\ \dfrac{\Gamma \circ \Gamma' \supset \Delta \circ \Delta_1', \phi(t)}{\Gamma \circ \Gamma' \supset \Delta \circ \Delta_1', \bigvee u\, \phi(u)}\end{array}$$

From derivations of $[\Gamma \supset \Delta, \theta]$ and $[\Gamma', \theta \supset \Delta_1', \phi(t)]$ we get a derivation of $(\Gamma', \theta \supset \Delta_1', \bigvee u\, \phi(u))$. Thus the same θ continues to work. The same argument applies when $\bigvee u\, \phi(u)$ is counted with Δ.

Now suppose the last rule in $\mathcal{D}$ is $(\bigvee \supset)$ with active formula $\bigvee u\, \phi(u)$, counted with $\Gamma' = \Gamma_1', \bigvee u\, \phi(u)$. Let j be the sort of u; then we must have $j \in J_o$. $\mathcal{D}$ has the form

$$\begin{array}{c}\vdots\\ (\bigvee \supset)\ \dfrac{\Gamma \circ \Gamma_1', \phi(u) \supset \Delta \circ \Delta'}{\Gamma \circ \Gamma_1', \bigvee u\, \phi(u) \supset \Delta \circ \Delta'}\end{array}$$

where w is of sort j and w is not free in $[\Gamma \circ \Gamma_1', \bigvee u\, \phi(u) \supset \Delta \circ \Delta']$. By inductive hypothesis find an A-interpolant θ for $[\Gamma \supset \Delta]$ and $[\Gamma_1', \phi(w) \supset \Delta']$

with respect to Rel . Now w may be free in θ ; write $\theta = \theta(w)$. We have A-derivations of $[\Gamma \supset \Delta, \theta(w)]$ and $[\Gamma_1', \emptyset(w), \theta(w) \supset \Delta']$. We can then make the following inferences:

$$(\supset \bigwedge)\ \frac{\Gamma \supset \Delta,\ \theta(w)}{\Gamma \supset \Delta,\ \bigwedge u\ \theta(u)} \qquad \begin{array}{c} (\bigwedge \supset)\ \dfrac{\Gamma_1',\ \emptyset(w),\ \theta(w) \supset \Delta'}{\Gamma_1',\ \emptyset(w),\ \bigwedge u\ \theta(u) \supset \Delta'} \\ (S, \bigvee\supset)\ \overline{\overline{\Gamma_1',\ \bigvee u\ \emptyset(u),\ \bigwedge u\ \theta(u) \supset \Delta'}} \end{array}$$

The first of these is permitted since $w \notin Fr(\Gamma \supset \Delta)$. Since $j \in J_0$, $\bigwedge u\ \theta(u)$ is quantifier-free/J_0 and hence provides the required interpolant. In case $\bigvee u\ \emptyset(u)$ is counted with Γ we proceed similarly, except now one must take $\bigwedge u\ \theta(u)$ as the new interpolant. This completes the proof.

Correction to §4.

In the statements 4.2, 4.3, on p. 4-3 either:

(a) omit Cn , or (b) change Sort to Tm-Sort.

In case (a) the arguments are unaffected. In case (b), note that since σ is relational Tm-Sort($\emptyset$) = {j: a variable or constant of sort j occurs in $\emptyset$}. Now the argument p. 4-7, Subcase IXa, must be modified if t is a constant which occurs in θ but not in $(\Gamma' \supset \Delta')$. Write $\theta = \theta(t)$, and let $w \in FV$ be of the same sort as u but not in $\mathfrak{D}$. Then we can derive $(\Gamma \supset \Delta,\ \theta(t))$ and $(\Gamma', \theta(w) \supset \Delta_1', \emptyset(t))$ and hence $(\Gamma \supset \Delta,\ \bigvee u\ \theta(u))$ and $(\Gamma',\ \bigvee u\ \theta(u) \supset \Delta_1',\ \bigvee u\ \emptyset(u))$. Subcase IXb p. 4-8 is modified similarly, in this case taking $\bigwedge u\ \theta(u)$ for the new interpolant.

These corrections do not affect the applications of 4.2 in §5.

§5. Applications to model theory.

5(a) Persistent and invariant formulas for extensions. Interpolation theorems have proved to be a powerful tool to derive certain model-theoretic results. Craig [Cr2] used his own theorem to get a quick proof of Beth's definability theorem [Be]. Lyndon used his sharpened version to characterize those properties expressed in $\mathcal{L}_{HF}$ which are preserved under homomorphism [Ly2]. Using his extensions of these interpolation theorems to $\mathcal{L}_{HC}$, Lopez-Escobar was able to derive the corresponding model-theoretic results for $\mathcal{L}_{HC}$ in the same way. Barwise's interpolation theorems permit the immediate generalization of these to the languages $\mathcal{L}_{A}$.

The problem which Lyndon solved belongs to a general area of model theory which has received considerable attention. The concern is to syntactically characterize those properties (expressed in a given language) which are preserved under, or persistent for, various relations between and operations on structures, e.g. under substructures, extensions, homomorphisms, direct products, unions of chains, etc. The first and paradigm of such results was obtained independently by Łos [Ls] and Tarski [Tr]: a formula $\emptyset$ of $\mathcal{L}_{HF}$ is preserved under (passage to) substructures if and only if there is a universal formula θ (of $\mathcal{L}_{HF}$) such that $\Vdash(\emptyset \leftrightarrow \theta)$. This immediately yields the characterization of the formulas of $\mathcal{L}_{HF}$ which are preserved under extensions, namely those equivalent to existential formulas. The arguments of Łos and Tarski were model-theoretic, principally applying compactness and the method of diagrams of models.

Subsequently model-theoretic arguments have been applied very successfully to recapture the interpolation theorems for $\mathcal{L}_{HF}$ and to solve many of the preservation problems. The works of Keisler [Ks1], [Ks2], [Ks3] alone give three quite distinct such developments, almost coextensive in their applications. However, because of the failure of compactness for $\mathcal{L}_{HC}$ and lack so far of any suitable generalization of the tool of ultraproducts, it has not been possible to extend the model-theoretic arguments to this and other infinitary languages.[20] Following the lead of [LE1], proof-theoretical arguments have received renewed attention.

---

[20] There is one exception that I know of ; cf. ftn. 21)

Malitz [Mℓ] generalized the Łoś-Tarski result to $\mathcal{L}_{HC}$ using a special interpolation theorem for universal formulas.[21] Curiously, his argument depended essentially on using infinitely long formulas and could not be applied to $\mathcal{L}_{HF}$. Barwise [Bw] easily extended it to the languages $\mathcal{L}_A$ for $A \neq HC$; with his strengthened interpolation theorem he was also able to obtain other related model-theoretical results (cf. [Bw], §4). The following presents a new argument which treats all of the $\mathcal{L}_A$ including $A = HF$ uniformly, by exploiting the many-sorted language. Moreover, it has turned out possible to adapt this method to obtain characterizations relative to another type of extension, the so-called <u>outer</u>(or $\in$-, or end-) extensions, for the same languages. This work, which generalizes the results reported in [F,K], will appear in [F4].[22]

Given two structures $\mathfrak{M}$ and $\mathfrak{M}'$ of signature σ,

$$\mathfrak{M} = (\langle M_j \rangle_{j\in J}, \langle R_i \rangle_{i\in I_0}, \langle a_i \rangle_{i\in I_1}, \langle F_i \rangle_{i\in I_2})$$

and

$$\mathfrak{M}' = (\langle M_j' \rangle_{j\in J}, \langle R_i' \rangle_{i\in I_0}, \langle a_i' \rangle_{i\in I_1}, \langle F_i' \rangle_{i\in I_2}),$$

we say that $\mathfrak{M}$ is an <u>extension/J_0</u> of $\mathfrak{M}'$ and that $\mathfrak{M}'$ is a <u>restriction/J_0</u> (or <u>substructure/J_0</u>) of $\mathfrak{M}$, and write $\mathfrak{M}' \subseteq_{J_0} \mathfrak{M}$ if (i) for each $j \in J$, $M_j' \subseteq M_j$, (ii) for each $j \in J_0$, $M_j' = M_j$, (iii) for each $i \in I_0$, $R_i' = R_i \cap (\bigcup M_j' [j \in J])^{k_i}$, (iv) for each $i \in I_1$, $a_i' = a_i$, (v) for each $i \in I_2$, and $(x_1,\ldots,x_{m_i}) \in \mathcal{D}(F_i')$, $F_i'(x_1,\ldots,x_{m_i}) = F_i(x_1,\ldots,x_{m_i})$. We call the elements of J_0 the <u>stationary sorts</u> of the extension. Let $M' = \bigcup M_j' [j \in J]$; we shall abbreviate (iii) by writing $R_i' = R_i \upharpoonright M'$. Let $\mathbb{M}$ be a collection of structures. Given a formula $\phi(u_1,\ldots,u_n)$ with just $u_1,\ldots,u_n$ free, we say that ϕ is <u>persistent for extensions/J_0 in</u> $\mathbb{M}$ if whenever $\mathfrak{M}'$, $\mathfrak{M}$ belong to $\mathbb{M}$ and $\mathfrak{M}' \subseteq_{J_0} \mathfrak{M}$ and $x_1,\ldots,x_n$ are in $\mathfrak{M}'$

---

[21] G. Choodnovsky reported in [Ch] that he found this result independently, and even more, its generalization to $\mathcal{L}_{\kappa,\kappa}$ for κ inaccessible. Some special compactness theorems for sets of universal sentences are also mentioned and said to be used in this work.

[22] Makkai's independent work [Mk] provides another uniform treatment of these various results. This is also not model-theoretic, but the techniques are somewhat different from those presented here.

and of the same sort as $u_1,\dots,u_n$ resp., then

$$\models_{\mathcal{M}'} \phi[x_1,\dots,x_n] \text{ implies } \models_{\mathcal{M}} \phi[x_1,\dots,x_n] .$$

ϕ is said to be <u>persistent for restrictions</u>/J_o <u>in</u> $\mathbb{M}$ if $\sim \phi$ is pers. for extensions/J_o in $\mathbb{M}$. ϕ is said to be <u>invariant for extensions</u>/J_o <u>in</u> $\mathbb{M}$ if it is persistent both for extensions/J_o and restrictions/J_o in $\mathbb{M}$. When $\mathbb{M}$ is the collection of all models of a set S, we read 'relative to S' instead of 'in $\mathbb{M}$'. <u>Throughout the following</u> S <u>is any set of</u> A-<u>sentences which, in case</u> $A \neq HF$, <u>is countable and</u> A-<u>rec. enum.</u>

5.1 <u>Theorem</u>. <u>An</u> A-<u>formula</u> ϕ <u>is persistent for extensions</u>/J_o <u>relative to</u> S <u>if and only if there is an</u> A-<u>formula</u> θ <u>which is existential</u>/J_o <u>for which</u> $S \Vdash (\phi \leftrightarrow \theta)$ <u>and</u> $Fr(\phi) \supseteq Fr(\theta)$.

<u>Proof</u>. It is easy to prove by induction on θ that if θ is existential/J_o (universal/J_o) then θ is persistent for extensions/J_o (restrictions/J_o). Then if θ is existential/J_o and $S \Vdash (\phi \leftrightarrow \theta)$ then ϕ is persistent for extensions/J_o rel. to S (we need not assume $Fr(\phi) \supseteq Fr(\theta)$ here).

For the converse, we wish first to argue that it is sufficient to prove the theorem for relational signatures, even without designated individuals. Consider the standard procedure for eliminating function symbols and constant symbols in favor of new relation symbols. Let ϕ_{Rel} be the A-formula associated with ϕ by this procedure. Let S_{Rel} consist of all Ψ_{Rel} for $\Psi \in S$, together with the statements expressing the functionality (on the appropriate sorts) of the new relations. Then, in case $A \neq HF$, S_{Rel} is also countable and A-rec. enum. If ϕ is persistent for extensions/J_o rel. to S then ϕ_{Rel} is persistent for extensions/J_o rel. to S_{Rel}. Hence by the result for purely relational signatures, there is an A-formula θ which is existential/J_o for which $S_{Rel} \Vdash (\phi_{Rel} \leftrightarrow \theta)$ and $Fr(\phi_{Rel}) \supseteq Fr(\theta)$. θ may involve the new relation symbols; however these can now be rewritten in terms of the original function and constant symbols without introducing new quantifiers, giving $S \Vdash (\phi \leftrightarrow \theta_1)$ for θ_1 existential/J_o . The reader may verify this in greater detail.

Thus we can now assume $I_1 = I_2 = 0$; we write $\sigma = (J, I_o, \langle k_i \rangle_{i \in I_o})$ and $\mathfrak{M} = (\langle M_j \rangle_{j \in J}, \langle R_i \rangle_{i \in I_o})$. We shall express the persistence of $\emptyset$ syntactically by considering an extended language $\mathcal{L}_A^+$ in which we can express the relation of restriction/J_o between a structure $\mathfrak{M}$ and a substructure determined by subsets $\langle M'_j \rangle_{j \in J}$. Let J' be in 1-1 correspondence with and disjoint from J ; for each $j \in J$, let j' be the corresponding element of J' . Let $J^+ = J \cup J'$, and $\sigma^+ = (J^+, I_o, \langle k_i \rangle_{i \in I_o})$. Given any $\mathfrak{M}$ of signature σ and $\mathfrak{M}' = (\langle M'_j \rangle_{j \in J}, \langle R'_i \rangle_{i \in I_o})$, we associate the σ^+-structure $[\mathfrak{M}, \mathfrak{M}'] = (\langle M_j^+ \rangle_{j \in J^+}, \langle R_i \rangle_{i \in I_o})$, where $M_j^+ = M_j$ and $M_{j'}^+ = M'_j$ for each $j \in J$. On the other hand, given any structure $\mathfrak{M}^+ = (\langle M_j^+ \rangle_{j \in J^+}, \langle R_i \rangle_{i \in I_o})$ of signature σ^+ , Let $M_j = M_j^+$ for $j \in J$, $M'_j = M_{j'}^+$ for $j \in J$ and $M' = \bigcup M'_j \; [j \in J]$. Then take $\mathfrak{M}^+ \upharpoonright J = (\langle M_j \rangle_{j \in J}, \langle R_i \rangle_{i \in I_o})$ and $\mathfrak{M}^+ \upharpoonright J' = (\langle M'_j \rangle_{j \in J}, \langle R_i \upharpoonright M' \rangle_{i \in I_o})$. Trivially, $\mathfrak{M}^+ = [\mathfrak{M}^+ \upharpoonright J, \mathfrak{M}^+ \upharpoonright J']$. Also, $\mathfrak{M}^+ \upharpoonright J' \subseteq_{J_o} \mathfrak{M}^+ \upharpoonright J$ if $M'_j \subseteq M_j$ for each $j \in J$ and $M'_j \supseteq M_j$ for each $j \in J_o$.

$\mathcal{L}_A^+$ is taken to be the language of σ^+-structures; it contains the language $\mathcal{L}_A$ of σ-structures directly as a sublanguage. Given a variable u of the form $v_{j,n}$ with $j \in J$ we shall write u' for the variable $v_{j',n}$. In the following, $u, w, u_1, w_1, \ldots$ range only over variables of $\mathcal{L}_A$. Given any formula Ψ of $\mathcal{L}_A$, let Ψ' be the result of replacing each <u>bound</u> variable u of Ψ by the corresponding variable u' ; $Fr(\Psi') = Fr(\Psi)$. Let S' be the set of sentences Ψ' for $\Psi \in S$.

Now define Ext to be the set consisting of the following sentences: (i) $\bigwedge u' \bigvee u \, (u' = u)$ for each sort of variable u , and (ii) $\bigwedge u \bigvee u' \, (u = u')$ for each sort of variable in J_o . Then $\mathfrak{M}^+$ is a model of Ext if and only if $\mathfrak{M}^+ \upharpoonright J' \subseteq_{J_o} \mathfrak{M}^+ \upharpoonright J$. Suppose $\mathfrak{M}^+$ is a model of Ext and $\mathfrak{M} = \mathfrak{M}^+ \upharpoonright J$, $\mathfrak{M}' = \mathfrak{M}^+ \upharpoonright J'$. Let $\Psi(w_1, \ldots, w_t)$ be any formula of $\mathcal{L}_A$ with just $w_1, \ldots, w_t$ free , and let $x_1, \ldots, x_t$ in $\mathfrak{M}$ be of the same sorts as $w_1, \ldots, w_t$ resp. Then we see by induction that

$$\models_{\mathfrak{M}^+} \Psi[x_1, \ldots, x_t] \text{ iff } \models_{\mathfrak{M}} \Psi[x_1, \ldots, x_t]$$

and, if $x_1, \ldots, x_t \in \mathfrak{M}'$ then

$$\models_{\mathfrak{M}^+} \Psi'[x_1,\ldots,x_t] \text{ iff } \models_{\mathfrak{M}'} \Psi[x_1,\ldots,x_t] .$$

In particular if $\mathfrak{M}^+$ is a model of $S \cup S'$ then both $\mathfrak{M}, \mathfrak{M}'$ are models of S.

Given $\emptyset(w_1,\ldots,w_n)$, with distinct free variables $w_1,\ldots,w_n$, which is persistent for extensions/J_o rel. to S, let $u_1,\ldots,u_n$ be distinct members of BV of corresponding sorts. Then it follows from the foregoing that

$$\text{Ext} \cup S \cup S' \cup \{\bigvee u_i'(u_i' = w_i): i = 1,\ldots,n\} \Vdash (\emptyset' \to \emptyset) .$$

Now $\text{Ext} \cup S \cup S'$ is also countable and A-rec. enum. if $A \neq HF$. Hence by the completeness-compactness theorem (3.9) (or the standard completeness-compactness theorem when $A = HF$), there are subsets Ext_1 of Ext and S_1 of S which belong to A for which we have a (cut-free) A-derivation of

$$(*) \qquad [\prod \text{Ext}_1 \wedge \prod S_1' \wedge \prod_{i=1}^{n} \bigvee u_i' \; (u_i' = w_i) \wedge \emptyset'] \to [\prod S_1 \to \emptyset] .$$

We now apply the interpolation theorem (4.2).[23] An A-interpolant θ can be chosen which satisfies (i) $\text{Fr}(\theta) \subseteq \text{Fr}(\emptyset) = \{w_1,\ldots,w_n\}$, (ii) $\text{Rel}(\theta) \subseteq \text{Rel}(\emptyset)$, (iii) every bound variable in θ is of sort in $\mathcal{L}_A$, and (iv) $\text{Un}(\theta) \subseteq J_o$. The reason for (iv) is that the only variables of $\mathcal{L}_A$ with an essentially universal occurrence in the hypothesis of (*) are just those which appear in Ext_1; by definition, the sort of each of these is in J_o. Hence θ is an existential/J_o formula of $\mathcal{L}_A$. Now from derivability of

$$[\prod \text{Ext}_1 \wedge \prod S_1' \wedge \prod_{i=1}^{n} \bigvee u_i' \; (u_i' = w_i) \wedge \emptyset'] \to \theta$$

---

[23] The implication (*) does not satisfy the hypothesis of (4.2) that $\text{Sort(concl.)} \cap \text{Un(hyp.)} \neq 0$ when $J_o = 0$. It also fails to satisfy the alternate hypothesis that $\text{Sort(hyp.)} \cap \text{Ex(concl.)} \neq 0$ when $\text{Ex}(\prod S_1 \to \emptyset) = 0$. In this case we consider instead of (*) the implication with conclusion modified to $[\prod S_1 \to \emptyset] \wedge \bigvee u(u = u)$, where u is any variable of $\mathcal{L}_A$. This does not affect the remainder of the argument.

and $\theta \to [\prod S_1 \to \emptyset]$ we easily get $\prod S_1 \Vdash (\emptyset \leftrightarrow \theta)$ and hence $S \Vdash (\emptyset \leftrightarrow \theta)$.

Note that only the completeness theorem (3.8) need be applied to obtain this result for $S \in A$, in particular for $S = 0$.

5.2 Theorem. Suppose every sentence in S is universal/J_o. Then an A-formula $\emptyset$ is invariant for extensions/J_o rel. to S if and only if there is an A-formula θ which is quantifier-free/J_o and for which $S \Vdash (\emptyset \leftrightarrow \theta)$. ($\theta$ can be chosen to satisfy $Fr(\theta) \subseteq Fr(\emptyset)$ if there is a constant term of each sort.)

Proof. Suppose $\emptyset$ is invariant for extensions/J_o rel. to S. By the preceding theorem we can find an A-formula θ_1 which is universal/J_o and an A-formula θ_2 which is existential/J_o such that $S \Vdash (\emptyset \leftrightarrow \theta_1)$ and $S \Vdash (\emptyset \leftrightarrow \theta_2)$. By completeness, compactness we get $S_1 \subseteq S$ with $S_1 \in A$ and a (cut-free) A-derivation of

$$(\prod S_1 \wedge \theta_1) \to \theta_2 .$$

The hypothesis of this implication is also universal/J_o. Hence by theorem (4.4) we can find an A-interpolant θ which is quantifier-free/J_o; this θ satisfies the conclusion of the theorem. (If there are constant terms of each sort, we substitute these for the variables in $Fr(\theta) - Fr(\emptyset)$ according to sort.)

For the applications in the next section, we now specialize to $J = \{0\}$, i.e. the case of a single-sorted language $\mathcal{L}_A$, and we take J_o to be empty. Let $Sub(S)$ be the collection $\mathbb{M}$ of all substructures of models of S. We obtain next a result for formulas which are invariant for extensions in $Sub(S)$. This can be derived from (5.2) in the case $A = HF$ if one uses the result that $\mathfrak{M} \in Sub(S)$ if and only if $\mathfrak{M}$ is a model of the set S_o of all universal sentences in $\mathcal{L}_{HF}$ which are consequences of S (cf., e.g., [R]). A generalization of this has been obtained by Barwise [Bw] but only to the case $A = HC$. However, the set of universal sentences in $\mathcal{L}_{HC}$ which are consequences of S is not countable, so the hypotheses of this section on S do not apply. It is thus necessary to make a separate argument.

5.3 Theorem. An A-formula $\emptyset$ is invariant for extensions in Sub(S) if and only if there is a quantifier-free A-formula θ such that $(\emptyset \leftrightarrow \theta)$ is valid in each member of Sub(S). (θ can be chosen with $Fr(\theta) \subseteq Fr(\emptyset)$ if there is at least one constant symbol.)

Proof. We shall essentially reduce this to the relational case again. However, one must be slightly more cautious here, since if there are function symbols, $\emptyset_{Rel}$ need not be invariant for extensions in $Sub(S_{Rel})$. For, a structure can be in $Sub(S_{Rel})$ without being of the form $\mathfrak{M}_{Rel}$ for some $\mathfrak{M} \in Sub(S)$. However, let $\overline{f_i}$ be the (m_i+1)-ary relation symbol introduced to correspond with f_i. Then a structure in $Sub(S_{Rel})$ is a model of the sentence $\prod_{i \in I_2} \bigwedge u_1 \ldots \bigwedge u_{m_i} \bigvee u \overline{f_i}(u_1,\ldots,u_{m_i},u)$ if and only if it is of the form $\mathfrak{M}_{Rel}$ for $\mathfrak{M} \in Sub(S)$; we denote this sentence by Clos.

Now extend the language by adjoining two new sorts of variables $u', w', u_1', w_1', \ldots$ (the "prime" sort) and $u^*, w^*, u_1^*, w_1^*, \ldots$ (the "star" sort). We follow the notation of the proof of (5.1). We now also take Ψ^* to be the result of replacing each bound variable u of Ψ by the corresponding variable u^*. Let $\emptyset(w_1,\ldots,w_n)$ be invariant for extensions in Sub(S). Let $\emptyset_1^*$ be $\emptyset_{Rel}^*(w_1',\ldots,w_n')$ and $\emptyset_1'$ be $\emptyset_{Rel}'(w_1',\ldots,w_n')$. Then it follows that

(1) $$S_{Rel} \cup \{\bigwedge u^* \bigvee u' (u^*=u'), \bigwedge u' \bigvee u (u'=u)\}$$

$$\cup \{Clos^*, Clos'\} \cup \{\prod_{i=1}^{n} \bigvee u_i^* (u_i^*=w_i')\} \Vdash (\emptyset_1^* \leftrightarrow \emptyset_1').$$

Hence by completeness, compactness we get some $S_1 \in A$ with $S_1 \subseteq S_{Rel}$ for which we have (cut-free) A-derivations of the following two implications:

(2) $$[\bigwedge u^* \bigvee u' (u^*=u') \wedge Clos^* \wedge \prod_{i=1}^{n} \bigvee u_i^*(u_i^*=w_i') \wedge \emptyset_1^*] \rightarrow$$

$$\rightarrow [\prod S_1 \wedge \bigwedge u' \bigvee u(u'=u) \wedge Clos' \rightarrow \emptyset_1']$$

and

(3) $[\prod S_1 \wedge \bigwedge u' \bigvee u(u'=u) \wedge \text{Clos}' \wedge \emptyset_1'] \rightarrow$

$$\rightarrow [\bigwedge u^* \bigvee u' (u^*=u') \wedge \text{Clos}^* \wedge \prod_{i=1}^{n} \bigvee u_i^* (u_i^*=w_i') \rightarrow \emptyset_1^*] .$$

Now we apply (4.2). In the case of (2) we get an A-interpolant, every free variable of which is among $w_1', \ldots, w_n'$, and every bound variable of which has prime sort. It thus has the form $\Psi'(w_1', \ldots, w_n')$ for some formula Ψ of the relational language got from $\mathcal{L}_A$. Moreover, the hypothesis of (2) is existential in the prime sorts, so Ψ is existential. It is then seen that $(\emptyset_{Rel} \rightarrow \Psi)$ holds in every structure which satisfies Clos and $(\Psi \rightarrow \emptyset_{Rel})$ holds in every structure in $\text{Sub}(S_{Rel})$ which satisfies Clos. Hence if we replace each occurrence $\overline{f}_i(u_1, \ldots, u_{m_i}, u)$ in Ψ by $f_i(u_1, \ldots, u_{m_i}) = u$, we obtain an existential A-formula Ψ_1 such that

(4) $\Vdash(\emptyset \rightarrow \Psi_1)$, and $(\Psi_1 \rightarrow \emptyset)$ is valid in every member of $\text{Sub}(S)$.

The common sorts in the hypothesis and conclusion of (3) are again the prime sorts; in this case the prime sorts have only an essentially universal bound occurrence in the conclusion. Hence, applying (4.2) to (3) and arguing as above, we conclude that there is a universal A-formula χ_1 such that

(5) $\Vdash(\chi_1 \rightarrow \emptyset)$, and $(\emptyset \rightarrow \chi_1)$ is valid in every member of $\text{Sub}(S)$.

By (4) and (5), $\Vdash(\chi_1 \rightarrow \Psi_1)$. Hence by (4.4) we can find a quantifier-free θ such that $\Vdash(\chi_1 \rightarrow \theta) \wedge (\theta \rightarrow \Psi_1)$. But then $(\emptyset \leftrightarrow \theta)$ is valid in every member of $\text{Sub}(S)$, q.e.d.

5(b) <u>Elimination of Quantifiers; general theory.</u> Throughout this section we shall assume for simplicity that $J = \{0\}$, i.e. that σ is a signature of single-sorted structures. However, as in the preceding section we shall make use of some extensions of the given language to many-sorted languages. We shall also continue to assume throughout that S <u>is a set of sentences in</u> $\mathcal{L}_A$ <u>which, in case</u> $A \neq HF$, is countable and A-recursively enumerable.

We say that S admits elimination of quantifiers in $\mathcal{L}_A$ if for every A-formula $\emptyset$ there exists a quantifier-free A-formula Ψ with $Fr(\Psi) \subseteq Fr(\emptyset)$ such that $S \Vdash (\emptyset \leftrightarrow \Psi)$. By the trivial part of (5.2) if S admits elimination of quantifiers in $\mathcal{L}_A$ then every A-formula is invariant for extensions rel. to S .[24] In particular, if this holds and $\mathfrak{M}, \mathfrak{M}'$ are models of S and $\mathfrak{M}' \subseteq \mathfrak{M}$ then $\mathfrak{M}$ and $\mathfrak{M}'$ satisfy exactly the same A-sentences. Consequently, if S has a prime model, i.e. a model $\mathfrak{M}_0$ which can be isomorphically embedded in every model $\mathfrak{M}$ of S , then S is complete in $\mathcal{L}_A$, i.e. for every A-sentence $\emptyset$ either $S \Vdash \emptyset$ or $S \Vdash \sim \emptyset$.

To obtain elimination of quantifiers for S it is sufficient to test a simple class of formulas. We say that $\emptyset$ is in Ex_1 if it has the form $\bigvee u\, \emptyset'$ where $\emptyset'$ is quantifier-free.

5.4 Lemma. Suppose that for each A-formula $\emptyset$ in Ex_1 there exists a quantifier-free A-formula Ψ with $Fr(\Psi) \subseteq Fr(\emptyset)$ and $S \Vdash (\emptyset \leftrightarrow \Psi)$. Then S admits elimination of quantifiers in $\mathcal{L}_A$.

Proof. Let $El(\emptyset, \Psi)$ hold if Ψ is quantifier-free, $Fr(\Psi) \subseteq Fr(\emptyset)$ and there exist $S_1 \in A$, $S_1 \subseteq S$ and an A-derivation $\mathcal{D}$ of $[\prod S_1 \rightarrow (\emptyset \leftrightarrow \Psi)]$. It can be seen that the relation El is $\sum_1^{(A)}$-definable. Then one proves by induction on A-formulas, using completeness, compactness (3.9) and the $\sum_1$-axiom of choice in A , that for every A-formula $\emptyset$ there exists Ψ with $El(\emptyset, \Psi)$.

In the case A = HF it is sufficient to test an even simpler class of formulas. For, each quantifier-free formula is equivalent to one in disjunctive normal form. Then by distributing the existential quantifier over the disjunction, it is sufficient to test formulas of the form $\bigvee u \prod K$ where each formula in K is atomic or the negation of an atomic formula. This is the classical method of proving elimination of quantifiers; cf. [K,K], Ch.4 for some typical illustrations. However, it can be shown that the disjunctive normal form theorem does not hold

[24] In the terminology of Robinson [R], for A = HF this conclusion is equivalent to saying that S is model-complete; in the terminology of Tarski-Vaught [T,V] , it is equivalent to saying that whenever $\mathfrak{M}, \mathfrak{M}'$ are models of S and $\mathfrak{M}' \subseteq \mathfrak{M}$ then $\mathfrak{M}$ is an elementary extension of $\mathfrak{M}'$.

in $\mathcal{L}_A$ for $A \neq HF$. Thus (5.4) cannot in general be improved.

In practice, the application of the classical method in $\mathcal{L}_{HF}$ is effective and leads to a decision method for telling whether a formula is a consequence of S. But the verification of this is in some cases extremely tedious and not very informative mathematically. This has led to the development of various model-theoretic methods which give the same consequences as to completeness or invariance for extensions, and in some cases even the fact of quantifier-elimination. These results make more efficient use of known mathematical information concerning models of S. We reconsider these here because they open up the possibility of corresponding results for the languages $\mathcal{L}_A$.

The simplest such result is the well-known one due to Vaught [V]: if the set S of $\mathcal{L}_{HF}$ sentences has only infinite models and is categorical in some infinite power (i.e. any two models of S of that power are isomorphic) then S is complete. Using the downward Skolem-Löwenheim theorem for sentences in $\mathcal{L}_{HC}$ (Scott [Sc1]), Barwise [Bw] (ch.4) has observed the following partial generalization and application of Vaught's theorem: if S is a countable set of HC-sentences with only infinite models, and S is categorical in power $\aleph_0$ then S is complete in $\mathcal{L}_{HC}$. (In fact, by a result of Scott [Sc1], the converse is also true.) For the application, take $S = ACF_n^\infty$ = the set of axioms for algebraically closed fields of infinite transcendence degree over the ground field of characteristic n (where n is 0 or a prime). By standard algebra, ACF_n^∞ has only infinite models and is categorical in power $\aleph_0$.

One very powerful method of obtaining more far-reaching results in $\mathcal{L}_{HF}$ is the use of ultraproducts (as developed initially in Keisler [Ks2] and Kochen [Ko2]). However, we don't know at present of any suitable generalization of this notion to $\mathcal{L}_A$ for $A \neq HF$, or even to $\mathcal{L}_{HC}$. The model-theoretic approach of Robinson presented in [R] provides another basis for the same results. Moreover, by use of a certain notion of relative model-completeness [R] also contains results giving sufficient conditions for eliminating quantifiers. A basic lemma for these is 5.3.1 in [R], proved by a somewhat complicated compactness argument. The theorem (5.5), to be obtained next, generalizes 5.3.1 to arbitrary $\mathcal{L}_A$; since the proof is direct, it may be of independent interest even for the case $A = HF$.

This has been used in [Bw], where a quantifier-elimination theorem for arbitrary $\mathcal{L}_A$ is stated. A still stronger result is obtained here (5.6); another source for it is the elegant reformulation of Robinson's results for $\mathcal{L}_{HF}$ in Shoenfield [Sh], Ch.5 and in the text of Kreisel-Krivine, [K,K], p. 125.

In the following, assume that S_0 is a set of A-sentences and, in case $A \neq HF$, that S_0 is countable and A-recursively enumerable. We say that a formula $\emptyset(w_1,\ldots,w_n)$ is (S,S_0)-invariant if whenever $\mathfrak{M},\mathfrak{M}'$ are models of S and $\mathfrak{M}^*$ is a model of S_0 and $\mathfrak{M}^* \subseteq \mathfrak{M}$ and $\mathfrak{M}^* \subseteq \mathfrak{M}'$ and $x_1,\ldots,x_n$ are in $\mathfrak{M}^*$ then

$$\models_{\mathfrak{M}} \emptyset[x_1,\ldots,x_n] \text{ if and only if } \models_{\mathfrak{M}'} \emptyset[x_1,\ldots,x_n] .$$ [25)]

5.5 Theorem. An A-formula $\emptyset$ is (S,S_0)-invariant if and only if there is an A-formula θ with $Fr(\theta) \subseteq Fr(\emptyset)$ such that whenever $\mathfrak{M}$ is a model of S, $\mathfrak{M}^*$ is a model of S_0, $\mathfrak{M}^* \subseteq \mathfrak{M}$ and $x_1,\ldots,x_n$ are in $\mathfrak{M}^*$ then

$$\models_{\mathfrak{M}} \emptyset[x_1,\ldots,x_n] \text{ if and only if } \models_{\mathfrak{M}^*} \theta[x_1,\ldots,x_n] .$$

Proof. It is trivial that $\emptyset$ is (S,S_0)-invariant if there exists such a formula θ. For the converse, one again shows first that it is sufficient to prove the result for the purely relational case; the details are left to the reader. The signature σ is thus assumed to be of the form $(\{0\}, I_0, \langle k_i\rangle_{i\in I_0})$; the structures of signature σ are of the form $\mathfrak{M} = (M, \langle R_i\rangle_{i\in I_0})$.

Extend the language $\mathcal{L}_A$ for structures of signature σ as follows: we adjoin two new sorts of variables $u', w', u_1', w_1', \ldots,$ and $u^*, w^*, u_1^*, w_1^*, \ldots$. We also adjoin new relation symbols r_i' of k_i arguments and r_i^* of k_i-arguments for each $i \in I_0$. The new language is referred to as $\mathcal{L}_A^+$, its signature as σ^+. Structures of signature σ^+ are of the form $\mathfrak{M}^+ = (\langle M, M', M^*\rangle, \langle R_i\rangle_{i\in I_0}, \langle R_i'\rangle_{i\in I_0}, \langle R_i^*\rangle_{i\in I_0})$. For each such $\mathfrak{M}^+$, let $\mathfrak{M}^+ \upharpoonright M = (M, \langle R_i \upharpoonright M\rangle_{i\in I_0})$, $\mathfrak{M}^+ \upharpoonright M' = (M', \langle R_i' \upharpoonright M'\rangle_{i\in I_0})$ and

---

25) For A = HF, if every model of S^* has an extension which is a model of S, this is equivalent to the notion of [R] : $\emptyset$ is invariant relative to S over S^*.

$\mathfrak{M}^+ \upharpoonright M^* = (M^*, \langle R_i^* \upharpoonright M^* \rangle_{i \in I_0})$. We define Ext to be the set consisting of all of the following sentences: (i) $\bigwedge u^* \bigvee u(u^*=u)$, (ii) for each $i \in I_0$, $\bigwedge u_1^* \ldots \bigwedge u_{k_i}^* [r_i(u_1^*,\ldots,u_{k_i}^*) \leftrightarrow r_i^*(u_1^*,\ldots,u_{k_i}^*)]$, and define Ext' to be the set consisting of the sentences (iii) $\bigwedge u^* \bigvee u'(u^*=u')$, (iv) for each $i \in I_0$, $\bigwedge u_1^* \ldots \bigwedge u_{k_i}^* [r_i'(u_1^*,\ldots,u_{k_i}^*) \leftrightarrow r_i^*(u_1^*,\ldots,u_{k_i}^*)]$. Then $\mathfrak{M}^+$ is a model of Ext if and only if $\mathfrak{M}^+ \upharpoonright M^* \subseteq \mathfrak{M}^+ \upharpoonright M$ and $\mathfrak{M}^+$ is a model of Ext' if and only if $\mathfrak{M}^+ \upharpoonright M^* \subseteq \mathfrak{M}^+ \upharpoonright M'$. For each formula Ψ, let Ψ' be the result of replacing each bound variable u of Ψ by the corresponding variable u', and each relation symbol r_i by the corresponding symbol r_i'; Ψ^* is defined similarly. Then for any set T, T' is taken to be the set of Ψ' for Ψ in T and T^* the set of Ψ^* for Ψ in T. If the free variables of Ψ are $w_1,\ldots,w_t$ and $x_1,\ldots,x_t \in M^* \subseteq M \cap M'$ then

$$\models_{\mathfrak{M}^+} \Psi[x_1,\ldots,x_t] \text{ iff } \models_{\mathfrak{M}^+ \upharpoonright M} \Psi[x_1,\ldots,x_t],$$

$$\models_{\mathfrak{M}^+} \Psi'[x_1,\ldots,x_t] \text{ iff } \models_{\mathfrak{M}^+ \upharpoonright M'} \Psi[x_1,\ldots,x_t], \text{ and}$$

$$\models_{\mathfrak{M}^+} \Psi^*[x_1,\ldots,x_t] \text{ iff } \models_{\mathfrak{M}^+ \upharpoonright M^*} \Psi[x_1,\ldots,x_t].$$

It follows that

$$\text{Ext} \cup \text{Ext}' \cup S \cup S' \cup S_0^* \cup \{\bigvee u_i^*(u_i^*=w_i): i = 1,\ldots,n\} \Vdash (\phi \rightarrow \phi').$$

To complete the proof, one shows by completeness, compactness, and the interpolation theorem (4.2) that there is a formula θ of $\mathcal{L}_A$ with $Fr(\theta) \subseteq Fr(\phi)$ such that

$$\text{Ext} \cup S \cup S_0^* \cup \{\bigvee u_i^*(u_i^* = w_i): i = 1,\ldots,n\} \Vdash (\phi \rightarrow \theta^*)$$

and $\text{Ext}' \cup S' \cup S_0^* \cup \{\bigvee u_i^*(u_i^* = w_i): i = 1,\ldots,n\} \Vdash (\theta^* \rightarrow \phi')$.

Thus θ satisfies the desired conclusion.

5.6 <u>Theorem</u>. <u>Assume $\mathcal{L}_A$ has at least one constant symbol</u>. <u>Then</u> S <u>admits elimination of quantifiers in $\mathcal{L}_A$ if and only if every A-formula in</u> Ex_1 <u>is</u> (S,0) <u>invariant</u>.[26]

<u>Proof</u>. Suppose S admits elimination of quantifiers in $\mathcal{L}_A$. Let $\phi(w_1,\dots,w_n)$ be any A-formula and $\Psi(w_1,\dots,w_n)$ any quantifier-free A-formula such that $S \Vdash (\phi \leftrightarrow \Psi)$. Suppose $\mathfrak{M}, \mathfrak{M}'$ are any models of S, $\mathfrak{M}^*$ any structure with $\mathfrak{M}^* \subseteq \mathfrak{M}$, $\mathfrak{M}^* \subseteq \mathfrak{M}'$, and $x_1,\dots,x_n$ are any elements of $\mathfrak{M}^*$. Then

$$\models_{\mathfrak{M}} \phi[x_1,\dots,x_n] \text{ iff } \models_{\mathfrak{M}} \Psi[x_1,\dots,x_n]$$

and

$$\models_{\mathfrak{M}} \Psi[x_1,\dots,x_n] \text{ iff } \models_{\mathfrak{M}^*} \Psi[x_1,\dots,x_n] .$$

The same holds for $\mathfrak{M}'$ instead of $\mathfrak{M}$, so

$$\models_{\mathfrak{M}} \phi[x_1,\dots,x_n] \text{ iff } \models_{\mathfrak{M}'} \phi[x_1,\dots,x_n] .$$

Now suppose that $\phi(w_1,\dots,w_n)$ is any A-formula in Ex_1 and that ϕ is (S,0) invariant. Then by (5.5) we can find on A-formula θ with $Fr(\theta) \subseteq Fr(\phi)$ such that whenever $\mathfrak{M}$ is a model of S and $\mathfrak{M}^* \subseteq \mathfrak{M}$ then

$$\models_{\mathfrak{M}} \phi[x_1,\dots,x_n] \text{ iff } \models_{\mathfrak{M}^*} \theta[x_1,\dots,x_n]$$

for all $x_1,\dots,x_n$ in $\mathfrak{M}^*$. We claim that θ is invariant for extensions in Sub(S). For suppose $\mathfrak{M}^* \subseteq \mathfrak{M}_1^* \subseteq \mathfrak{M}$, where $\mathfrak{M}$ is a model of S. Then by the foregoing, we also have $\models_{\mathfrak{M}} \phi[x_1,\dots,x_n]$ iff $\models_{\mathfrak{M}_1^*} \theta[x_1,\dots,x_n]$. Hence we can conclude by (5.3) that there is a quantifier-free A-formula Ψ with $Fr(\Psi) \subseteq Fr(\theta)$ such that $(\theta \leftrightarrow \Psi)$ is valid in every member of Sub(S). Since $(\phi \leftrightarrow \theta)$ is valid in every model of S, it follows that $S \Vdash (\phi \leftrightarrow \Psi)$. This argument applies to any Ex_1 A-formula ϕ, hence S admits elimination of quantifiers in $\mathcal{L}_A$ by (5.4).

---

[26] This result and the remaining results of §5 were found subsequent to the presentation of the lectures.

5.7 Corollary. Assume $\mathcal{L}_A$ has at least one constant symbol. Suppose that (i) S_o is a set of universal sentences, (ii) every model of S is a model of S_o, and (iii) every A-formula in Ex_1 is (S,S_o)-invariant. Then S admits elimination of quantifiers in $\mathcal{L}_A$.

Proof. By (i), (ii), if $\mathfrak{M}^* \subseteq \mathfrak{M}$ and $\mathfrak{M}$ is a model of S then $\mathfrak{M}^*$ is a model of S_o. Hence any formula which is (S,S_o)-invariant is also $(S,0)$-invariant.

This improves the quantifier-elimination theorem stated in [Bw], Ch.4, by removing several of the hypotheses, principally that every model of S_o can be extended to a model of S. Note that (5.7) immediately implies the non-trivial part of (5.6), and includes as a special case the result of [K,K], p. 125.

We now consider several algebraic conditions which are useful in simplifying the work for various S to show that every A-formula in Ex_1 is $(S,0)$-invariant.

S is said to satisfy the isomorphism lifting condition if whenever $\mathfrak{M}_o$, $\mathfrak{M}_1$ are models of S and $\mathfrak{M}_o^* \subseteq \mathfrak{M}_o$, $\mathfrak{M}_1^* \subseteq \mathfrak{M}_1$ and $\mathfrak{M}_o^* \cong_H \mathfrak{M}_1^*$ (i.e. $H: M_o^* \to M_1^*$ establishes the isomorphism) then we can find $\overline{\mathfrak{M}}_o$, $\overline{\mathfrak{M}}_1$ and $\overline{H}$ such that $\overline{\mathfrak{M}}_o$, $\overline{\mathfrak{M}}_1$ are models of S, $\mathfrak{M}_i^* \subseteq \overline{\mathfrak{M}}_i \subseteq \mathfrak{M}_i$ for $i = 0,1$, $\overline{\mathfrak{M}}_o \cong_{\overline{H}} \overline{\mathfrak{M}}_1$ and $\overline{H}$ extends H.

Given any structure $\mathfrak{M}$ and subset X of $\mathfrak{M}$, let $Gen_{\mathfrak{M}}(X)$ be the substructure of $\mathfrak{M}$ generated by X; we shall omit the subscript $\mathfrak{M}$ when there is no confusion. S is said to satisfy the simple extension condition if whenever $\mathfrak{M}, \mathfrak{M}'$ are models of S, $X = \{x_1,\ldots,x_n\}$ is a subset of $\mathfrak{M}$, $X' = \{x_1',\ldots,x_n'\}$ is a subset of $\mathfrak{M}'$, and $Gen(X) \cong_H Gen(X')$ with $H(x_i) = x_i'$ for each i, then for any $y \in \mathfrak{M}$ there exists $y' \in \mathfrak{M}'$ and $\overline{H}$ such that $Gen(X \cup \{y\}) \cong_{\overline{H}} Gen(X' \cup \{y'\})$, $\overline{H}$ extends H, and $\overline{H}(y) = y'$. S is said to satisfy the simple retraction condition if this conclusion holds whenever it is also assumed that $\mathfrak{M}' \subseteq \mathfrak{M}$ and $x_i = x_i'$ for each i.

We say that S is relatively categorical in power $\aleph_o$ if whenever $\mathfrak{M}, \mathfrak{M}'$ are countable models of S, $X = \{x_1,\ldots,x_n\}$ is a subset of $\mathfrak{M}$, $X' = \{x_1',\ldots,x_n'\}$ is a subset of $\mathfrak{M}'$ and $Gen(X) \cong_H Gen(X')$ with $H(x_i) = x_i'$ for each i then there exists $\overline{H}$ extending H such that $\mathfrak{M} \cong_{\overline{H}} \mathfrak{M}'$. Note that if S is relatively categorical in power $\aleph_o$ and is such that whenever $\mathfrak{M}, \mathfrak{M}'$ are countable models of S we have $\mathrm{Gen}_{\mathfrak{M}}(0) \cong \mathrm{Gen}_{\mathfrak{M}'}(0)$, then S is categorical in power $\aleph_0$.

The following is left to the reader.

5.8 Theorem. Consider the following conditions:

$(i)^A$ Every A-formula in Ex_1 is $(S,0)$-invariant.

$(ii)^A$ S satisfies the isomorphism lifting condition and every A-formula in Ex_1 is invariant for extensions relative to S .

(iii) S satisfies the isomorphism lifting condition and the simple retraction condition.

Then $(ii)^A$ implies $(i)^A$, (iii) implies $(ii)^A$ and $(ii)^{HC}$ implies (iii) .

The first implication here combined with (5.6) gives a generalization of the quantifier-elimination theorem of [Sh], Ch.5.

5.9 Theorem. The following conditions are equivalent, assuming $\mathcal{L}_A$ has at least one constant symbol.

(i) S admits elimination of quantifiers in $\mathcal{L}_{A_1}$ for each $A_1 \supseteq A$.

(ii) Every (HC-)formula in Ex_1 is $(S,0)$ invariant.

(iii) S satisfies the simple extension condition.

(iv) S is relatively categorical in power $\aleph_0$.

Proof. The equivalence of (i),(ii),(iii) is straightforward. The proof that (iii) implies (iv) is by a back-and-forth argument like Cantor's argument for the categoricity in power $\aleph_0$ of dense orderings without first and last element. To show that (iv) implies (iii), one makes use of the downward Skolem-Löwenheim theorem [Sc] in the following form: if $\mathfrak{M}$ is a model of S and X is a finite subset of $\mathfrak{M}$ then there is a countable model $\mathfrak{M}_0$ of S with $X \subseteq \mathfrak{M}_0 \subseteq \mathfrak{M}$. (This application is permitted by our general assumption that S is countable in case $A \neq HF$.) Then given $\mathfrak{M}, \mathfrak{M}'$, X , X' and H satisfying the hypothesis of the simple extension condition for S , and given y in $\mathfrak{M}$, let

$X \cup \{y\} \subseteq \mathcal{M}_o \subseteq \mathcal{M}$, $X' \subseteq \mathcal{M}'_o \subseteq \mathcal{M}'$ with $\mathcal{M}_o$, $\mathcal{M}'_o$ countable models of S . By relative categoricity, the isomorphism H can be extended to an isomorphism $\mathcal{M}_o \cong_H \mathcal{M}'_o$; then take $y' = H(y)$.

Theoretically speaking, the preceding theorem shows that we cannot expect interesting consequences of elimination of quantifiers results independent of A which cannot already be obtained by suitable categoricity arguments. However, in practice it may be more convenient to verify (5.9)(iii) or the sufficient condition (5.8)(iii) .

5(c) Elimination of quantifiers; some applications.

We conclude with some examples illustrating the use of (5.6), (5.8) and (5.9).

I. ACF_n^∞ admits elimination of quantifiers in every $\mathcal{L}_A (A \neq HF)$. Here we can verify (5.8)(iii) . For the simple retraction condition, if X is a finite subset of $\mathcal{M}' \subseteq \mathcal{M}$, where $\mathcal{M}, \mathcal{M}'$ are both models of ACF_n^∞ and y is in $\mathcal{M}$ but not in $\mathcal{M}'$, choose y' in $\mathcal{M}'$ which is transcendental over X . The conclusion for arbitrary $\mathcal{L}_A$, $A \neq HF$, is a result of the fact that ACF_n^∞ belongs to HH , which is the smallest such A .

It follows that every formula of $\mathcal{L}_{HC}$ is invariant for extensions in algebraically closed fields of infinite transcendence degree. This is a very strong metamathematical formulation of what is sometimes called Lefschetz' transfer principle, for extending results of algebraic geometry in the complex numbers to arbitrary fields satisfying ACF_o^∞ . It is easily seen to include the transfer principle obtained by Scott and Tarski [S,T] formulated in terms of weak 2nd order logic (finite formulas with quantification over finite sets of individuals). At first sight it is not theoretically comparable with the transfer result described by Kochen [Ko2] for some special higher order finite formulas. However, it is not difficult to see that the result presented here can be extended to corresponding higher order HC-formulas in the same way. (cf. [Wl], Ch.X for a discussion of how the informal principle is realized in practice.)

II. Let B be the set of axioms (in $\mathcal{L}_{HF}$) for Boolean algebras $\mathcal{M} = (M,+,\cdot,-,0,1)$. Let $u \leq w$ abbreviate $u+w = w$, and let at(u) be the formula

$u \neq 0 \wedge \bigwedge w(w \leq u \rightarrow w = 0 \vee w = u)$. Adjoin a new unary predicate symbol r_n for each n . Let AtB^∞ consist of B together with

(i) $\bigwedge u\{r_n(u) \leftrightarrow \bigvee u_1 \dots \bigvee u_n\ [\prod_{i \leq n} (at(u_i) \wedge u_i \leq u) \wedge \prod_{i<j \leq n} (u_i \neq u_j)]\}$

for each $n > 0$, (ii) $\bigwedge u(u \neq 0 \rightarrow r_1(u))$ and

(iii) $\bigwedge u\ [\prod_{0<n<\omega} r_n(u) \rightarrow \bigvee u_1 \bigvee u_2 (u = u_1 + u_2 \wedge u_1 \cdot u_2 = 0 \wedge \prod_{0<n<\omega} r_n(u_1) \wedge r_n(u_2))]$.

(ii) expresses that the algebra is atomistic, and (iii) that every element containing infinitely many atoms is the sum of a pair of disjoint elements with the same property. The conjunction of AtB^∞ is of simple recursive structure so certainly AtB^∞ is an HH-rec. enum. set of HH-sentences. The result here is that AtB^∞ <u>admits elimination of quantifiers in every $\mathcal{L}_A$ $(A \neq HF)$</u> . For, it can be seen that AtB^∞ satisfies the simple extension condition. Given $\mathfrak{M}, \mathfrak{M}'$, X , X' , y satisfying the hypotheses of that condition, a y' can be chosen with $Gen(x \cup \{y\}) \cong Gen(X' \cup \{y'\})$ by examining the "size" of the intersection of y and $\bar{y}$ with each element of the standard partition of 1 associated with the members $x_1,\dots,x_n$ of X . Note that if $\mathfrak{M}, \mathfrak{M}'$ are any two models of AtB^∞ with $\models_{\mathfrak{M}} r_n(1)$ iff $\models_{\mathfrak{M}'} r_n(1')$ for every n , then $Gen_{\mathfrak{M}}(0) \cong Gen_{\mathfrak{M}'}(0)$. Hence $AtB^\infty \cup \{\prod_{0<n<\omega} r_n(1)\}$ is categorical in power $\aleph_0$.

III. The results of the preceding two examples are independent of the choice of A , as long as $A \neq HF$. We now consider examples where the result holds only for one $A \neq HF$. In general here we have to make use of a wider syntactic framework than admitted in §2(a). For example, we want to be able to specify that for a certain A-recursive set N , the language $\mathcal{L}_A$ has unary function symbols f_a for each $a \in N$, and similarly for functions symbols of other arguments, and for relation symbols and constants. The basic syntactic notions are still seen to be A-recursive as in (2.6), (4.1), and all the results of §§3-5 apply without change to such $\mathcal{L}_A$.

The problem dealt with here is connected with the general problem to characterize the true A-sentences of $(\omega_1,<)$. We sketch a solution for the cases that $O(A) < \omega_1$, where $O(A)$ is the least ordinal not in A .

The basic language in $\mathcal{L}_A$ has a binary relation symbol, for which we use the symbol $<$. We extend this language by adding a constant symbol $\bar{0}$ and unary function symbols Sc_α and Pd_α for each $\alpha < O(A)$. The set of $\alpha < O(A)$ is A-recursive, so the remarks above concerning $\mathcal{L}_A$ apply here. Let SO consist of the HF-sentences expressing that $<$ is a simple-ordering relation. Then take WO_A to consist of SO together with all of the following A-sentences ;

(i) $\bigwedge u \sim (u < 0)$, (ii) $\bigwedge u\, [Sc_0(u) = u]$,

(iii) $\bigwedge u \bigwedge w\, [w < Sc_\alpha(u) \leftrightarrow w < u \vee \sum_{\beta<\alpha} w = Sc_\beta(u)]$, for each $\alpha < O(A)$,

(iv) $\bigwedge u \bigwedge w\, [u = Sc_\alpha(w) \rightarrow Pd_\alpha(u) \neq u]$, for each $\alpha < O(A)$,

(v) $\bigwedge u\, [Pd_\alpha(u) \neq u \rightarrow u = Sc_\alpha\, (Pd_\alpha(u))]$, for each $\alpha < O(A)$,

(vi) $\bigwedge u \bigwedge w\, [w < Pd_\alpha(u) \rightarrow u \neq Sc_\alpha(w)]$, for each $\alpha < O(A)$,

(vii) $\bigwedge u_1 \ldots \bigwedge u_k \sim \prod_{t\in T} \sum_{t'\in T}(t' < t)$, for each non-empty set T of terms where $T \in A$ and $Fr(t) \subseteq \{u_1,\ldots,u_k\}$ for each $t \in T$.

It can be seen that WO_A is A-rec. enum. Then WO_A <u>admits elimination of quantifiers in</u> $\mathcal{L}_A$. To see this, note first that WO_A consists entirely of universal sentences. Hence the class of models of WO_A is closed under substructures, and the isomorphism lifting condition is trivially satisfied. We claim that every A-formula in Ex_1 is invariant for extensions in models of WO_A . Suppose given such a formula $\bigvee u\, \phi(w_1,\ldots,w_n,u)$ and given two such models $\mathfrak{M}' \subseteq \mathfrak{M}$. First, there is an upper bound $\alpha_0 < O(A)$ to the ordinals α such that a symbol Sc_α or Pd_α occurs in ϕ ; one can assume that α_0 is closed under ordinal addition. Now suppose given $x_1,\ldots,x_n$ in $\mathfrak{M}'$ and y in $\mathfrak{M}$ but not in $\mathfrak{M}'$, such that $\models_{\mathfrak{M}} \phi[x_1,\ldots,x_n,y]$. It is sufficient to find y' in $\mathfrak{M}'$ such that $\{Val_{\mathfrak{M},x_1,\ldots,x_n,y}(t) : t \in Tm(\phi)\} \cong_H \{Val_{\mathfrak{M}',x_1,\ldots,x_n,y'}(t) : t \in Tm(\phi)\}$ with $H(x_i) = x_i$ $(1 \leq i \leq n)$ and $H(y) = y'$. Note that each term of ϕ contains at most one of the variables $w_1,\ldots,w_n,u$. Let T be the set of terms in ϕ in which u occurs. By (vi) there must be a least among the elements $Val_{\mathfrak{M},y}(t)$ for $t \in T$; call this y_0 . Let $y_1 = \min(y,y_0)$; then $y = Sc_\beta(y_1)$ for some $\beta < \alpha_0$. Now if $z = \sup(x_i : x_i < y)$ or $z = 0$ if $y < x_i$ for each i , we can take $y' = Sc_{\alpha_0+\beta}(z)$.

A prime model of WO_A is the set of ordinals $O(A)$ with $Sc_\alpha(x) = x + \alpha$ and $Pd_\alpha(x) =$ least y s.t. $y + \alpha = x$, if there is such a y, otherwise $Pd_\alpha(x) = x$. It follows that WO_A is complete in $\mathcal{L}_A$. Every well-ordered structure $(M, <)$ with order-type $O(A) \cdot \gamma$ for some $\gamma \neq 0$ can be regarded as a model of WO_A : in particular this holds for $(\omega_1, <)$. Note that the HC-sentence $\bigvee u \prod_{\alpha < O(A)} [Sc_\alpha(\bar{0}) < u]$ is not decided by WO_A.

The argument above is analogous to familiar arguments for $\mathcal{L}_{HF}$ where in verifying the invariance for extensions of HF-formulas in Ex_1 one need only consider a finite part of $Gen(X \cup \{y\})$.

<u>Addendum</u>. Karp [Kp2, end of Bibliography] studies the following notions. Let $\mathcal{L}_V$ be the language with conjunctions and disjunctions of unlimited cardinality but only finite quantification, i.e. the formulas of $\mathcal{L}_V$ are built up like those of $\mathcal{L}_{HC}$ but now $\sum K$ and $\prod K$ can be formed for any non-empty set K of formulas. Attention is again restricted to those formulas which contain only finitely many free variables. Define the quantifier-rank $qr(\phi)$ of a formula ϕ by: $qr(\phi) = 0$ for ϕ atomic, $qr(\sim\phi) = qr(\phi)$, $qr(\sum K) = qr(\prod K) = \sup(qr(\phi): \phi \in K)$, and $qr(\bigvee u\, \phi) = qr(\bigwedge u\, \phi) = qr(\phi) + 1$. Let $\mathfrak{M} \equiv_\delta \mathfrak{M}'$ hold if for every sentence ϕ of $\mathcal{L}_V$ with $qr(\phi) < \delta$, $\models_{\mathfrak{M}} \phi$ iff $\models_{\mathfrak{M}'} \phi$, and $\mathfrak{M} \equiv \mathfrak{M}'$ if $\mathfrak{M} \equiv_\delta \mathfrak{M}'$ for every δ.

The main results of [Kp2] give necessary and sufficient conditions for $\mathfrak{M} \equiv_\delta \mathfrak{M}'$, resp. $\mathfrak{M} \equiv \mathfrak{M}'$ to hold, in terms of sets of isomorphisms between finitely generated substructures of $\mathfrak{M}, \mathfrak{M}'$. Suppose S is a countable set of HC-sentences and that there is at least one constant symbol in $\mathcal{L}_V$. It follows immediately from (the easy half) of Theorem 2(l.c.) that if S satisfies the simple extension condition then any two models $\mathfrak{M}, \mathfrak{M}'$ of S with $Gen_{\mathfrak{M}}(0) \cong Gen_{\mathfrak{M}'}(0)$ are $\equiv$; we can thus obtain the same conclusion if S satisfies any of the conditions 5.9(i)-(iv).

This surprisingly strong result can be explained by first proving an even stronger result: <u>if S admits elimination of quantifiers in $\mathcal{L}_{HC}$ then it admits elimination of quantifiers in $\mathcal{L}_V$</u>. The argument rests on the following two observations. (i) Every quantifier-free formula of $\mathcal{L}_V$ is equivalent to one in disjunctive normal form; (ii) there are only countably many atomic formulas (since the signature σ is countable), so any conjunction of atomic formulas and their negations is the conjunction of a countable set. Now we claim that <u>if S admits elimination of quantifiers in $\mathcal{L}_{HC}$ then for each formula $\emptyset$ of $\mathcal{L}_V$ there is a (quantifier-free) formula Ψ of $\mathcal{L}_{HC}$ with</u> $Fr(\emptyset) = Fr(\Psi)$ <u>and</u> $S \Vdash (\emptyset \leftrightarrow \Psi)$. For by the preceding, given an enumeration $\theta_0, \ldots, \theta_n, \ldots$ of all atomic formulas with $Fr(\theta_n) \subseteq Fr(\emptyset)$ we can find $P_1 \subseteq \mathcal{P}(\omega)$, $P_2 \subseteq \mathcal{P}(\omega)$ such that

$S \Vdash [\emptyset \leftrightarrow \sum_{X \in P_1} (\prod_{n \in X} \theta_n \wedge \prod_{n \notin X} \sim \theta_n)]$ and $S \Vdash [\sim \emptyset \leftrightarrow \sum_{Y \in P_2} (\prod_{n \in Y} \theta_n \wedge \prod_{n \notin Y} \sim \theta_n)]$.

Thus both $\emptyset$ and $\sim \emptyset$ define generalized analytic sets in the sense of [Bw], Ch.6, from which the existence of the desired Ψ follows (essentially by a suitable application of the interpolation theorem). Thus, with this hypothesis on S , $\mathcal{L}_V$ provides no greater expressive power than $\mathcal{L}_{HC}$. Of course, this is not the case in general.

Theorem 3 of [Kp2] bears on III above and in one respect yields a somewhat stronger result: suppose δ is closed under ordinal exponentiation; then for any ordered structure $\mathfrak{M} = (M, <)$ with least element, $(\delta, <) \equiv_\delta (\delta, <) \circ \mathfrak{M}$ (ordered product). The hypothesis applies to $\delta = O(A)$ for any A satisfying the conditions considered here; moreover, it is seen that if $\emptyset$ is a formula of $\mathcal{L}_A$ then $qr(\emptyset) < O(A)$.

§6. Constructive consistency proofs; provable well-orderings of formal theories.

As discussed in §1(b), systems of Gentzen-type with infinitary rules have been used successfully to give a new and basically clearer approach to constructive consistency proofs of certain formal theories. Moreover, their use has elicited the role of well-orderings and transfinite induction in such proofs. The only general exposition of this approach is Schütte's text [Sch1]. References to more recent work of interest will be found in the following.

Besides the usual rules for the propositional connectives and certain cases of quantification, [Sch1] makes use of various special rules for quantification and a special rule $(\bar{S})$ called progressionschluss. The former rules vary from theory to theory, e.g. the ω-rule in the case of number theory, certain comprehension rules in the case of ramified analysis, etc. The rule $(\bar{S})$ is used to simplify proofs which give upper bounds for provable well-orderings.

An important simplification was made by Tait [Tt], who showed how to associate with the formulas ϕ of the various formal theories which had been treated by the above methods certain formulas ϕ^* in the propositional calculus with countably long disjunctions and conjunctions, i.e. the quantifier-free part of $\mathcal{L}_{HC}$. Then using a cut-elimination theorem for this calculus, he showed how to extract consistency proofs and the known bounds. A rule of type $(\bar{S})$ is still used for the latter in [Tt].

We consider here a bit more general approach, taking the full language of $\mathcal{L}_{HC}$ for its framework. It is also shown how to treat provable well-orderings directly, without use of any new rules. This treatment is illustrated in detail for a formal system Z of number theory. The propositional reduction of [Tt] still does the main work here. However, the formulation including quantifiers can be used to give additional information for certain parts of ramified analysis described below.

Let $\mathcal{N} = (\omega, 0, ', +, \cdot)$, where $x' = x + 1$. The corresponding language $\mathcal{L}_{HC}$ has a constant symbol c_o, and function symbols f_o(unary), and f_1, f_2 (both binary). We write t' for $f_o(t)$, $(t_1 + t_2)$ for $f_1(t_1,t_2)$ and $(t_1 \cdot t_2)$ for $f_2(t_1,t_2)$. $\bar{n}$ is defined inductively by : $\bar{0}$ is c_o and $\overline{n+1}$ is $\bar{n}'$. A closed term is one which does not contain any variables. If t is a closed term, Val(t) is the value of t in $\mathcal{N}$ ($\mathrm{Val}_{\mathcal{N},x}(t)$ for any assignment x). We take

T to be the sentence $(\bar{0} = \bar{0})$ and $\bot = \sim T$. Z_o is the set consisting of the following sentences in $\mathcal{L}_{HF}$: $\bigwedge u(u' \neq \bar{0})$, $\bigwedge u_1 \bigwedge u_2(u_1' = u_2' \rightarrow u_1 = u_2)$,
$\bigwedge u_1 \bigwedge u_2 (u_1 + \bar{0} = u_1 \wedge u_1 + u_2' = (u_1 + u_2)')$,
$\bigwedge u_1 \bigwedge u_2 (u_1 \cdot \bar{0} = \bar{0} \wedge u_1 \cdot u_2' = (u_1 \cdot u_2) + u_1)$.

We shall also consider certain extensions of this basic language. In particular, to deal with questions of provable well-orderings without higher order quantifiers, it is convenient to adjoin a unary relation symbol r_o (sometimes called a <u>free predicate variable</u>). We shall write $p(t)$ instead of $r_o(t)$, and $\mathcal{L}_{HC}(p)$ for this extended language; $\mathcal{L}_A(p)$ is the language of A-formulas in $\mathcal{L}_{HC}(p)$. $\phi(p)$ or $\phi(p; u_1,\ldots,u_n)$ indicates that the symbol p may occur in ϕ . Given $\phi(p)$ and $\Psi(u)$, we denote by $\phi(\hat{u}\,\Psi(u))$ the result of replacing each occurrence of an atomic formula $p(t)$ in ϕ by $\Psi(t)$. Another kind of extension of the language to be considered later is that obtained by adjunction of a second sort of variable, intended to range over subsets of ω , and a binary relation symbol ε .

Let $I_\omega(p)$ be the sentence $[p(\bar{0}) \wedge \bigwedge u(p(u) \rightarrow p(u')) \rightarrow \bigwedge u\, p(u)]$. Then take $Ind_\omega(p)$ to be the set consisting of all the formulas $I_\omega(\hat{u}\,\Psi(u))$ for every formula Ψ of $\mathcal{L}_{HF}(p)$, and Ind_ω the subset of formulas in $Ind_\omega(p)$ which do not contain the symbol p . In more usual terms, Ind_ω consists of the instances of the scheme of induction in the natural numbers which are expressed in the basic language. We take $Z = Z_o \cup Ind_\omega$ and $Z(p) = Z_o \cup Ind_\omega(p)$. If $Z(p) \vdash \phi$ then ϕ is valid in every structure $\mathfrak{N}(P) = (\omega, P, 0, ', +, \cdot)$ where $P \subseteq \omega$. It is easily proved by induction that if $Z(p) \vdash \phi(p)$ and $\Psi(u)$ does not contain p then $Z \vdash \phi(\hat{u}\,\Psi(u))$. In particular, if ϕ does not contain p and $Z(p) \vdash \phi$ then $Z \vdash \phi$ or, as it is usually expressed, $Z(p)$ <u>is a conservative extension of</u> Z . (This applies to any sublanguage $\mathcal{L}_A$ of $\mathcal{L}_{HC}$.)

Let Ax_ω be the sentence $\bigwedge u \sum_{n<\omega} (u = \bar{n})$. We take $Z^{(\infty)} = Z_o \cup \{Ax_\omega\}$; $Z^{(\infty)}(p)$ is the same as $Z^{(\infty)}$, except regarded as a set of sentences in $\mathcal{L}_{HC}(p)$. Ax_ω has a simple recursive structure, so it is certainly in $\mathcal{L}_{HH}$. Since $Z^{(\infty)}$ is a finite set, we can also regard it as a finite sequence. The following gives closure under the ω-rule of derivability from $Z^{(\infty)}(Z^{(\infty)}(p))$, and thence derivability of Z $(Z(p))$; it also provides some useful bounds.

6.1 <u>Lemma</u>

(i) <u>Suppose for each</u> n , $\mathcal{D}_n$ <u>is a derivation of</u> $[\Gamma \supset \Delta, \phi(\bar{n})]$ <u>from</u> $Z^{(\infty)}$; <u>let</u> $\alpha = \sup_n od(\mathcal{D}_n)$, $\rho = \sup_n \rho(\mathcal{D}_n)$ <u>and</u> $\nu = od(\phi)$. <u>Then we can form a derivation</u> $\mathcal{D}$ <u>of</u> $[\Gamma \supset \Delta , \bigwedge u\, \phi(u)]$ <u>from</u> $Z^{(\infty)}$ <u>with</u> $od(\mathcal{D}) < \max(\alpha,\nu) + \omega$ <u>and</u> $\rho(\mathcal{D}) = \max(\rho, \nu + 2)$.

(ii) <u>With each</u> HF-<u>derivation</u> $\mathcal{D}$ <u>of a sequent</u> $(\Gamma \supset \Delta)$ <u>from</u> Z <u>can be associated a derivation</u> $\mathcal{D}'$ <u>of</u> $(\Gamma \supset \Delta)$ <u>from</u> $Z^{(\infty)}$ <u>with</u> $od(\mathcal{D}') < \omega\cdot 2$ <u>and</u> $\rho(\mathcal{D}') < \omega$.

(iii) <u>The results</u> (i) <u>and</u> (ii) <u>hold for</u> $Z^{(\infty)}(p)$ <u>and</u> $Z(p)$ <u>instead of</u> $Z^{(\infty)}$, Z , <u>resp</u>.

<u>Proof</u>. (i) First infer $[\Gamma \supset \Delta, \prod_{n<\omega} \phi(\bar{n})]$. It is seen by induction on ϕ that we can derive $[u=\bar{n}, \phi(\bar{n}) \supset \phi(u)]$ for any n in ν steps. Then infer $[u=\bar{n}, \prod_{m<\omega} \phi(\bar{m}) \supset \phi(u)]$. Hence by cut applied to $\prod_{m<\omega} \phi(\bar{m})$, we obtain $[\Gamma, u=\bar{n} \supset \Delta, \phi(u)]$ for each n . But then we can infer $[\Gamma, \sum_{n<\omega}(u=\bar{n}) \supset \Delta, \phi(u)]$, and thence $[\Gamma, \bigwedge u \sum_{n<\omega} (u=\bar{n}) \supset \Delta, \phi(u)]$. The conclusion is obtained by applying cut once more, to Ax_ω . (ii) Note first that for each axiom $I_\omega(\hat{u}\, \Psi(u))$ with Ψ an HF-formula, the sequent $0 \supset [\Psi(0) \wedge \bigwedge u(\Psi(u) \rightarrow \Psi(u')) \rightarrow \Psi(\bar{n})]$ has a derivation $\mathcal{D}_n$ with $od(\mathcal{D}_n) < \omega$ and $\rho(\mathcal{D}_n) = od(\Psi) + 1$. Then by (i), each axiom of $(0 \supset Z)$ has a derivation bounded as required. Now given an HF-derivation $\mathcal{D}$ of $(\Gamma \supset \Delta)$ from Z , form the new derivation $\mathcal{D}'$ by placing above each axiom of Z its derivation from $Z^{(\infty)}$ just described.

Given a derivation of $(\Gamma \supset \Delta)$ from $Z^{(\infty)}$ we obtain a derivation of $(Z^{(\infty)}, \Gamma \supset \Delta)$ of the same length with no new cuts (return to the argument for the Deduction Theorem). Thus we immediately obtain the following from (3.7). <u>If</u> $(\Gamma \supset \Delta)$ <u>has an</u> HF-<u>derivation from</u> Z <u>then we can find a cut-free derivation</u> $\mathcal{D}^*$ <u>of</u> $(Z^{(\infty)}, \Gamma \supset \Delta)$ <u>with</u> $od(\mathcal{D}^*) < \epsilon_0$. An analysis of this argument can then be given to provide a constructive consistency proof of Z (in $\mathcal{L}_{HF}$) , with transfinite induction up to ϵ_0 being the only non-finitist principle involved.

To make such an analysis more perspicuous and to treat the provable well-orderings of Z simply, we now make use of the obvious association of a quantifier-free sentence ϕ^* with each sentence ϕ of $\mathcal{L}_{HC}(p)$. Namely, $(\bigwedge u\, \phi(u))^* = \prod_{n<\omega}(\phi(\bar{n}))^*$, $(\bigvee u\, \phi(u))^* = \sum_{n<\omega}(\phi(\bar{n}))^*$. For simplicity, when ϕ is an atomic sentence $(t_1=t_2)$

we take ϕ^* to be $\top$ if $\mathrm{Val}(t_1) = \mathrm{Val}(t_2)$ and $\perp$ if $\mathrm{Val}(t_1) \neq \mathrm{Val}(t_2)$; if ϕ is an atomic sentence $p(t)$ and $\mathrm{Val}(t) = n$ we take ϕ^* to be $p(\bar{n})$. Of course $(\sim \phi)^* = \sim \phi^*$, $(\sum_{\phi \in K} \phi)^* = \sum_{\phi \in K} \phi^*$ and $(\prod_{\phi \in K} \phi)^* = \prod_{\phi \in K} \phi^*$. It follows directly that $\mathrm{od}(\phi^*) = \mathrm{od}(\phi)$. If $\phi(w_1,\ldots,w_k)$ is a formula with free variables among $w_1,\ldots,w_k$, we write $\phi^*(\bar{n}_1,\ldots,\bar{n}_k)$ for $(\phi(\bar{n}_1,\ldots,\bar{n}_k))^*$. For a sequence $\Gamma(w_1,\ldots,w_k)$ with free variables among $w_1,\ldots,w_k$, $\Gamma^*(\bar{n}_1,\ldots,\bar{n}_k)$ is the sequence with terms $\phi^*(\bar{n}_1,\ldots,\bar{n}_k)$ for each term ϕ of Γ .

6.2 <u>Lemma</u>. <u>Suppose the free variables of</u> $(\Gamma \supset \Delta)$ <u>are</u> $w_1,\ldots,w_k$ <u>and that</u> $\mathcal{D}$ <u>is a derivation of</u> $\Gamma(w_1,\ldots,w_k) \supset \Delta(w_1,\ldots,w_k)$ <u>from</u> $Z^{(\infty)}(p)$. <u>Then for each</u> $n_1,\ldots,n_k$ <u>we can find a derivation</u> $\mathcal{D}'$ <u>of</u> $[\Gamma^*(\bar{n}_1,\ldots,\bar{n}_k) \supset \Delta^*(\bar{n}_1,\ldots,\bar{n}_k)]$ <u>with</u> $\mathrm{od}(\mathcal{D}') \leq \omega + \mathrm{od}(\mathcal{D})$ $\rho(\mathcal{D}') = \rho(\mathcal{D})$.

<u>Proof</u>. Consider, for example, the member $[0 \supset \bigwedge u \sum_{n<\omega}(u = \bar{n})]$ of $Z^{(\infty)}$. The associated propositional sequent is $[0 \supset \prod_{m<\omega} \sum_{n<\omega} T_{m,n}]$ where $T_{m,n} = \top$ if $m = n$, and $= \perp$ if $m \neq n$. This has a cut-free derivation of length 3 . Suppose given a derivation $\mathcal{D}$ of the form

$$\begin{array}{c} \vdots\, \mathcal{D}_0 \\ (\supset \bigwedge)\ \dfrac{\Gamma(w_1,\ldots,w_k) \supset \Delta(w_1,\ldots,w_k),\ \phi(w_1,\ldots,w_k,w)}{\Gamma(w_1,\ldots,w_k) \supset \Delta(w_1,\ldots,w_k),\ \bigwedge u \phi(w_1,\ldots,w_k,u)}\ , \end{array}$$

where w is not free in the conclusion. Then, for each $n_1,\ldots,n_k,m$ we get a derivation $\mathcal{D}'_{0,m}$ of

$$[\Gamma^*(\bar{n}_1,\ldots,\bar{n}_k) \supset \Delta^*(\bar{n}_1,\ldots,\bar{n}_k),\ \phi^*(\bar{n}_1,\ldots,\bar{n}_k,\bar{m})]\ .$$

with $\mathrm{od}(\mathcal{D}'_{0,m}) \leq \omega + \mathrm{od}(\mathcal{D}_0)$, $\rho(\mathcal{D}'_{0,m}) = \rho(\mathcal{D}_0)$. Then we can infer $[\Gamma^*(\bar{n}_1,\ldots,\bar{n}_k) \supset \Delta^*(\bar{n}_1,\ldots,\bar{n}_k)\ ,\ \prod_{m<\omega} \phi^*(\bar{n}_1,\ldots,\bar{n}_k,\bar{m})]$, with length of derivation $\leq \omega + \mathrm{od}(\mathcal{D}_0) + 1$ and same cut-rank. The remainder of the argument is left to the reader.

We now apply (3.7) to (6.1(ii)) and (6.2) .

6.3 <u>Corollary</u>. <u>Suppose</u> $(\Gamma \supset \Delta)$ <u>consists of HF-sentences and that</u> $Z \vdash (\Gamma \supset \Delta)$. <u>Then we can find a cut-free derivation</u> $\mathcal{D}$ <u>of</u> $(\Gamma^* \supset \Delta^*)$ <u>with</u> $\mathrm{od}(\mathcal{D}) < \epsilon_0$. <u>The</u>

<u>same holds with</u> Z(p) <u>instead of</u> Z .

We immediately conclude that Z <u>is consistent</u>, i.e. it is not the case that $Z \vdash (0 \supset 0)$. For, given any cut-free derivation $\mathfrak{D}$ of any sequent $(\Gamma_1 \supset \Delta_1)$, we prove by induction on $\mathfrak{D}$ that $\Gamma_1 \neq 0$ or $\Delta_1 \neq 0$. The difference between this consistency proof and a model-theoretic proof is that the latter makes use of the non-constructive notion of validity in $\mathfrak{N}$. However, the present proof is not yet seen to be constructive, since it involves the set-theoretical apparatus associated with infinite formulas, infinite derivations, and ordinals. But the proof, and the work on which it depends, can be seen to be constructive by re-examining these notions in a suitable light. The reader with some expert knowledge can find reasonable indications as to how this is to be done in several scattered sources, in particular [Kr3], §6 and [Tk2], §1. Unfortunately, there is no detailed exposition available which is suitable for the general reader, nor is there space here to provide such. For convenience, we simply repeat some of the main ideas involved.

First, abstract ordinals are replaced by certain specific <u>recursive well-ordering relations</u> $\preccurlyeq$ with field $\{a: a \preccurlyeq a\}$ contained in ω. Let a,b range over this field: we write $a \prec b$ if $a \preccurlyeq b$ and $a \neq b$. The ordinal corresponding to any initial segment $\{b: b \prec a\}$ is denoted by $|a|$. With respect to any such ordering we can deal with a class of formulas and derivations which are recursively represented by natural numbers, with order and rank bounded by elements of the field of $\preccurlyeq$. We could take here a coding which follows the set-theoretical representation of §2(d), except that now also applications of infinitary operations or rules are to be accompanied by explicit enumerations. Thus, for example, $2^a \cdot 3^6 \cdot 5^e$ represents $\sum_{n<\omega} \phi_n$ if for each n, $\{e\}(n) = 2^{a_n} \cdot 3^{b_n} \cdot 5^{c_n}$ represents ϕ_n with $a_n \prec a$.[27] Such clauses will insure that if f represents ϕ then $od(\phi) \leq |(f)_0|$. If f represents ϕ we can effectively extract from f a number representing the subformula of ϕ lying at any specified point along any specified branch in the tree of ϕ ; f also reveals what operations are used to build ϕ at each node. It can then be seen that there is a completely elementary predicate, Fmla(f), such that

---

27) [LE2] gives a similar coding, but without consideration of accompanying constructive ordinal bounds.

We use here the notation from recursion theory of [Kl] .

Fmla(f) holds if and only if f represents some formula relative to $\preccurlyeq$.[28] Numbers which represent sequents are obtained in the obvious way from numbers which represent formulas.

Similarly, derivations $\mathcal{D}$ are to be represented by numbers d which completely described the tree of $\mathcal{D}$ and such that $od(\mathcal{D}) \leq |(d)_o|$, $\rho(\mathcal{D}) \leq |(d)_1|$. Thus if $(d)_1 = 0$, d represents a cut-free derivation. We shall take it that the coding is such that if d represents $\mathcal{D}$ then $(d)_2$ represents the conclusion of $\mathcal{D}$. Again, one obtains a completely elementary predicate Dern(d)[28] which holds if and only if d represents some derivation relative to $\preccurlyeq$.

The next step is to see what elementary statements can be made which correspond to the basic cut-elimination results (3.3)-(3.7). For this purpose, it is necessary to deal with relations $\preccurlyeq$ satisfying certain closure conditions. A function $\varphi(\alpha_1,\ldots,\alpha_n)$ of ordinals is said to be <u>recursively representable on</u> $\preccurlyeq$ if we have a partial recursive function $F(a_1,\ldots,a_n)$ such that for each $a_1,\ldots,a_n$ (in the field of $\preccurlyeq$), $F(a_1,\ldots,a_n)$ is in the field of $\preccurlyeq$ and $|F(a_1,\ldots,a_n)| = \varphi(|a_1|,\ldots,|a_n|)$. For example, the functions $(\alpha_1+\alpha_2)$ and ω^{α_1} are recursively representable on the natural recursive well-ordering $\preccurlyeq_o$ of ω with order-type ϵ_o ; we write $(a_1 \oplus a_2)$, w^{a_1} for the corresponding number-theoretic functions. This can be embedded in a natural recursive well-ordering $\preccurlyeq_1$ of order type Γ_o on which the function $\kappa^{(\alpha_2)}(\alpha_1)$ is recursively representable; cf. [Ff2] or [Sch2] .

Take $|0| = 0$, $|1| = 1$, $|w| = \omega$ in $\preccurlyeq_o$. Regard the predicates Fmla(f), Dern(d) as given relative to $\preccurlyeq_o$. We are thus led to consider the following statements corresponding to (3.6), (3.7):

(E3.6) <u>If</u> Dern(d) <u>and</u> $(d)_o = a$, $(d)_1 \preccurlyeq_o b \oplus 1$, <u>then there exists</u> d' <u>such that</u> $(d')_2 = (d)_2$, $(d')_1 \preccurlyeq_o b$ <u>and</u> $(d')_o \preccurlyeq_o w^{w \oplus a}$.

(E3.7) <u>If</u> Dern(d) <u>and</u> $(d)_1 \prec w$ <u>then there exists</u> d' <u>such that</u> $(d')_2 = (d)_2$ <u>and</u> $(d')_1 = 0$.

Then reconsideration of the arguments of (3.6), (3.7) leads one to obtain construc-

---

[28] To be precise, the predicate Fmla(f) can be taken to be in the Π^o_2 level of arithmetical hierarchy, i.e. expressible in the form (x)(Ey)R(f,x,y) with R primitive recursive. The same applies to the predicate Dern(d) introduced next.

tive proofs of (E3.6), (E3.7). In each case, one finds a partial recursive function G such that for any d satisfying the hypothesis, $d' = G(d)$ satisfies the conclusion. Indeed, the definition of the desired G is set up in advance by following out the cases in the proofs of (3.6), (3.7) (like those in (3.3), (3.4)). Then the argument for the present results is carried through by applying transfinite induction to the relation $\preccurlyeq_0$, in each case to a predicate of the form: if d satisfies the hypothesis then $G(d)$ is defined and satisfies the conclusion.

One can similarly obtain constructive results corresponding to (3.3)-(3.5) by using the relation $\preccurlyeq_1$.

Next turning specifically to Z, we consider throughout formulas and derivations represented relative to $\preccurlyeq_0$. Now it is very easy to formulate elementary versions (E6.1)-(E6.3) of (6.1)-(6.3) and to obtain constructive proofs for the former directly from (E3.7) and the arguments for the latter. In this way we obtain a constructive consistency proof for Z. The only essentially non-combinatorial part of this proof consists in the application of transfinite induction with respect to $\preccurlyeq_0$. For stronger theories for which one has consistency proofs via infinitary systems such as to be described below, the general method of constructivization is the same, only the ordering used may be different.

If we take the preceding for granted, it follows by Gödel's second underivability theorem that certain specific elementary instances of transfinite induction w. r. to $\preccurlyeq_0$ are not derivable in Z. We shall now establish by direct means for a wide class of definable well-orderings $\preccurlyeq$ that if the order type of the relation is at least ϵ_0, then the <u>general</u> principle of transfinite induction with respect to $\preccurlyeq$ cannot be derived in Z.

Suppose given any formula $\Lambda(w_1,w_2)$ of $\mathcal{L}_{HF}$ with two free variables w_1,w_2. Since we are interested in the situation that this defines an ordering relation, we write $(w_1 \preccurlyeq w_2)$ instead of $\Lambda(w_1,w_2)$. We then write $(w_1 \prec w_2)$ for $(w_1 \preccurlyeq w_2) \wedge (w_1 \neq w_2)$.

Let $SO_{\preccurlyeq}$ be the conjunction of the sentences expressing that this relation is a simple ordering (of its field) namely, $\bigwedge u_1 \bigwedge u_2 (u_1 \preccurlyeq u_2 \wedge u_2 \preccurlyeq u_1 \rightarrow u_1 = u_2)$,
$\bigwedge u_1 \bigwedge u_2 \bigwedge u_3 (u_1 \preccurlyeq u_2 \wedge u_2 \preccurlyeq u_3 \rightarrow u_1 \preccurlyeq u_3)$,
$\bigwedge u_1 \bigwedge u_2 (u_1 \preccurlyeq u_1 \wedge u_2 \preccurlyeq u_2 \rightarrow u_1 \preccurlyeq u_2 \vee u_2 \preccurlyeq u_1)$.

Then the general principle of transfinite induction w.r. to $\preccurlyeq$ is expressed by the sentence $TI_{\preccurlyeq}(p)$ which is the conjunction of $SO_{\preccurlyeq}$ and

$$\bigwedge u\,[u \preccurlyeq u \wedge \bigwedge u_1(u_1 \prec u \rightarrow p(u_1)) \rightarrow p(u)] \rightarrow \bigwedge u(u \preccurlyeq u \rightarrow p(u)) .$$

If $Z(p) \vdash TI_{\preccurlyeq}(p)$ then the formula $(w_1 \preccurlyeq w_2)$ defines a well-ordering relation, since $TI_{\preccurlyeq}(p)$ holds in $\mathfrak{N}(P)$ for all $P \subseteq \omega$. In this case, we denote by $|\preccurlyeq|$ the ordinal to which this relation is isomorphic.

6.4 <u>Theorem</u>. <u>If</u> $Z(p) \vdash TI_{\preccurlyeq}(p)$ <u>then</u> $|\preccurlyeq| < \epsilon_0$.

<u>Proof</u>. If $\preccurlyeq'$ is provably isomorphic to $\preccurlyeq$ in Z then also $Z(p) \vdash TI_{\preccurlyeq'}(p)$. If $TI_{\preccurlyeq_1}(p)$ and $TI_{\preccurlyeq_2}(p)$ are both provable in $Z(p)$ we can find provably isomorphic $\preccurlyeq_1'$, $\preccurlyeq_2'$ with disjoint fields. Then the relation $\preccurlyeq^+$ which is the ordered sum of $\preccurlyeq_1'$, $\preccurlyeq_2'$ has $TI_{\preccurlyeq^+}(p)$ provable in $Z(p)$. In particular, if we take $\preccurlyeq_2$ with $|\preccurlyeq_2| = \omega+1$, $\preccurlyeq^+$ is a relation with provably infinite field and with a final element. For the result, we can thus assume now that $\preccurlyeq$ has the same properties. Now we can define a relation $\preccurlyeq_1$ which is provably isomorphic to $\preccurlyeq$ and which has field equal to ω . Mathematically speaking, simply enumerate the field $\{x: x \preccurlyeq x\}$ of $\preccurlyeq$ in increasing $<$ order: for each n, $F(n) =$ (least x) $[x \preccurlyeq x$ and $F(i) < x$ for all $i < n]$. Then take $m \preccurlyeq_1 n$ iff $F(m) \preccurlyeq F(n)$. Thus $\preccurlyeq_1$ is recursive in $\preccurlyeq$. The relation $F(n) = x$ holds just in case there exists a finite sequence $x_1,\ldots,x_n$ such that for each $i \leq n$, $(x_i \preccurlyeq x_i)$ and such that for no $y < x_1$ and no y with $x_i < y < x_{i+1}$ do we have $(y \preccurlyeq y)$. Using the representation of finite sequences in Z due to Gödel ([Kl], Ch. IX) this relation and thence $\preccurlyeq_1$ is explicitly definable in Z .

Without loss of generality, we can assume now that the field of $\preccurlyeq$ is ω ; trivially, 0 can be taken to be the least element in the $\preccurlyeq$ relation. Moreover, we may assume $(w_1 \preccurlyeq w_2)$ expressed in prenex form

$\bigwedge u_1 \bigvee v_1 \ldots \bigwedge u_k \bigvee v_k\, \Psi(w_1,w_2,u_1,\ldots,u_k,v_1,\ldots,v_k)$ (Ψ quantifier-free). It is seen by successive applications of the ω rule (6.1(i)) that there is a $q < \omega$ such that for any n,m with $(n \preccurlyeq m)$ there is a derivation $\mathcal{D}_{n,m}$ of $\overline{n} \preccurlyeq \overline{m}$ from $Z^{(\infty)}$ with $od(\mathcal{D}_{n,m}) < \omega$, $\rho(\mathcal{D}_{n,m}) < q$. Then the same obviously holds for $\overline{m} \prec \overline{n}$

when $m \prec n$, and for $\sim (m \prec n)$ when $m \not\prec n$.[29]

Suppose $Z(p) \vdash TI_{\preccurlyeq}(p)$. By (6.2) there is a derivation of order $< \omega.3$ and finite cut-rank of $(TI_{\preccurlyeq}(p))^*$, i.e. of

$$\prod_{n<\omega} \{(\bar{n} \preccurlyeq \bar{n})^* \wedge [\prod_{m<\omega} ((\bar{m} \prec \bar{n})^* \rightarrow p(\bar{m})) \rightarrow p(\bar{n})]\} \rightarrow \prod_{n<\omega} p(\bar{n}) .$$

It is seen by combining this with the preceding derivations $\mathcal{D}_{n,m}$ that there is then a derivation of

$$p(\bar{0}) \wedge \prod_{0<n<\omega} [\prod_{m \prec n} p(\bar{m}) \rightarrow p(\bar{n})] \rightarrow \prod_{n<\omega} p(\bar{n})$$

with order $< \omega^2$ (certainly) and also of bounded cut-rank. Hence, for each k there is a cut-free derivation $\mathcal{D}$ of $\{p(\bar{0}), \prod_{0<n<\omega}[\prod_{m \prec n} p(\bar{m}) \rightarrow p(\bar{n})] \supset p(\bar{k})\}$ with $od(\mathcal{D}) < \epsilon_0$. This reduces the matter to the proof of the following.

6.5 **Lemma**. *Suppose $\preccurlyeq$ is any well-ordering of ω , with least element 0 . If $\mathcal{D}$ is a cut-free derivation of*

$$p(\bar{0}) , \prod_{0<n<\omega} [\prod_{m \prec n} p(\bar{m}) \rightarrow p(\bar{n})] \supset p(\bar{k}) ,$$

then $|k| \leq od(\mathcal{D})$.

The proof of this is left to the reader; it is a non-trivial exercise in analyzing possibilities in cut-free derivations.

To complete the proof of the theorem, take k to be the final element in the ordering $\preccurlyeq$. Then $|k| < \epsilon_0$, and hence $|\preccurlyeq| < \epsilon_0$.

The result (6.4) is best possible. For, as shown by Gentzen, for each proper initial segment $\preccurlyeq_0 \upharpoonright n$ of the natural ordering of type ϵ_0 we have $Z(p) \vdash TI_{\preccurlyeq_0 \upharpoonright n}(p)$; cf. [Sch1] for a simple proof.

We now briefly sketch the extension of this treatment to the system R_0 of arithmetical analysis. In this case $\mathcal{L}_{HC}$ is taken to be the number-theoretical language extended by adding a second sort of variable called *set variables*, for

[29] If one wishes to avoid the non-constructive applications of the ω-rule here, the hypothesis of the theorem can be restricted to relations $\preccurlyeq$ for which such derivations $\mathcal{D}_{n,m}$ exist. In particular, this is easily seen to apply to recursive relations $\preccurlyeq$.

which we use the letters X,Y,..., and a binary relation symbol ε . Then $\mathcal{L}_{HC}(p)$ is determined as before. Ind_ω (p) is now taken to consist of all formulas $I_\omega(\hat{u}\ \Psi(u))$ in this extended language $\mathcal{L}_{HC}$ (p) , and Ind_ω the subset of formulas in Ind_ω (p) which do not contain the symbol p . The sentence $\bigwedge X \bigwedge Y [\bigwedge u(u\varepsilon X \leftrightarrow u\varepsilon Y) \rightarrow X = Y]$ is called the axiom of extensionality, denoted by Ext.

A formula of $\mathcal{L}_{HC}$ (p) is said to be arithmetical if it contains no bound set variables; it is said to be strictly arithmetical if it contains no set variables at all. In the following, $\mathcal{O}\!\mathcal{l}, \mathcal{L}$ range over arithmetical formulas. By CA(p) we mean the set consisting of all formulas

$$\bigvee X \bigwedge u\ [u\varepsilon X \leftrightarrow \mathcal{O}\!\mathcal{l}(u)]$$

where $\mathcal{O}\!\mathcal{l}$ is an arithmetical formula of $\mathcal{L}_{HF}$ (p) . CA is the subset of CA(p) consisting of formulas which do not contain p . The members of CA are the instances of the relative arithmetic comprehension axiom. We take $R_o(p) = Z_o \cup Ind_\omega(p) \cup \{Ext\} \cup CA(p)$, and R_o to be the subset $Z_o \cup Ind_\omega \cup \{Ext\} \cup CA$. One now has the following form of conservative extension result: if $R_o(p) \vdash \phi(p)$ and p is not in $\mathcal{L}(u)$ and $\mathcal{L}$ is arithmetical then $R_o \vdash \phi(\hat{u}\mathcal{L}(u))$. Given $\phi(X)$, write $\phi(p)$ for the result of substituting p(t) for each occurrence $(t \in X)$ in ϕ ; this explains the notation $\phi(\hat{u}\ \Psi(u))$ below. Note that $R_o(p) \vdash \phi(X)$ if and only if $R_o(p) \vdash \phi(p)$.

For any $P \subseteq \omega$, let Ar(P) be the collection of all subsets of ω which are arithmetically definable from P , i.e. definable in $\mathfrak{N}(P)$. Then $(\omega, Ar(P); P, 0, ', +, \cdot)$ is a model of $R_o(p)$. In particular, for P = 0 and Ar = Ar(0) , $(\omega, Ar; 0, ', +, \cdot)$ is a model of R_o .

As before, we have associated infinitary systems $R_o^{(\infty)} = Z_o \cup \{Ax_\omega, Ext, CA\}$, $R_o^{(\infty)}(p) = Z_o \cup \{Ax_\omega, Ext, CA(p)\}$. Then we can obtain a result like (6.1) for the systems R_o, $R_o^{(\infty)}$ and $R_o(p)$, $R_o^{(\infty)}(p)$, using the same bounds.

The first main difference comes with the associated infinitary propositional sentences. We consider first the language without p . We can give a simple enumeration $\mathcal{O}\!\mathcal{l}_n(u)$, n = 0,1,2,... of all (strictly) arithmetical formulas of $\mathcal{L}_{HF}$

with just one free variable u . The operation $\emptyset^*$ is defined for every sentence of $\mathcal{L}_{HC}$ as follows. For atomic sentences, numerical quantifiers, $\sim$, $\sum$ and $\prod$, this operation is treated as before. We now take $(\bigwedge X\, \emptyset(X))^* = \prod_{n<\omega}[\emptyset(\hat{u}\,\mathfrak{A}_n(u))]^*$ and $(\bigvee X\, \emptyset(X))^* = \sum_{n<\omega}[\emptyset(\hat{u}\,\mathfrak{A}_n(u))]^*$. Then it is seen that for any $\emptyset$, $od(\emptyset^*) \leq \omega + od(\emptyset)$.

The next step corresponds to (6.2). One shows that if $\mathcal{D}$ is a derivation of $\Gamma(w,\ldots,Y,\ldots) \supset \Delta(w,\ldots,Y,\ldots)$ from $R_0^{(\infty)}$ then for any $n,\ldots,m,\ldots$ we can find a derivation $\mathcal{D}'$ (from the empty set) of

$$\Gamma^*(\bar{n},\ldots,\hat{u}\,\mathfrak{A}_m(u),\ldots) \supset \Delta^*(\bar{n},\ldots,\hat{u}\,\mathfrak{A}_m(u),\ldots)$$

with $od(\mathcal{D}') \leq \omega + od(\mathcal{D})$ and $\rho(\mathcal{D}') \leq \omega + \rho(\mathcal{D})$. In particular, by the extension of (6.1) to R_0 , if $(\Gamma \supset \Delta)$ consists of HF-sentences and $R_0 \vdash (\Gamma \supset \Delta)$, there is a derivation $\mathcal{D}$ of $(\Gamma^* \supset \Delta^*)$ with $od(\mathcal{D}) < \omega.2$ and $\rho(\mathcal{D}) < \omega.2$. But then by the exercise (ii) at the end of §3(b), there is a cut-free derivation $\mathcal{D}$ of $(\Gamma^* \supset \Delta^*)$ with $od(\mathcal{D}) < \epsilon_{\epsilon_0}$. In this way we obtain a consistency proof of R_0 by transfinite induction up to ϵ_{ϵ_0} .

We can obtain similar results for derivability from $R_0(p)$ by using the following modification of $\emptyset^*$. We now construe $\mathfrak{A}_n(u) = \mathfrak{A}_n(p;u)$ to be an enumeration of all (strictly) arithmetical formulas of $\mathcal{L}_{HF}(p)$. Then the same bounds can be obtained. Thus if $(\Gamma \supset \Delta)$ consists of sentences from $\mathcal{L}_{HF}(p)$, and $R_0(p) \vdash (\Gamma \supset \Delta)$ there is a cut-free derivation $\mathcal{D}$ of $(\Gamma^* \supset \Delta^*)$ with $od(\mathcal{D}) < \epsilon_{\epsilon_0}$.

To treat the provable well-orderings of R_0 , it is not necessary to pass to $R_0(p)$. For, from earlier observations, the following are equivalent for a formula $(w_1 \preccurlyeq w_2)$ of $\mathcal{L}_{HF}$: $R_0(p) \vdash TI_{\preccurlyeq}(p)$, $R_0(p) \vdash TI_{\preccurlyeq}(X)$, $R_0 \vdash TI_{\preccurlyeq}(X)$. However, the use of the symbol p is still of simplifying value. Again the result will apply to any arithmetical formula, or more generally to any formula $(w_1 \preccurlyeq w_2)$ such that for some $\alpha < \epsilon_0$ and $q < \omega$ there is for each n,m with $n \preccurlyeq m$ a derivation $\mathcal{D}_{n,m}$ of $(\bar{n} \preccurlyeq \bar{m})$ from $R_0^{(\infty)}$ with $od(\mathcal{D}_{n,m}) \leq \alpha$, $\rho(\mathcal{D}_{n,m}) \leq \omega + q$. Assuming this, and reducing to the case of an ordering with field ω as in (6.4), we can argue just as done there that if $R_0(p) \vdash TI_{\preccurlyeq}(p)$ then for each k there is a cut-free derivation $\mathcal{D}$ of

$$p(\bar{0}) \, , \prod_{0<n<\omega}\left[\prod_{m \prec n} p(m) \rightarrow p(n)\right] \supset p(k)$$

with $od(\mathcal{D}) < \epsilon_{\epsilon_0}$. But then we can apply (6.5) without change to conclude that $|k| < \epsilon_{\epsilon_0}$. Hence, as before , $|\preccurlyeq| < \epsilon_{\epsilon_0}$. Once more, it is not difficult to show that this result is best possible.

The proof-theoretical methods discussed here yield much more information about the formal systems treated (including those to be discussed below) than is contained in the statements of consistency and the results about provable well-orderings. This is emphasized by Kreisel [Kr3] , where much of the additional information is related to the notion of what is called a reflection principle. Indeed this is the central notion for a general formulation in [Kr3] of what is to be accomplished in extensions of Hilbert's program. It is insufficient to say that the main purpose is to prove consistency of various S by transfinite induction with respect to certain recursive well-ordering relations. For, one can easily make up a primitive recursive well-ordering $\preccurlyeq$ of order-type ω such that the consistency of Z follows in a trivial way from the statement that this ordering has a least element. On the other hand, Kreisel [Kr4] has shown that under quite general circumstances if one can deduce certain instances of the reflection principle for S from $TI_{\preccurlyeq}(p)$ then $|\preccurlyeq|$ is an upper bound for the provable well-orderings of S . Moreover, the consistency of S is equivalent to a particularly simple instace of the reflection principle for S . Though more informative, we have not taken up this approach here, because this would involve bringing in the arithmetization (formalization in Z) of provability in formal systems, which would require more development.

The discussion of R_0 above contains the basic ideas for a proof-theoretical study of systems of ramified analysis R_α , $R_\alpha^{(\infty)}$ of finite and transfinite rank α in the framework of $\mathcal{L}_{HC}$, and which recaptures the known results [Sch1] . In connection with the study of the informal notion of predicativity, Schütte [Sch2], [Sch3] and the author [Ff1], [Ff3] considered autonomous sequences of ramified systems. The autonomous ordinals α are inductively defined essentially as follows: if β is autonomous and $\beta < \alpha$ and we have a relation $\preccurlyeq$ defined in R_β with $R_\beta(p) \vdash TI_{\preccurlyeq}(p)$ and $|\preccurlyeq| = \alpha$ then α is autonomous. It was shown in the papers mentioned that the autonomous ordinals α are just those with $\alpha < \Gamma_0$.

In [Ff1], I also considered some non-ramified subsystems of classical analysis H_α, and established reductions of these systems to the ramified systems and vice versa. The result that Γ_0 is an upper bound for the ordinals autonomous with respect to the systems H_α could then be extracted from the result for the ramified systems. Modifying ideas of Tait [Tt], it is shown in [Ff3] how to obtain this bound directly from results concerning $\mathcal{L}_{HC}$. It turns out advantageous there to consider systems of arithmetical analysis which are infinitary in still a further sense: Given any ordinal γ, take CA_γ to consist of all formulas

$$\bigvee X \bigwedge u\, [u \varepsilon X \leftrightarrow \mathfrak{A}(u)]$$

for $\mathfrak{A}$ any arithmetical formula of $\mathcal{L}_{HC}$ with $od(\mathfrak{A}) < \gamma$. Then it can be shown that all theorems of the autonomous H_α are derivable from the system with axioms CA_{Γ_0} (and the other axioms of $R_0^{(\infty)}$). The quantifiers have a very useful role here. For each $\mathfrak{A}(u, w_1, \ldots, w_n, Y_1, \ldots, Y_m)$ we can introduce a function symbol $f_{\mathfrak{A}}$ and modify the above comprehension axiom to

$$\bigwedge u[u \varepsilon\, f_{\mathfrak{A}}(w_1, \ldots, w_n, Y_1, \ldots, Y_m) \leftrightarrow \mathfrak{A}(u, w_1, \ldots, w_n, Y_1, \ldots, Y_m)]\ .$$

By extensionality, this is a conservative extension procedure. In this way, all the axioms of this infinitary system of arithmetical analysis become universal, in the sense of §4. Then we can apply the interpolation theorem (4.9) whenever an existential formula is derived in this system. Exact statements of results concerning autonomous systems and more details concerning the application of proof theory in $\mathcal{L}_{HC}$ indicated here are to be found in [Ff3].

A much stronger subsystem of 2nd-order classical analysis than the ones mentioned so far has as principal axioms the instances of the comprehension schema given by Π^1_1-formulas; we refer to it as Π^1_1-analysis. Takeuti [Tk1], [Tk2] has obtained a constructive consistency proof of a system S which contains Π^1_1-analysis, but is not far removed from it. This is the strongest system that we know of at the time of writing for which such a proof has been obtained. Takeuti studies derivability in S via a certain 2nd-order extension of Gentzen's sequential predicate calculus. In [Tk2], the ω-rule is adjoined to get a full cut-elimination theorem. The proof is by transfinite induction with respect to a certain ordering called a system of ordinal diagrams. Every initial segment of this ordering can be proved to be well-ordered in S , and it is shown that every provable well-ordering of S can be effectively embedded in it. However, in contrast to the orderings used in the treatment of the ramified systems, it is difficult to understand what leads one to the use of ordinal diagrams in Takeuti's work. If an alternative method of assignment of ordinal notations can be found which gives the same ordinals but does not have this ad hoc character, this may remove one obstacle to the treatment of substantially stronger systems.

The other known approaches to proof theory have also not succeeded in proceeding constructively beyond this point. It may be that proof theory has reached its limits as a tool in uncovering the constructive content of classical mathematics. Even if not, it may be that Gentzen's particular approach cannot be significantly extended. However, at the moment it seems to me possible that some quite fresh way of extending it, comparable to the introduction of infinitary systems, can open the way ahead.

Bibliography

[Ba] Bachmann, H., Transfinite Zahlen, Springer, Berlin (1955)

[Bw] Barwise, K. J., Infinitary logic and admissible sets, Dissertation, Stanford University (1967).

[Be] Beth, E. W., On Padoa's method in the theory of definition, Indag. Math., v. 15 (1953), pp. 330-339.

[Ch] Choodnovsky, G., Some results in theory of infinitely long expressions, (abstract) Proc. 3d Intl. Cong. for Logic, Methodology and Philos. of Science, Amsterdam, 1967 (to appear).

[Cr1] Craig, W., Linear reasoning. A new form of the Herbrand-Gentzen theorem, Journ. Symbolic Logic, v. 22 (1957), pp. 250-268.

[Cr2] ________, Three uses of the Herbrand-Gentzen theorem in relating model theory and proof theory, Journ. Symbolic Logic, v. 22 (1957), pp. 269-285.

[Ff1] Feferman, S., Systems of predicative analysis, Journ. Symbolic Logic, v. 29 (1964) pp. 1-30.

[Ff2] ________, Systems of predicative analysis, II: representations of ordinals, Journ. Symbolic Logic (to appear).

[Ff3] ________, Autonomous transfinite progressions and the extent of predicative mathematics, Proc. 3d Intl. Cong. for Logic, Methodology and Philos. of Science, Amsterdam, 1967 (to appear).

[Ff4] ________, Persistent and invariant formulas for outer extensions, Compos. Math. (to appear).

[F,K] Feferman, S. and G. Kreisel, Persistent and invariant formulas relative to theories of higher order, (Research Announcement) Bull. Amer. Math. Soc. v. 72 (1966), pp. 480-485.

[Fr] Fraïssé, R., Une notion de récursivité relative, in Infinitistic Methods (Proc. Symp. founds. maths, Warsaw, 1959) Pergamon, Oxford (1961), pp. 323-328.

[Gn1] Gentzen, G., Untersuchungen über das logische Schließen, Math. Zeitschr. v. 39 (1934) pp. 176-210, 405-431.

[Gn2] ________, Die Widerspruchsfreiheit der reinen Zahlentheorie, Math. Annalen, v. 112 (1936), pp. 493-565.

[Gn3] ________, Neue Fassung des Widerspruchsfreiheitsbeweis fur die reine Zahlentheorie, Forsch. zur Logik und zur Grundlegung der exacten Wissenschaften, no. 4 (1938), Leipzig pp. 19-44.

[Gn4] ________, Beweisbarkeit und Unbeweisbarkeit von Anfangsfallen der transfiniten Induktion in der reinen Zahlentheorie, Math. Annalen, v. 119 (1943) pp. 140-161.

[Gd1] Gödel, K., Über formal unentscheidbare Satze der Principia Mathematica und verwandter Systeme I. Monats. Math., Physik v. 38 (1931) pp. 173-198 (transl. in [vH], pp. 596-616).

[Gd2] ________, Über eine bisher noch nicht benützte Erweiterung des finiten Standpunktes, Dialectica, v. 12 (1958), pp. 280-287.

[G,M,R] Grzegorczyk, A., A. Mostowski and C. Ryll-Nardzewski, The classical and the ω-complete arithmetic, Journ. Symbolic Logic, v. 23 (1958), pp. 188-206.

[vH] van Heijenoort, J. (ed.) From Frege to Gödel (A source book in mathematical logic, 1879-1931), Harvard Univ. Press, Cambridge (1967).

[Hn] Henkin, L., An extension of the Craig-Lyndon interpolation theorem, Journ. Symbolic Logic, v. 28 (1963), pp. 201-216.

[Hr] Herbrand, J. Recherches sur la théorie de la demonstration, Trav. Soc. Sciences, Lettres de Varsovie, Cl. III, no. 33 (1930), 128 pp. (transl. in part in [vH], pp. 525-581).

[Hℓ] Hilbert, D. Über das unendlichen, Math. Annalen, v. 95 (1926), pp. 161-190 (transl. in [vH], pp. 367-392).

[H,B] Hilbert, D. and P. Bernays, Grundlagen der Mathematik, vol. 2, Springer, Berlin (1939).

[Kp] Karp, C., Languages with expressions of infinite length, North-Holland, Amsterdam (1964).

[Ks1] Keisler, H. J., Theory of models with generalized atomic formulas, Journ. Symbolic Logic, v. 25 (1960), pp. 1-26.

[Ks2] ________, Ultraproducts and elementary classes, Indag. Math., v. 23 (1961), pp. 477-495.

[Ks3] ________, Some applications of infinitely long formulas, Journ. Symbolic Logic, v. 30 (1965), pp. 339-349.

[Kl] Kleene, S. C., Introduction to metamathematics, Van Nostrand, Princeton (1952).

[Ko1] Kochen, S., Completeness of algebraic systems in higher order calculi, in Summaries of talks presented at the summer institute for symbolic logic, Cornell University 1957, 2nd edition, Inst. for Def. Analyses (1960), pp. 370-376.

[Ko2] ________, Ultraproducts in the theory of models, Annals Math., v. 74 (1961) pp. 231-261.

[Kr1] Kreisel, G., Model-theoretic invariants: applications to recursive and hyperarithmetic operations, in The theory of models (eds. Addison, Henkin, Tarski), North-Holland, Amsterdam (1965), pp. 190-205.

[Kr2] ________, Relative recursiveness in metarecursion theory (abstract) Journ. Symbolic Logic, v. 33 (1969), p. 442.

[Kr3] ________, A survey of proof theory, Journ. Symbolic Logic (to appear).

[Kr4] ________, So-called consistency proofs by means of transfinite induction (provable ordinals) (abstract for Leeds meeting ASL,1967), Journ. Symbolic Logic (to appear).

[K,K] Kreisel, G. and J. L. Krivine, Elements of mathematical logic (model theory), North-Holland, Amsterdam (1967).

[K,S] Kreisel, G. and G. Sacks, Metarecursion theory, Journ. Symbolic Logic, v. 30 pp. 318-338.

[Kk] Kripke, S., Transfinite recursion on admissible ordinals, I, II, (abstracts) Journ. Symbolic Logic v. 29 (1964), pp. 161-162.

[Ku] Kunen, K., Implicit definability and infinitary languages, Journ. Symbolic Logic (to appear).

[Lv] Lévy, A., A hierarchy of formulas in set theory, Memoirs Amer. Math. Soc., no. 57 (1965).

[L-E1] Lopez-Escobar, E. G. K., An interpolation theorem for denumerably long formulas, Fund. Math. v. 57 (1965), pp. 253-272.

[L-E2] ________, Remarks on an infinitary language with constructive formulas, Journ. Symbolic Logic v. 32 (1967), pp. 305-318.

[Lr] Lorenzen, P., Algebraische und logistische Untersuchungen über freie Verbände, Journ. Symbolic Logic, v. 16 (1951) pp. 81-106 .

[Ls] Łos, J., On the extending of models, I, Fund. Math. v. 42 (1955), pp. 38-54.

[Ly1] Lyndon, R., An interpolation theorem in the predicate calculus, Pacif. Journ. Math., v. 9 (1959), pp. 129-142.

[Ly2] ________, Properties preserved under homomorphism, Pacif. Journ. Math. v. 9 (1959), pp. 253-272.

[Mh] Maehara, S., On the interpolation theorem of Craig (Japanese), Sûgaku v. 12 (1960/61), pp. 235-237.

[Mk] Makkai, M., Preservation theorems for the logic with denumerable conjunctions and disjunctions, to appear in Journal Symbolic Logic.

[Ml] Malitz, J. I., Problems in the model theory of infinite languages, Dissertation, University of California at Berkeley (1965).

[Mn] Mendelson, E., Introduction to mathematical logic, Van Nostrand, Princeton (1964).

[N] Novikov, P. S., Inconsistencies of certain logical calculi (Russian) in Infinitistic methods (Proc. symp. founds. maths., Warsaw, 1959), Pergamon, Oxford, 1961, pp. 71-74.

[O] Oberschelp, A., Untersuchungen zur mehrsortigen Quantorenlogik, Math. Annalen, v. 145 (1962), pp. 297-333.

[Pl] Platek, R., Foundations of recursion theory, Dissertation, Stanford University (1966).

[Pr] Prawitz, D., Natural deduction. A proof theoretical study. Acta Univ. Stock., Stockholm studies in philosophy, no. 3 (1965).

[R] Robinson, A., Introduction to model theory and to the metamathematics of algebra, North-Holland, Amsterdam (1963).

[Sch1] Schütte, K., Beweistheorie, Springer, Berlin (1960).

[Sch2] ________, Predicative well-orderings, in Formal systems and recursive functions (eds. Crossley, Dummett), North-Holland, Amsterdam (1963), pp. 279-302.

[Sch3] ________, Eine grenze fur die Beweisbarkeit der Transfiniten Induktion in der verzweigten Typenlogik, Arch. math. Logik, Grundl., v. 7 (1965) pp. 45-60.

[Sch4] ________, Der Interpolationssatz der intuitionistischen Prädikatenlogik, Math. Annalen, v. 148 (1962), pp. 192-200.

[Sc1] Scott, D., Logic with denumerable long formulas and finite strings of quantifiers in The theory of models (eds. Addison, Henkin, Tarski), North-Holland,

Amsterdam (1965), pp. 32-341.

[Sc2] ________, Interpolation theorems, Proc. 3d Intl. Cong. for Logic, Methodology and Philos. of Science, Amsterdam 1967 (to appear).

[S,T] Scott, D. and A. Tarski, Extension principles for algebraically closed fields, (abstract) Notices Amer. Math. Soc., v. 5 (1958), pp. 778-779.

[Sh] Shoenfield, J. R., *Mathematical Logic*, Addison-Wesley, Reading (1967).

[Sm] Smullyan, R. M., A unifying principle in quantification theory, in *The theory of models* (eds. Addison, Henkin, Tarski), North-Holland, Amsterdam, 1965, pp. 433-434.

[Tt] Tait, W. W., Normal derivability in classical logic, Journ. Symbolic Logic (to appear).

[Tk1] Takeuti, G., Consistency proofs of subsystems of classical analysis, Annals Math., v. 86 (1967), pp. 299-348.

[Tk2] ________, Π_1^1-comprehension axioms and ω-rule, Proc. Leeds 1967 Inst. in Logic (this volume).

[Tr] Tarski, A. Contributions to the theory of models, Indag. Math. v. 16 (1954), Part I pp. 572-581, Part II pp. 582-588.

[T,V] Tarski, A. and R. L. Vaught, Arithmetical extensions of relational systems, Compos. Math. v. 13 (1957), pp. 81-102.

[V] Vaught, R. L., Applications of the Löwenheim-Skolem-Tarski theorem to problems of completeness and decidability. Indag. Math., v. 16 (1954), pp. 467-472.

[Wg] Wang, H., Logic of many sorted theories, Journ. Symbolic Logic, v. 17 (1952), pp. 105-116.

[Wl] Weil, A., *Foundations of algebraic geometry*, Amer. Math. Soc. Colloq. Publs. v. 29, Rev. Ed. (1962).

[Kp2] Karp, C., Finite-quantifier equivalence, in *The theory of models* (eds. Addison, Henkin, Tarski), North-Holland, Amsterdam, 1965, pp. 407-412.

Partitions and Models

by Michael Morley

It is well known that if an elementary theory has an infinite model, then it has models of every infinite power. To produce such models having particular non-elementary properties is considerably more difficult. This paper is an exposition of some results that can be obtained by fairly simple methods, namely, the use of cardinality arguments and partition theorems. No results presented here are new but many of the proofs are. These may be of interest even to those already familiar with the results.

We have chosen not to discuss more complex methods of constructing models, in particular, the method of ultraproducts. (Two excellent review articles [16] and [4], have appeared recently on the use of ultraproducts in constructing models.)

Many beautiful results of model theory have been omitted simply because they did not fit in with our exposition.

0. Preliminaries

We shall work in naive set theory with the axiom of choice · Parts of Section 7 and all of Section 11 presuppose a knowledge of axiomatic set theory. Ordinal numbers are so defined that each ordinal is equal to the set of preceding ordinals: $\alpha = \{\beta; \beta \prec \alpha\}$. Cardinals are those ordinals which cannot be mapped one-one onto any smaller ordinal. Ordinals are denoted by α, β, γ, with the first infinite ordinal denoted by ω. Cardinals are denoted by K, λ, μ, with the understanding that K and λ are always infinite cardinals. The first cardinal greater than K is K^+. The cardinality of the smallest set of ordinals cofinal with K is $\operatorname{cf} K$. A cardinal is __regular__ if $\operatorname{cf} K = K$. The cardinality of a set X is denoted by $|X|$ and the set of all subsets of X by $P(X)$. For each infinite cardinal K and ordinal $\alpha, 2(K, \alpha)$ is defined inductively by $2(K, 0) = K$, $2(K, \alpha+1) = |P(2(K, \alpha))|$ and $2(K, \alpha) = \sup 2(K, \beta)$, $\beta \prec \alpha$, for limit ordinals α. In particular, $\beth \alpha = 2(\omega, \alpha)$. The generalized continuum hypothesis may be stated as $\aleph_\alpha = \beth_\alpha$ for all α.

A __relation structure__ $\circledA = \langle A, R_i \rangle_{i \in I}$ is a set A with an indexed set R_i, $i \in I$, of finitary relations, operations, and distinguished elements. The structure $\circledB = \langle B, S_i \rangle_{i \in I}$ is similar to $\circledA$ if correspondingly indexed relations and operations are of the same kind. The notions of __isomorphism__, __subsystem__ and __monomorphism__ (= isomorphism into) are all defined in the obvious way for similar systems. In particular, a subsystem is always assumed closed under all the operations and to contain all the distinguished elements. To denote that $\circledA$ is a subsystem of $\circledB$, we wrote $\circledA \subset \circledB$.

Corresponding to each similarity type of structure there is an __elementary language__ (as distinct from non-elementary languages appearing later). This will have as symbols:

(i) A countable set of variables v_0, v_1, $\cdots$,

(ii) relation, function, and individual symbols corresponding to the relations, operations, and distinguished elements of the structure,

(iii) a binary relation symbol $=$

(iv) logical connectives $\wedge$, $\vee$, $\sim$, $\rightarrow$, $\leftrightarrow$ (and, or, not, implies, if and only if),

(v) for each variable v universal and existensional quantifiers (v), $(\exists v)$.

Terms are defined as usual by induction:

(i) a variable is a term,

(ii) an individual symbol is a term,

(iii) if R is an n-ary operation symbol and $t_1, \cdots, t_n$ are terms then $R(t_1, \cdots, t_n)$ is a term.

Terms not containing a variable are constant terms.

Formulas are defined as usual by:

(i) if R is an n-ary relation symbol and $t_1, \cdots, t_n$ are terms, then $R(t_1, \cdots, t_n)$ is a formula,

(ii) if φ_1 and φ_2 are formulas, then $\varphi_1 \wedge \varphi_2$, $\varphi_1 \vee \varphi_2$, $\sim \varphi_1$, $\varphi_1 \rightarrow \varphi_2$ and $\varphi_1 \leftrightarrow \varphi_2$ are also formulas.

(iii) if φ is a formula and v a variable, then $(\exists v)\varphi$ and $(v)\varphi$ are formulas.

A formula with no free (i.e. unquantified) variables is a sentence.

If L is the language corresponding to the similarity type of Ⓐ, then Ⓐ is called an L-structure.

The reader is assumed to understand the notions of satisfaction of a formula in Ⓐ and of validity of a sentence in Ⓐ. If the sequence $a_o, \cdots, a_n$ of elements of A satisfies the formulas $\varphi(v_o, \cdots, v_n)$ in Ⓐ, we write Ⓐ $\models \varphi(a_o, \cdots, a_n)$. If the sentence φ is valid in Ⓐ, we write Ⓐ $\models \varphi$. The set of all sentences valid in Ⓐ is called the theory of Ⓐ, written Th (A). A theory T is a set of sentences. Ⓐ is said to be a model of a theory T if every sentence of T is valid in Ⓐ. A substructure Ⓐ $\subset$ Ⓑ is an elementary subsystem (Ⓐ $\prec$ Ⓑ)

if every sequence of A satisfies the same set of formulas in $\mathfrak{A}$ as in $\mathfrak{B}$. Similarly for elementary monomorphism.

Given a relation structure $\mathfrak{A}$, we can form another relation structure by taking every element of A as a distinguished element. The resulting system is denoted by $\langle A, a\rangle_{a\in A}$. $Th(\langle A,a\rangle_{a\in A})$ is called the complete diagram of $\mathfrak{A}$, written $D(A)$.

1. Skolem Functions

Let L be an elementary language. Corresponding to each formula $\varphi(v_o,\cdots,v_n)$ of L, we can add an n-ary function symbol f_φ (the Skolem function) to L to form a new language L'. (If $n = 0$, take f_φ to be an individual constant.) Define the langauges $L_n{}^*$ and L^* inductively by setting $L_o{}^* = L$, $L^*_{n+1} = (L_n{}^*)'$, $L^* = \bigcup_{n\in\omega} L_n{}^*$. For each f_φ there is a defining sentence:

$$(v_1)\cdots(v_n)((\exists v_o)\varphi(v_o,\cdots,v_n) \rightarrow \varphi(f_\varphi(v_1,\cdots,v_n),\ v_1,\cdots,v_n)).$$

An L^*-structure is a Skolem model if every defining sentence is valid in it. If T is a theory in L, its Skolem closure is $T^* = T \cup$ {defining sentences of L^*}. With the use of the axiom of choice, Skolem functions can be defined in any L-structure in such a way as to make it a Skolem model. Hence, for any formula φ of L, we have $T \models \varphi$ if and only if $T^* \models \varphi$.

THEOREM 1.1. For every formula φ in L^* there is a formula φ' in L^* without quantifiers such that φ and φ' are equivalent in all Skolem models.

Proof: By induction on the length of φ using the defining sentence to eliminate quantifiers.

COROLLARY 1.2. Every subsystem of a Skolem model is an elementary subsystem and every monomorphism of one Skolem model into another is an elementary monomorphism.

COROLLARY 1.3. In a Skolem model Ⓐ, the elements represented by constant terms form an elementary subsystem called the Skolem hull of Ⓐ.

We shall repeatedly use Corollary 1.2 to show that there are models of a theory T having some given property. In each case we will extend T* to a larger theory T_o such that (i) the compactness theorem may be used to show T_o consistent and (ii) the Skolem hull of any model of T_o has the desired properties. The following two lemmas are sometimes useful in the process.

LEMMA 1.4. Let $t_o, \cdots, t_n$ be constant terms in L*. There is a formula $D(v_o, \cdots, v_n)$ in L such that for any L-structure Ⓐ there is a way to define Skolem functions so as to make $a_i = t_i$ if and only if Ⓐ$\models D(a_o, \cdots, a_n)$.

Proof: We give only a hint. The proof proceeds by replacing the Skolem functions by their defining equations. The formula $D(v_o, \cdots, v_n)$ is obtained as a complex amalgam of these defining equations.

An application of Lemma 1.4 gives:

LEMMA 1.5. Let $\varphi(v_o, \cdots, v_n)$ be a formula in L, $t_o, \cdots, t_n$ constant terms L* and T a theory in L. Then $T^* \models \varphi(t_o, \cdots, t_n)$ if and only if $T \models D(v_o, \cdots, v_n) \to \varphi(v_o, \cdots, v_n)$ where D is related to the terms $t_o, \cdots, t_n$ as in Lemma 1.4.

COROLLARY 1.6. If Ⓐ is an infinite L-structure, then Ⓐ has:

(i) elementary subsystems in every power λ, $|A| \geq \lambda \geq |L|$.

(ii) elementary extensions in every power $\lambda \geq |A| + |L|$.

Proof: (i) follows immediately for Corollary 1.2. To prove (ii), add a set Y of power λ of new individual symbols to the language. Let

$T_1 = D(A) \cup \{y_1 \neq y_2$ for every pair of distinct symbols y_1, y_2 in $Y\}$.

Then T_1 is consistent (use the compactness theorem) and the Skolem hull of T_1 has the desired property.

COROLLARY 1.7. If T is a theory in L with an infinite model, then T has model in every power $\lambda \geq |L|$.

Corollary 1.6 completely describes the situation when $|A| \geq |L|$. When $|A| < |L|$ the situation is somewhat more subtle.

THEOREM 1.8. If $|A| = \lambda \geq \omega$, then Ⓐ has elementary extensions in every power $K \geq \lambda^{\omega}$.

Proof: Extend Ⓐ to a Skolem model Ⓐ* and let $T = Th(A^*, a)_{a \in A}$. Let $X \subseteq A$ be a countably infinite set. Each term $t(v_0, \cdots, v_n)$ defines a function $t: X^{n+1} \to A$. Say two terms are equivalent if they define the same function. There are at most λ^{ω} equivalence classes. Extend T to a theory T' by adding K new individual symbols c_α, $\alpha < K$, and the following sentences:

(i) $c_\alpha \neq a \quad \alpha < K, a \in A$

(ii) $c_\alpha \neq c_\beta \quad \alpha \neq \beta$

(iii) $t_1(c_{\alpha_1}, \cdots, c_{\alpha_n}) = t_2(c_{\alpha_1}, \cdots, c_{\alpha_n})$ whenever t_1 is equivalent to t_2 and $\alpha_1, \cdots, \alpha_n < K$.

The theory T' is consistent since every finite subset of T' may be satisfied by elements of X. The Skolem hull of T' is an elementary extension of Ⓐ of power $K \geq \lambda^{\omega}$.

The example of [2] shows that the restriction $K \geq \lambda^{\omega}$ is necessary (at least for reasonable size cardinals).

2. Omitting Types

Let T be a theory in the elementary langauge L and Σ a set of formulas in L, each formula having no free variable other than v_o. An element a of an L-structure Ⓐ is of type Σ if it satisfies every formula of Σ. Ⓐ realizes Σ if it has an element of type Σ ; otherwise it omits Σ. It follows from the compactness theorem that a necessary and sufficient condition for a theory T to have a model realizing Σ is that it have models realizing every finite $\Sigma_1 \subseteq \Sigma$. The condition for omitting a type is more complex.

THEOREM 2.1. Suppose T a theory and Σ a type in a countable language L. A necessary and sufficient condition that there be a model of T omitting Σ is that there be a consistent theory $T' \supset T$ such that : (#) for every formula $\varphi(v_o)$ if $T' \models \varphi(v_o) \to \sigma(v_o)$ for all $\sigma \in \Sigma$, then $T' \models \sim(\exists v_o)\varphi(v_o)$.

Remark: If T is complete, then T' = T and the condition (#) simply says Σ is not implied by any single consistent formula. In any event, since (#) is preserved under intersections, there is a least $T_1 \supset T$ with property (#).

Proof: (i) Necessity. Suppose Ⓐ is a model of T which omits Σ. Let T' = Th(A).

(ii) Sufficiency. Suppose $T' \supset T$, T' is consistent and has property (#). Extend T' to its Skolem closure T*. Let $\{t_n; n \in \omega\}$ be an enumeration of the constant terms with each term preceded by all its subterms. We prove by induction the existence of a sequence $\sigma_n(v_o)$ in Σ with $T^* \cup (\sim\sigma_n(t_n); n \in \omega)$ consistent.

Suppose σ_i has been chosen for $i < n$ and $T^* \cup \{\sim \sigma_i(t_i), i < n\}$ is consistent. I assert that there is a way to choose σ_n so that $T^* \cup \{\sim \sigma_i(t_i), i \leq n\}$ is consistent. Otherwise, for all $\sigma \in \Sigma$,

$$T^* \models \sim \sigma_o(t_o) \wedge \cdots \wedge \sim \sigma_{n-1}(t_{n-1}) \to \sigma(t_n).$$

By Lemma 1.4, there is a formula $D(v_o,\cdots,v_n)$ such that, for all $\sigma \varepsilon \Sigma$

$$T' \models (v_o) \cdots (v_n)(D(v_o,\cdots,v_n) \rightarrow (\sim \sigma_o(v_o) \wedge \cdots \wedge \sim \sigma_{n-1}(v_{n-1}) \rightarrow \sigma(v_n)).$$

Then by assumption

$$T' \models \sim (\exists v_o \cdots \exists v_n)(D(v_o,\cdots,v_n) \wedge \sim \sigma_o(v_o) \wedge \cdots \wedge \sim \sigma_{n-1}(v_{n-1}))$$

contrary to the induction hypothesis that $T^* \cup (\sim \sigma_i(t_i);\ i < n)$ is consistent.

The examples of [10] show that the assumption that the language be countable is essential for Theorem 2.1 to hold. Results known for uncountable languages are considerably less satisfactory, but see [1].

Omitting types is a subject closely related to what is often called ω-logic. An ω-logic is formed by adding to an elementary language new variables w_o, w_1, $\cdots$ which in each model of ω-logic will range over ω and new constants Δ_o, Δ_1, $\cdots$ which correspond to the elements of ω. A model Ⓐ for ω-logic is a set of the form $A \cup \omega$ together with relations, functions, etc., on it. In interpreting the formulas, the variables w_o, w_1, $\cdots$ are always assumed to range only over ω and v_o, v_1, $\cdots$ only over A.

Suppose T is a theory in a countable elementary language and Σ a type of element. Let σ_o, σ_1, $\cdots$ be an enumeration of Σ. Extend T to a theory T_1 in ω-logic by adding a new binary relation R and the sentences

$$(v_o)(R(\Delta_n, v_o) \leftrightarrow \sigma_n(v_o))$$
$$(v_o)(\exists w_o) \sim R(w_o, v_o).$$

Clearly:

THEOREM 2.2 With T, Σ_1 and T_1 as above there is a natural one-one correspondence between models of T which omit Σ and models of T_1.

Conversely, let T_1 be a theory in ω-logic. Form a theory T in elementary logic by adding:

(i) Two new unary relation symbols A and W,

(ii) the axioms:

$$(v_o)(A(v_o) \vee W(v_o))$$
$$W(\Delta_n) \quad \text{for each} \quad n \in \omega.$$

Then replace each sentence φ in T_1 by a sentence φ' formed by replacing the variables in φ by variables relativized to A and W thusly:

(i) v_i is replaced by v_{2i}, w_i by v_{2i+1}

(ii) $(\exists v_i)$ is replaced by $(\exists v_{2i})(A(v_{2i}) \wedge \cdots)$
(v_i) by $(v_{2i})(A(v_{2i}) \rightarrow \cdots)$

(iii) $(\exists w_i)$ is replaced by $(\exists v_{2i+1})(W(v_{2i}) \wedge \cdots)$
(w_i) by $(v_{2i+1})(W(v_{2i+1}) \rightarrow \cdots)$

Finally, let Σ be the set of formulas $\{W(v_o),\ v_o \neq \Delta_o,\ v_1 \neq \Delta_1,\ \cdots\}$ Then:

THEOREM 2.3. With T, Σ, and T_1 as above, there is a natural one-one correspondence between models of T omitting Σ and models of T_1.

Thus ω-logic and omitting types are essentially two ways of looking at the same problem. The result in ω-logic corresponding to Theorem 2.1 is often called the ω-completeness theorem (see [18]).

Another way to view omitting types is in terms of infinitely long formulas. If L is an elementary language, define the language $\hat{L}$ by adding to the formation rules for formulas:

(iv) if C is any countable set of formulas, then $\bigwedge_{\varphi \in C} \varphi$ and $\bigvee_{\varphi \in C} \varphi$ are formulas.

These formulas are to be interpreted as asserting respectively that every $\varphi \in C$ holds and some $\varphi \in C$ holds. Notice that if $\bigwedge_{\varphi \in C} \varphi$ or $\bigvee_{\varphi \in C} \varphi$ is to appear as part of a sentence, i.e a formula with no free variables, then the number of distinct free variables appearing in $\wedge\varphi$ or $\vee\varphi$ must be finite. It is obvious that:

THEOREM 2.4. If T is a theory and Σ a type in a countable elementary language L, then Ⓐ is a model of T omitting Σ if and only if Ⓐ is a model of "$(\bigwedge_{\varphi\in T}\varphi)\wedge\sim(\exists\, v_o)(\bigwedge_{\varphi\in\Sigma}\varphi)$".

Conversely, suppose ψ is a sentence in $\hat{L}$. As remarked earlier, any subformula of ψ can have only a finite number of free variables. Suppose $\{\varphi_o,\varphi_1,\cdots\}$ is a set of formulas with free variables among $v_o,\cdots,v_k$. Introduce a new relation symbol R and write (in ω-logic) the formula

$$R(\Delta_n,v_o,\cdots,v_k)\rightarrow\varphi_n(v_o,\cdots,v_k).$$

Then the conjunction of the φ_i's may be replaced by $(w_o)R(w_o,v_o,\cdots,v_k)$ and the disjunction by $(\exists\, w_o)R(w_o,v_o,\cdots,v_k)$, In this fashion the sentence ψ becomes equivalent to a theory in ω-logic and hence by Theorem 2.3.

THEOREM 2.5. Let ψ be a sentence in $\hat{L}$. There is a theory T and a type Σ in a countable elementary language $L_1\supset L$ such that a natural one-one correspondence exists between models of ψ and models of T which omit Σ.

3. Two Cardinal Theorems

Let T be a theory in a countable elementary language L and U a particular unary relation symbol in L. Say an L-structure Ⓐ is of <u>type</u> (K,λ) if Ⓐ is of power K and the interpretation of U in Ⓐ, U^A, is a subset of A of power λ. We shall be interested in theorems of the following form: If a theory has a model of type (K,λ), then it possesses a model of type (K_1,λ_1). Such theorems are written in the abbreviated form $(K,\lambda)\rightarrow(K_1,\lambda_1)$ The following theorem is immediate from the Lowenheim-Skolem Theorem.

THEOREM 3.1. For all $K_1, (K,\lambda) \to (K_1,K_1)$. And for all $K \geq K_1 \geq \lambda$, $(K,\lambda) \to (K_1,\lambda)$.

A slightly deeper result follows.

THEOREM 3.2. $(K,\lambda) \to (K_1,\lambda_1)$ for all $K \geq K_1 \geq \lambda_1 \geq \lambda^{\omega}$.

Proof: According to the preceding theorem, it suffices to show that $(K,\lambda) \to (K,\lambda_1)$ for every $K \geq \lambda_1 \geq \lambda^{\omega}$. The proof is similar to the proof of Theorem 1.8. Let Ⓐ be a model of T of type (K,λ) and let X be a countably infinite subset of U. Let $T_1 = Th(A,a)_{a\epsilon A}$ and T_1^* its Skolem closure. Say two terms $t_1(v_o,\cdots,v_n)$ and $t_2(v_o,\cdots,v_n)$ are equivalent if for every $x_o,\cdots,x_n$ in X, either (i) $t_1(x_o, x_1,\cdots,x_n) = t_2(x_o,x_1,\cdots,x_n)$ or (ii) neither $t_1(x_o,\cdots,x_n)$ nor $t_2(x_o,\cdots,x_n)$ is in U. There are at most λ^W equivalence classes of terms. Extended T_1^* to T_2 by adding to T_1^* a set Y of power λ_1 of new individual symbols and axioms asserting:

(i) the elements of Y are distinct elements of U

(ii) for equivalent terms t_1 and t_2 and elements $y_o,\cdots,y_n$ in Y,
$U(t_1(y_o,\cdots,y_n)) \wedge U(t_2(y_o,\cdots,y_n)) \to t_1(y_o,\cdots;y_n) = t_2(y_o,\cdots,y_n)$.

The compactness theorem yields the consistency of T_2 and the Skolem hull of a model of T is of type (K,λ_1).

THEOREM 3.3. Let T be a theory in a countable language. If T has a (K,λ) model, $K > \lambda$, then it has an (ω_1,ω) model.

Proof: The proof uses the notion of omitting type and Theorems 2.1.

Theorem 3.1 implies that T has a (λ^+,λ) model. Let Ⓐ be such. Well order Ⓐ by a binary relation $<$. Let $T' = Th(Ⓐ, <)$ and let Ⓑ be any countable model of T'. We claim that Ⓑ has a proper elementary extension Ⓒ with $U^{Ⓑ} = U^{Ⓒ}$.

Consider the theory T_1,

$$T_1 = T' \cup D(B) \cup \{c > b;\ \text{all } b \in B\}.$$

Let $\Sigma = \{U(v_o),\ v_o \neq a_o,\ v_o \neq a_1,\ v_o \neq a_2,\ \cdots\}$ where $\{a_i\}$, $i \in \omega$, is an enumeration of U^B. To prove that Ⓑ has an elementary extension Ⓒ with $U^B = U^C$, it suffices to show that T_1 has a model omitting the type Σ. Recall that property (#) of Theorem 2.1 gives a necessary and sufficient condition for a theory to have a model omitting a type. The theory T_1 possesses the property (#) in a strong sense. We prove that for any formula $\varphi(v_o)$ such that $(\exists v_o)\varphi(v_o)$ is consistent with T_1, then not for all $\sigma \in \Sigma$ does $T_1 \models \varphi(v_o) \rightarrow \sigma(v_o)$

Assume $\varphi(v_o) \rightarrow \sigma(v_o)$, all $\sigma \in \Sigma$. Then $\varphi(v_o)$ has the form

$$\varphi(v_o) = \varphi(v_o, b_1, \cdots, b_n, c)$$

where $\varphi(v_o, v_1, \cdots, v_n, v_{n+1})$ is a formula of T' and $b_1, \cdots, b_n$ are individual symbols representing elements of B. Assume $(\exists v_o)\varphi(v_o, b_1, \cdots, b_n, c)$ is a sentence $\psi(c)$ consistent with T_1. The assumption that $\psi(c)$ is consistent with T_1 implies the existence in B of a cofinal sequence $\{d_j\}$, $j \in \omega$, for which $T_1 \models \psi(d_j)$. For otherwise the sentence $(v)(v > b \rightarrow \sim\psi(v))$, for some element $b \in B$, would be a sentence of $D(B)$, leading to the contradiction $T_1 \models \sim \psi(c)$. The cofinality of the sequence $\{d_j\}$ is expressed as

$$T_1 \models (t)(\exists v_{n+1})(\exists v_o)(v_{n+1} > t \rightarrow \varphi(v_o, b_1, \cdots, b_n, v_{n+1}).$$

We claim that it is a theorem of $T' = Th(A, \prec)$ and, a fortiori, of T_1 that

$$T' \models (v_1) \cdots (v_n)[((\exists v_o)(t)(\exists v_{n+1})(v_{n+1} > t \wedge \varphi(v_o, v_1, \cdots, v_n, v_{n+1}) \wedge U(v_o)$$

$$\vee (t)(\exists v_{n+1})(\exists v_o)(v_{n+1} > t \wedge \varphi(v_o, v_1, \cdots, v_{n+1}) \wedge \sim U(v_o))].$$

For examine cofinal sequences in the (λ^+, λ) model $(Ⓐ, <)$. Note that λ is not cofinal in λ^+. Therefore, any cofinal sequence $\{d_j\}$ on which $\exists v_o \varphi(v_o, v_1, \cdots, v_n, d_j)$ is valid contains a cofinal subsequence $\{d'_j\}$ for which $\varphi(a_o, v_1, \cdots, v_n, d'_j) \wedge U(a_o)$ holds for some fixed element a_o of U or $(\exists v_o)\varphi(v_o, \cdots, v_n, d'_j) \wedge \sim U(v_o)$ holds. This fact, as stated above, is a theorem of T_1. Thus if $T_1 \models \varphi(v_o) \rightarrow U(v_o)$, then $\varphi(v_o) \wedge (v_o = a_i)$ for some $a_i \in U^{Ⓑ}$ is consistent with T_1. Therefore, $(\exists v_o)\varphi(v_o)$ is inconsistent if $T_1 \models \varphi(v_o) \rightarrow \sigma(v_o)$, all $\sigma \in \Sigma$.

Iteration of the procedure leading from Ⓑ to Ⓒ ω_1 times will yield an (ω_1, ω) model.

An examination of the proof of Theorem 3.3 leads to several other results.

COROLLARY 3.4. Suppose $K > \lambda$ and Ⓐ a (K, λ) model. We can find Ⓑ, Ⓒ with Ⓑ $\prec$ Ⓐ and Ⓑ $\prec$ Ⓒ, $|B| = \omega$, $|C| = \omega_1$, and the extension of Ⓑ to Ⓒ leaves U fixed.

In Section 2, we showed that ω-models may be correlated with models in which a set W has only certain specific elements. Setting $U' = U \cup W$ gives:

THEOREM 3.5. Both Theorem 3.3 and Corollary 3.4 are true for ω-logic.

The construction of an elementary extension of Ⓑ in Theorem 3.3 leaving U fixed depends only on the fact that a certain sentence is a theorem of T'. Thus a theory T will have an (ω_1, ω) model if and only if this sentence can be consistently added to it. The compactness theorem gives:

Theorem 3.6. The class of theories with (ω_1, ω) models is ω-compact, i.e. a countable theory has such a model if and only if every finite subset of it does.

This compactness result may be extended to other pairs (K, λ).

THEOREM 3.7. If $\lambda^{\omega} = \lambda$, the class of theories having (K, λ) models is ω-compact.

<u>Proof</u>: Since the case where $K = \lambda$ is trivial, assume $K > \lambda$. Suppose T a theory for which every finite subtheory has a (K, λ) model. We may also assume that T itself is a Skolem theory. Let $T_0, T_1, \cdots$ be an enumeration of its finite subtheories and $\mathfrak{A}_0, \mathfrak{A}_1, \cdots$ their corresponding (K, λ) models. No loss of generality results from assuming that the $\mathfrak{A}_i$ are Skolem models all with the same universe A and $U \subset A$ is the same set for each $\mathfrak{A}_i$. Let $\mathfrak{A}_j{}' = \langle A_j, a\rangle_{a \in A}$. Let L' be the language corresponding to the models $\mathfrak{A}_j{}'$, i.e. the elementary language of the theory T with individual symbols <u>a</u> corresponding to the elements of A added to it. Say two constant terms t_1 and t_2 are equivalent (in the language L') if for every i,

$$\mathfrak{A}_i{}' \models U(t_1) \wedge U(t_2) \rightarrow t_1 = t_2.$$

Note there are $\lambda = \lambda^{\omega}$ equivalence classes. Let

$$T_0 = T \cup \{U(t_1) \wedge U(t_2) \rightarrow t_1 = t_2; \text{ for all } t_1 \text{ equivalent to } t_2\}.$$

T_0 is consistent and the Skolem hull of any model to T_0 is a (K, λ) model of T.

The following theorem is proved by methods more complex than those used in this paper.

THEOREM 3.8. If $\lambda < K$ implies $K^{\lambda} = K$ or, <u>a fortiori</u>, if K is regular and the generalized continuum hypothesis holds, then $(\omega_1, \omega) \rightarrow (K^{+}, K)$.

Closely allied with two cardinal problems are the notions of K-like models and of generalized quantifiers. Suppose Ⓐ is a relation structure with a particular distinguished binary relation E. The <u>field</u> of E is, by definition, the set of elements satisfying $(\exists v_1)(v_o E v_1 \vee v_1 E v_o)$. A is said to be <u>partly ordered</u> (not to be confused with partially ordered) if E linearly orders its field. Ⓐ is <u>K-like</u> if the field of E has power K but every initial segment has power $\prec K$.

With every elementary language L, there is an associated <u>generalized quantifier</u> language L_Q formed by adding a new quantifier symbol Q and a new formation rule for formulas:

if ψ is a formula and v_i a variable, then $(Q v_i)\psi$ is a formula.

To interpret a formula of L_Q, shoose a cardinal K and interpret $(Q v_i)$ to mean "there exist at least K v_i's such that $\cdots$". Call this the <u>Q_K-interpretation</u>. Similarly, if T is a theory in L_Q, then A is a <u>Q_K-model</u> of T if it satisfies the sentences of T under the Q_K-interpretation.

THEOREM 3.9. If T is a theory in L, then there is a theory T' in L_Q such that Ⓐ is a K-like model of T if and only if Ⓐ is a Q_K-model of T'.

<u>Proof</u>: Simply add to T a sentence asserting that E is a linear ordering of its field and the sentence

$$(Qv_o)(\exists v_1)(v_o E v_1 \vee v_1 E v_o) \wedge (v_1) \sim (Qv_o)(v_o E v_1).$$

Conversely

THEOREM 3.10. Let T' be a theory in L_Q. There is a theory T in a language $L' \supset L$ such that:

i) a natural many-one correspondence associates each K-like model of T with a Q_K model of T'

ii) each Q_K model of T' corresponds in a natural way with a non-empty set of K-like models of T.

Proof: Here is a sketch of the basic idea of the proof. Suppose a K-like linear ordering is given. It is possible to state that a set X has power K by asserting the existence of a one-one map of the field of the linear ordering into X and to state that X has power K by asserting the existence of a one-one map of X into an initial segment of the field. By induction on the length of a formula, each occurence of a generalized quantifier may be replaced by a function symbol.

Now to relate the foregoing to two cardinal theorems. First, note that a K-like model has a K-like elementary subsystem of power K. Second, a K^+-like linear ordering must have a particular initial segment of power K and one-one maps of its initial segments into it. Add a unary relation symbol U for this particular initial segment and a ternary relation symbol F with F(x,y,z) to mean "y corresponds to z under the one-one map of the initial segment determined by x into the set U". In this fashion, a correspondence is set up between each theory T with a K^+-like model and a theory T' with a(K^+,K) model. Then:

THEOREM 3.11. Let T be a theory in L_Q.

(i) If T has a Q_{K^+}-model, T has a Q_{ω_1} model.

(ii) (Assuming the generalized continuum hypothesis) if T has a Q_{ω_1}-model, then for regular K, T has a Q_{K^+} model.

To indicate the necessity of some of the hypothesis in Theorem 3.11, we point out that it is easy to find a theory in L_Q which has a Q_K-model if and only if K is a successor cardinal.

Theorem 3.6 yields:

THEOREM 3.12. Q_{ω_1} is ω-compact, i.e. a countable theory in L_Q has a Q_{ω_1}-model if and only if every finite subtheory of it has a Q_{ω_1}-model.

This theorem can be improved to:

THEOREM 3.13. If $\mathrm{cf}\, K > \omega$ and $\lambda < K$ implies $\lambda^{\omega} < K$, then Q_K is ω-compact.

<u>Proof</u>: According to Theorem 3.9, it suffices to show that a countable theory T in L has a K-like model if and only if every finite subtheory of it does. Let $T_0, T_1, \cdots$ be an enumeration of the finite subtheories of T and suppose $\mathfrak{A}_n$ a K-like model of T_n. Denote by E the distinguished binary relation which provides a K-like ordering on its field. Add a set $\{c_\alpha; \alpha < K\}$ of individual symbols to represent the elements of the field of E in A_n. Let $\mathfrak{A}_n' = \langle A_n, c_\alpha \rangle_{\alpha<K}$. No less of generality results from assuming that each $\mathfrak{A}_n'$ is a Skolem model and that the same individual symbols are used to denote the field of E in the various A_n's (though the symbols may satisfy different formulas in different models).

Let t_1 and t_2 be arbitrary constant terms. Extend the theory T to the theory T_0 by adding all instances of sentences of the following form that hold in all the models $\mathfrak{A}_n'$:

(i) t_1 is in the field of E

(ii) $t_1 < c_\alpha \wedge t_2 < c_\alpha \rightarrow t_1 = t_2$

(iii) $t_1 < c_\alpha$

The Skolem hull of a model of T_0 is a K-like model of T. For the number of unequal terms less than c_α will be at most a countable product of λ's each less than K. But by assumption on K this product is less than K. On the otherhand, each constant term t in the field of E will be less than some c_α by the assumption of $K \prec \omega$ and the condition (iii).

Finally, consider Q_ω. An ω-like linear ordering is in fact isomorphic to ω. Hence, for every theory T in elementary logic there is a corresponding theory T' in ω-logic whose models correspond to the ω-like models of T. Similarly, there is a correspondence between theories in L_Q having Q_ω-models and theories in ω-logic.

4. Partition Theorems

So far, the theorems of this study have been largely based on the fact that if a set of large cardinality is partitioned into a small number of classes, some one class will have large cardinality. The analogous result for n-ary relations - that a set X of large cardinality with X^n partitioned into a small number of classes contains a subset Y of large cardinality with Y^n lying entirely within one of the partition classes - is false. For consider $n = 2$, let X be linearly ordered and partition X^2 into the classes $C_1 = \{(x_1, x_2); x_1 \prec x_2\}$ and $C_2 = X^2 - C_1$.

Instead of X^n, consider the notion $X^{(n)} = \{y; y \subseteq X$ and $|y| = n\}$. Notice that if X has some given linear order, then $X^{(n)}$ may be represented as the set of n-tuples $\{(x_0, \cdots, x_{n-1}); x_0 \prec \cdots \prec x_{n-1}\}$. This convention will generally be followed.

Let X be a set with $|X| = K$. Partition $X^{(n)}$ into disjoint classes. A set $Y \subseteq X$ with $Y^{(n)}$ included entirely within one partition class is said to be <u>homogenous</u> for the partition. We are interested in theorems which give information about the cardinality of homogenous sets. If, for any partition of $X^{(n)}$ into μ classes, there is a set Y homogenous for the partition with $|Y| = \lambda$, we write "$K \to (\lambda)^n_\mu$".

Partition theorems of this type are closely related to ramification systems. A <u>ramification system</u> $(P, <)$ is a partially ordered set with a least element all of whose initial segments (i.e. the set of elements $<$ some element) are well ordered. The order of an element p of P is the ordinal α of the order type of the initial segment determined by it.

<u>Example</u>: Suppose P is a set of (well ordered) sequences of o's and 1's with $p \in P$ if and only if every initial segment of p is in P. If $p' < p$ is interpreted as p' is an initial segment of p, the result is a ramification system. The order of each p is its length. Systems of this type are called <u>binary ramification systems</u>.

Suppose X is a set and $X^{(2)} = C_o \cup C_1$, $C_o \cap C_1 = \phi$. Suppose further X is well ordered by $<$. Corresponding to X is a binary ramification system satisfying the conditions,

(i) there is a one-one map f of P onto X

(ii) $p_1 < p_2 \rightarrow f(p_1) < f(p_2)$

(iii) for $p_1 < p_2$, p_1 of length α, and the αth element of $p_2 = j (j = 0,1)$, we have $(f(p_1), f(p_2)) \in C_j$.

We construct P by induction. To the least element x_o of X, assign the empty sequence. Suppose for $x \in X$, every element of X less than x has been assigned a binary sequence. The sequence p corresponding to x is defined by the following prescription:

(a) The zeroth element of the sequence p is 0 or 1 depending on whether (x_o, x) is in C_o or C_1.

(b) Suppose the first α elements of the sequence p have been assigned. Call this sequence p_α. If no $y < x$ has been assigned p_α, let $f(p_\alpha) = x$.

(c) If for some $y < x$, $f(p_\alpha) = y$, let the αth element of p be 0 or 1 according as (y,x) is in C_o or C_1.

The ramification system constructed this way obviously satisfies the specified conditions (i), (ii), (iii). Essentially the same induction shows that any ramification system satisfying these conditions is isomorphic to P.

Assume the binary ramification system satisfying (i), (ii), (iii) contains a linearly ordered subsystem of power λ. The initial segments of a sequence p of length λ correspond to the linearly ordered subsystem. Define $Y_i, (i = 0,1)$ to be the set $\{f(q);\ q$ an initial segment of p for which the least element of $p - q = i\}$. By (iii), $Y_i{}^2 \subset C_i$. Either $|Y_0| = \lambda$ or $|Y_1| = \lambda$. Thus :

THEOREM 4.1. If every binary ramification system of power K has a linearly ordered subset of power λ, then $K \to (\lambda)^2_2$.

A similar argument gives:

THEOREM 4.2. Given μ and λ with $\mu < \mathrm{cf}\,\lambda$. Suppose every ramification system of power K which has (for every ordinal α) at most $\mu^{|\alpha|}$ elements or order α must contain a linearly ordered subset of power λ. Then $K \to (\lambda)^2_\mu$.

COROLLARY 4.3. $\omega \to (\omega)^2_m$ for all finite m.

COROLLARY 4.4. $(2^K)^+ \to (K^+)^2_K$.

If λ is regular, Theorem 4.1 has a converse.

THEOREM 4.5. If λ is regular and $K \to (\lambda)^2_2$, then every binary ramification system of power K has a linearly ordered subset of power λ.

Proof: Let P be a binary ramification system of power K. Let R_1 be the lexicographic ordering on P and R_2 a well ordering of P of order type K. C_0 is the class of pairs on which R_1 and R_2 agree and C_1 the class on which the orderings disagree. If $\kappa \to (\lambda)^2_2$ holds then the ordering R_1 must have either an increasing or decreasing sequence of length λ. The regularity of λ implies that P must have a linearly ordered subset of power λ.

COROLLARY 4.6. Not $2^\kappa \to (\kappa^+)^2_2$.

The proof of Theorem 4.1 involved the construction of a ramification system such that on linearly ordered subsets the class of a pair is determined by the first element alone. This process can be generalized to partitions of $X^{(n+1)}$. On linearly ordered subsets the class of an n+1-sequence depends only on its first n elements. Thus:

LEMMA 4.7. Let $(X, \prec)$ be a well ordered set and let $X^{(n+1)} = \bigcup C_i (i \in I)$ be a partition of $X^{(n+1)}$ into disjoint sets. There is a ramification system $(P, \prec)$, unique up to isomorphism, satisfying:

(i) P is mapped one-one onto X by a function f,

(ii) $p_1 \prec p_2$ implies $f(p_1) \prec f(p_2)$

(iii) $p_1 \prec p_2 \cdots p_n \prec q \prec r$ implies $(f(p_1),\cdots,f(p_n), f(q))$ and $(f(p_1),\cdots,f(p_n), f(r))$ are in the same class.

(iv) If q and r are incomparable, there is an n-tuple $p_1 \prec p_2 \cdots p_n \prec q$ and $p_n \prec r$ with $(f(p_1),\cdots,f(p_n), f(q))$ and $f(p_1),\cdots,f(p_n), f(r))$ in different classes.

Proof: By induction on X as before.

THEOREM 4.8. Suppose μ, λ, κ are as in Theorem 4.2. Let X be a well ordered set of power κ with $X^{(n+1)}$ partitioned into at most μ classes. Then there exists a subset $Y \subset X$ of power λ such that the class of each n+1-tuple in Y is determined by its first n elements.

Proof: Similar to the proof of Theorem 4.1 using μ-ary sequences in place of binary sequences.

THEOREM 4.9. $\omega \to (\omega)^{n+1}_m$ for all n, $m \in \omega$.

Proof: Using Theorem 4.8, iterate the procedure which leads from Theorem 4.2 to Corollary 4.3 n times.

THEOREM 4.10. $(2(k,n))^+ \to (K^+)^{n+1}_K$.

Proof: Using Theorem 4.8, iterate the procedure which leads from Theorem 4.2 to Corollary 4.4 n-times.

The following negative result concludes this section.

THEOREM 4.11. If $\text{cf } K < K$, then not $K \to (K)^2_2$.

Proof: Let $X \subset K$ be a cofinal subset of power $< K$. Partition $K^{(2)}$ into two classes depending on whether there is or is not an element of X between them. For no subset Y of power K can $Y^{(2)}$ be entirely in one class.

5. The Ehrenfeucht-Mostowski Theorem

For a given theory T, we raise the question whether it is possible to find a model $\textcircled{A}$ of T with an infinite set $X \subseteq A$ such that not only are all the elements of X alike in the sense that they all satisfy the same formulas, but also all n-tuples are alike. In general, this is not possible. For example, in a linearly ordered set, half the pairs satisfy $v_0 < v_1$ and the other half do not. Rather surprisingly, linear orderings are about the strongest obstacle.

THEOREM 5.1. Suppose T a theory in a countable elementary language L which has an infinite model and $(X, <)$ a linearly ordered set. There is a model $\textcircled{A}$ of T with $X \subseteq A$ with the property that for every formula

$\varphi(v_o,\cdots,v_n)$ and every $x_o \prec\cdots\prec x_n$ and $y_o\prec\cdots\prec y_n$ in X, Ⓐ $\models \varphi(x_o,\cdots,x_n)$ if and only if Ⓐ $\models \varphi(y_o,\cdots,y_n)$.

Remarks: The set X is said to be homogenous for Ⓐ. If X is infinite, the set of formulas satisfied by properly ordered sequences from X is called an Ehrenfeucht-Mostowski (E. -M.) set. It follows from the compactness theorem that for any E.-M. set Φ and any linear ordering $(Y, \prec)$, there is a model containing $(Y, \prec)$ for which Φ is an E.-M. set.

Proof: Let Ⓑ be an infinite model of T and $\varphi_o, \varphi_1, \cdots$ an enumeration of the formulas of L. Impose a linear ordering $\prec$ on B. (In general, this ordering is unrelated to the relations of Ⓑ.) Suppose φ_o has the free variables $v_o,\cdots,v_n$. The formula φ_o determines a partition of $B^{(n+1)}$ into two classes depending on whether $(b_o,\cdots,b_n)$ when properly ordered by $\prec$ satisfies φ_o or not. By Theorem 4.9, there is an infinite subset $B_o \subset B$ homogenous for this partition. Let φ_1 have free variables $v_o,\cdots,v_k$. Then φ_1 determines a partition of $B_o^{(k+1)}$ into two classes with an infinite subset $B_1 \subset B_o$ homogenous for the partition. Construct a B_i for each $i \in \omega$ in a similar fashion. Enlarge L by adding a new constant x for each $x \in X$ and enlarge T by adding to it for each formula $\varphi_i(v_o,\cdots,v_k)$ and each $x_o\prec\cdots\prec x_k$ in X the sentence $\varphi_i(x_o,\cdots,x_k)$ or $\sim\varphi_i(x_o,\cdots,x_k)$ depending on which sentence properly ordered sequences of B_i satisfy. The resulting theory is consistent and has a model which satisfies the conclusions of the theorem. The same proof yields:

COROLLARY 5.2. If $\varphi(v)$ is satisfied by an infinite set of elements in some model of T, then it is possible to take all elements of X to satisfy it. If $\varphi(v_o,v)$ linearly orderes an infinite set in some infinite model of T, then the ordering of X may be taken to agree with φ.

A particularly interesting case arises when T is a Skolem theory. In that case, according to Corollary 1.3 we can talk about the elementary submodel generated by X.

THEOREM 5.3. Let Ⓐ and Ⓑ be models of the Skolem theory T^* with linearly ordered (by $\prec$) subsets X and Y respectively. Suppose X and Y are homogenous with respect to the same E.-M. set Φ. Then any one-one order preserving map $f:X \to Y$ induces an elementary map of the model generated by X into the model generated by Y. The induced map is onto if and only if f is onto.

Proof: The map f induces a mapping of the term $t(x_0, \cdots, x_n)$ to $t(f(x_0), \cdots, f(x_n))$. Since both X and Y satisfy the same E.-M. set Φ, this map must preserve the validity of all formulas. All the elements in the models generated by X and Y are represented by such terms t. Hence f is defined. Obviously, if X maps onto Y; then the terms involving the elements of X map onto the terms involving the elements of Y. Conversely, suppose f map X onto a proper subset $Y_0 \subset Y$. Let $y \in Y - Y_0$. I assert that y is not in the model generated by Y_0 and hence not in the model generated by the image of X. For suppose y were an element of the model generated by Y_0. As none of the elements of Y_0 are equal to y, y must equal a term $t(y_0, \cdots, y_m)$ where the y_i are elements of Y_0. Because Y_0 is homogenous with respect to the E.-M. set Φ, Φ must contain the formula

$$(*) v_n = t(v_0, \cdots, v_{n-1}, v_{n+1}, \cdots, v_{n+j})$$

$$= t(v_0, \cdots, v_{n-1}, v_{\ell+1}, \cdots, v_{\ell+j}) = v_\ell .$$

where $\ell \neq n$. It follows from (*) that if y were a member of the model generate by Y_0, then at most a finite number of elements of any linear ordering homogenous for Φ could be distinct. But this leads to a contradiction, since by definition an E.-M. set must be realized by an infinite linear ordering.

Given a Skolem theory T^*, a linear ordering τ, and an E.-M. set Φ of T^*, there is a model of T^* generated by τ in which the properly ordered sequences of τ satisfy the formulas Φ. Denote this model, which is by the preceding theorem unique up to isomorphism, by $M(\Phi, \tau)$.

COROLLARY 5.4. Every theory with an infinite model has models with large automorphism groups; for example, a countable model with automorphism group of power 2^{ω} .

<u>Proof</u>: Let τ have order type of the rationals.

COROLLARY 5.5. Every theory with an infinite model has a model which can be mapped elementarily onto a proper submodel.

<u>Proof</u>: Let $\tau = \omega$.

COROLLARY 5.6. If a theory has an infinite model, then it has models in every power which realize only a countable number of distinct types. Further, it has models Ⓐ in every infinite power which realize only a countable number of types in $Th(A,a)_{a \varepsilon W}$, for every countable $W \subseteq A$.

<u>Proof</u>: (i) It is easy to see that $M(\Phi,\tau)$ realizes only a countable number of types. (ii) Take τ to be any well ordering. Let $W \subseteq M(\Phi,\tau)$ be countable. Then W is contained in the submodel generated by some countable subset τ_o of τ. Any $a \varepsilon M(\Phi,\tau)$ must be of the form $t(x_o,\cdots,x_n)$ where $t(v_o,\cdots,v_n)$ is a term and $x_o,\cdots,x_n$ are elements of τ. Its type with respect to τ_o and, <u>a fortiori</u>, with respect to W, is determined by t and by the order relation of $x_o,\cdots,x_n$ with respect to τ_o. There are only a countable number of terms and since τ_o is well ordered and countable, there are only a countable number of ways to interpolate a finite set into it.

THEOREM 5.7. If $K > \omega$ and T is <u>K-categorical</u>, that is, all models of T of power K are isomorphic, then no formula $\varphi(v_o, v_1)$ linearly orders any infinite subset of any model of T.

<u>Proof</u>: Suppose otherwise. Let $(X, <)$ be a linearly ordered subset of power ω_1 containing a dense countable subset W. Extend T by adding

individual symbols x to represent the elements of X and the sentence $\varphi(x_1,x_2)$ if and only if $x_1 > x_2$. Let this extension of T be the theory T_1. T_1 is consistent by the compactness theorem and has models in every power $\geq \omega_1$. Each of these, considered as a model of T_1, realizes an uncountable number of types with respect to W since each element of X realizes a different cut in W. Thus, in every power we have a model of T non-isomorphic to the one given by Corollary 5.3 (ii).

6. Some Hanf Number Calculations

Let P be some property of models of a theory T. Then there are two possibilities: either T has arbitrarily large models with property P or there is a least cardinal K(T) such that no model of power $\geq$ K has property P. Suppose θ is a set of theories and $K = \sup K(T)$, $(T \in \theta)$. It follows that a theory $T \in \theta$ has arbitrary large models with property P if it has such models of power K. The cardinal K defined in this manner is called the Hanf number of θ. In this section, we shall carry out the computation of some Hanf numbers.

THEOREM 6.1. Suppose T is a theory and Σ a type in a countable elementary language. If for every $\alpha < \omega_1$, T has a model of power $> \beth_\alpha$ which omits Σ, then T has models omitting Σ in every infinite power.

Remark: It can be shown by examples that for every $\alpha < \omega_1$, there is a T and a Σ such that T has a model omitting Σ of power $\beth_\alpha$ but none of higher power. Theorem 6.1 may be stated succinctly as: the Hanf number for omitting types is $\beth_{\omega_1}$.

Proof: Assume T is a Skolem theory. Under the hypothesis of the theorem, we prove the existence of an E.-M. set Φ such that for every linearly ordered set X, $M(\Phi,X)$ omits Σ. Let $\varphi_0, \varphi_1, \cdots$ be a enumeration of the formulas of T; $t_0, t_1, \cdots$ an enumeration of the Skolem terms of T and

$\sigma_0, \sigma_1, \cdots$ an enumeration of Σ. We define inductively a sequence of formulas ψ_n such that ψ_{2n} will be φ_n or $\sim\varphi_n$ and ψ_{2n+1} will be $\sim\sigma_i(t_n)$ for some $i \in \omega$ and such that for each $\alpha < \omega_1$ and each $n \in \omega$:

(*) there is a model A_α of T which omits Σ and a linearly ordered set $(X_\alpha, <)$ of power $\geq \beth_\alpha$ included in A_α such that for all $m < n$, $\psi_m(x_1, \cdots, x_{h(m)})$ holds for $x_1 < \cdots < x_{h(m)}$ in X_α where $h(m)$ is the number of free variables in ψ_m.

Note that for $n = 0$, this says no more than the hypothesis of the theorem.

Suppose ψ_m defined for all $m < n$. Let X_α and A_α be the set and model respectively satisfying (*).

Case 1. $n = 2r + 1$ is odd. Assume t_r has $h(r)$ free variables. Since $A_{\alpha+h(r)}$ omits Σ, for every $x_1 \cdots x_h$ in X_α there is an $i \in \omega$ with $\sim\sigma_i(t_r(x_1, \cdots, x_h))$. Choosing the least such i, we have a partition of $X^{(h)}_{\alpha+h}$ into ω classes. By Theorem 4.10, there is a $Y \subseteq X_{\alpha+h}$, $|Y| \geq \beth_\alpha$, Y homogenous, i.e. for every $x_1 < \cdots < x_h$ in Y, $\sim\sigma_i(t_r(x_1, \cdots, x_h))$. The choice of i here depends on α. But since there are only ω possible choices of i, there must be at least one value of i which corresponds to a set of α's cofinal with ω_1. For this i, let $\psi_n = \sim\sigma_i(t_r)$.

Case 2. n is even. Similar to above but easier.

The models $M(\Psi, X)$ with $\Psi = \{\psi_n\}$ satisfy the theorem.

COROLLARY 6.2. If a countable set of sentences in ω-logic has models of power $\beth_\alpha$ for each $\alpha \in \omega_1$, then it has models in every infinite power.

COROLLARY 6.8. Similarly for countable sets of L_Q sentences having Q_ω models.

COROLLARY 6.4. Similarly for single sentences in $\hat{L}$.

Using the method of proof of the preceding theorem, we can obtain a two cardinal theorem.

THEOREM 6.5. Suppose T is a theory in a countable language such that for every $n \in \omega$, there is a (κ,λ) model with $\kappa = 2(\lambda,n)$. Then for every κ there is a (κ,ω) model.

<u>Proof</u>: Recall that a (κ,λ) model of a theory with a unary relation U is a model A of T with $|A| = \kappa$ and $|U^A| = \lambda$ We shall construct an E.-M. set Ψ such that in $M(\Psi,X)$ the set U is countable.

The construction is parallel to the one in the proof of Theorem 6.1 except now ψ_{2n+1} is either $\sim U(t_n)$ or $t_n(i_1,\cdots,i_{h(n)}) = t_n(j_1,\cdots,j_{h(n)})$ whenever $i_1 < \cdots < i_{h(n)}$ and $j_1 < \cdots < j_{h(n)}$. The inductive definition is the same with "for each $\alpha < \omega_1$" replaced by "for each $\alpha < \omega$" Since in this case we have only two choices at each step and two is not cofinal with ω, the proof still works.

COROLLARY 6.6. If $2^{\omega} = \omega_1$, then any theory T satisfying the hypothesis of Theorem 6.5 has models of type (κ,λ) for all $\kappa \geq \lambda$.

<u>Proof</u>: Immediate from Theorem 3.2. The hypothesis $2^{\omega} = \omega_1$ is not necessary and could be removed by a more complex argument.

THEOREM 6.7. Suppose T is a countable set of sentences in L_Q and Q_κ is ω-compact (in particular, $\kappa = \omega_1$ or κ satisfies the hypothesis of Theorem 3.12). If for each $n \in \omega$, T has Q_κ models of power $\geq 2(\kappa,n)$, then T has Q_κ models in every power $\geq \kappa$.

Proof: By Theorem 3.9, it is sufficient to prove the equivalent result for theories T (in an elementary language) with K-like partly ordered models. Consider a model Ⓐ with $E(v_0,v_1)$ a binary relation which orders its field in a K-like fashion. Let $t(v_1,\cdots,v_{n+k})$ be a term and $a_1,\cdots,a_n$ and $a_1',\cdots,a_n'$ n-tuples in A. Say $(a_1,\cdots,a_n)$ is t-equivalent to $(a_1',\cdots,a_n')$ if for every k-tuple $(b_1,\cdots,b_k)$ in the field of E, either $t(a_1,\cdots,a_n, b_1,\cdots,b_k) = t(a_1',\cdots,a_n', b_1,\cdots,b_k)$ or neither $t(a_1,\cdots,a_n, b_1,\cdots,b_k)$ nor $t(a_1,\cdots,a_n, b_1,\cdots,b_k)$ is in the field of E. A term t determines at most 2^K t-equivalence classes. A finite set of terms $\{t_1,\cdots,t_k\}$ also determines at most 2^K $\{t_1,\cdots,t_k\}$-equivalence classes. Thus, from the assumption that there is a K-like model of power $\geq 2(k,n)$, it follows that for every finite set of terms, $\{t_1,\cdots,t_k\}$ there is an infinite linearly ordered set embedded in a K-like model such that all properly ordered sequences of it are t_i-equivalent for $i = 1,\cdots,k$. Note the fact that properly ordered sets are t_i-equivalent is expressible in the elementary language. Further, Q_K is ω-compact. Hence there is a K-like model containing an infinite ordered set X such that all properly ordered sequences are t_i-equivalent for all terms t_i. Let Ⓐ be such a model and Ⓑ the elementary submodel generated by {field of E} ∪ X. I assert that it is possible to find arbitrarily large extensions of Ⓑ which do not add any element to the field of E. Simply embed the linear ordering X in a larger one Y and add axioms asserting that all the properly ordered sequences of Y are t-equivalent to the sequences in X for every term t. This new theory is consistent by the compactness theorem. But it says that every element of the field of E in the model generated by {field of E} ∪ Y was already in Ⓑ.

7. Models of Number Theory and Set Theory

Let Ⓐ be a partly ordered Skolem model which is ω-like and let $T = Th(A)$. By Corollary 5.2, we may find an E.-M. set Ψ for T which in particular contains $v_0 \, E \, v_1$ where E is the relation which ω-like orders a subset of A. Indeed, it follows from the proof of Theorem 5.1 that Ψ may be constructed using only infinite subsets of the field of E. Such an E.-M. set Ψ must contain the following formulas:

(i) $v_o \, E \, v_1$

(ii) $\sim (v_{n+1} \, E \, t(v_1, \cdots, v_n))$ for each term t.

(iii) $t(v_1, \cdots, v_n) \, E \, v_k \rightarrow t(v_1, \cdots, v_n) = t(v_1, \cdots, v_k, v_{n+1}, \cdots, v_{2n-k})$

for all $k < n$.

The reason (ii) and (iii) must be included is that the negation of any such formula would imply that some initial segment of the field E would be infinite and therefore the well ordering could not be ω-like. Let $(X, <)$ be a well ordered set of type K, K an uncountable cardinal. We assert the $M(\Psi, X)$ is K-like. For (i) implies that the field of E has power K, (ii) that X is cofinal in the field of E, and (iii) that the number of elements preceding an element of X has power $< K$. We have thus proved

THEOREM 7.1. If T has an ω-like model, then it has a K-like model for every infinite cardinal K.

COROLLARY 7.2. If T is a countable set of sentences in L_Q and T has a Q_ω-model, then T has a Q_K-model for every uncountable cardinal K.

One may obtain some results about models of number theory by noting that Ramsey's Theorem, Theorem 4.9, is itself a theorem of number theory. By this we mean the following: Let $A = (w, +, x, <, o, 1, R_i)_{i \in I}$ be a model in which $+, x, <, o, 1$ have their usual meaning and I is countable. Let $T = Th(A)$.

THEOREM 7.3. Let $\varphi(v_1, \cdots, v_n)$ be a formula and suppose $T \models$"φ determines a mapping of properly ordered n-tuples from some cofinal subset into an initial segment". There is then a formula $\psi(v_o)$ with the property $T \models$ "the set of elements satisfying ψ is cofinal and on that set φ maps all properly ordered n-tuples onto the same element".

Proof: The proof, which is omitted, is a tedious verification that each step in the proof of Theorem 4.9 (Ramsey's Theorem) may be carried out by finite induction.

Theorems 7.2, 7.3, and 7.4, below actually apply to a larger class of theories. We may let $T = Th(A)$ where $\textcircled{A} = (A, +, x, o, 1, R_i)_{i \in I}$ and $\textcircled{A}$ satisfies

(i)   a particular set T_o of axioms of arithmetic which guarantee that inductively defined sets are explicitly definable sets.

(ii)   the axiom scheme of induction, that is, for any formula $\varphi(v)$,
$$\varphi(o) \wedge (v)(\varphi(v) \rightarrow \varphi(v+1)) \rightarrow (v)\varphi(v).$$

In (i), T_o is the set of axioms needed to prove Ramsey's Theorem as a formal theorem of $Th(A)$.

Theorem 7.4 about models of number theory and Theorem 7.5 about countable models of Z.-F. set theory plus the axiom of choice will be proved together. Models of set theory are partly ordered sets with the ordinals playing the rule of the linearly ordered set. The similarity between the two theorems arises from the fact that Ramsey's Theorem is a theorem of formal number theory and the Erdős-Rado Theorem (Theorem 4.9) is a theorem of Z.-F. plus the axiom of choice.

THEOREM 7.4. Suppose $\textcircled{A}$ is a model of number theory, i.e. a theory satisfying (i) and (ii) above. Let $A' = \langle A, a \rangle_{a \in A}$ and $T = Th(A')$. Then there is an E.-M. set Ψ for T such that for every uncountable cardinal $K > |A|$:

(i) $M(\Psi, K)$ is K-like

(ii) $M(\Psi, K) > \textcircled{A}$ and further $M(\Psi, K)$ is an end extension of $\textcircled{A}$, i.e. $\textcircled{A}$ is an initial segment of $M(\Psi, K)$.

THEOREM 7.5. Suppose $\textcircled{A}$ is a countable model of Z.-F. plus the axiom of choice. Let $A' = \langle A, a \rangle_{a \in A}$ and let $T = Th(A')$. Then there is an E.-M. set Ψ for T such that for linearly ordered sets X, $M(\Psi, X)$ is an <u>end</u>

extension of Ⓐ', i.e. the ordinals of Ⓐ' are an initial segment of the ordinals of $M(\Psi,X)$.

Proof: There is no loss in generality in assuming T is a Skolem theory. In number theory, we take the Skolem function to pick out the least element satisfying the given formula. The resulting theory satisfies (i) and (ii). For set theory the problem is more subtle. We note that on any set of the model we can pick out the Skolem function on that set using the axiom of choice.

The proofs are modifications of the Ehrenfeucht-Mostowski Theorem 5.1. The problem here is to guarantee that $M(\Psi,X)$ is an end extension of Ⓐ. In number theory, we choose Ψ so that its finite subsets are satisfied by properly ordered n-tuples of definable cofinal sequences of Ⓐ. In set theory, finite subsets of Ψ are satisfied by properly sequences of ordinals of arbitrarily high cardinality.

Case (i) Ⓐ is countable model of number theory. We use Ramsey's Theorem as a formal theorem of number theory.

Enumerate the formulas of $T: \varphi_o, \varphi_1, \cdots$. If φ_o has free variables $x_o, \cdots, x_k$, then φ_o induces a partition of $A^{(k+1)}$ into two disjoint subsets. According to Ramsey's Theorem there is a definable cofinal sequence on which φ_o or $\sim\varphi_o$ is satisfied by the properly ordered k+1-tuples of the sequence. If the formula obtained is of the form $t(v_o,\cdots,v_k) < a_1$ for a Skolem term t and element a_1 of A, then we choose a an element A for which $t(v_o,\cdots,v_k) = a$ is satisfied on a definable cofinal sequence. Since t maps a cofinal sequence into an initial segment, namely the elements $< a_1$, and this segment is formally finite such an a must exist according to Ramsey's Theorem.

It is easy to see how to continue the process by finite induction. At each step a definable cofinal subsequence of the previously defined cofinal sequence is obtained. The E.-M. set obtained satisfies the theorem.

Case (ii). Ⓐ is a countable model of set theory. Using the Erdős-Rado Theorem, we find an E.M. set Ψ such that every finite subset of Ψ is satisfied by properly ordered sequences of ordinals of arbitrarily high cardinality (in the sense of the model). Notice we are not justified in asserting that there is a single cofinal sequence with this property. At each stage of the finite induction, if the formula $t(v_o, \cdots, v_n) < a$ is obtained, we find a constant a_1 for which $t(v_o, \cdots, v_n) = a$ is satisfied. Such an a_1 must exist according to the Erdős-Rado Theorem.

Case (iii) Ⓐ is an uncountable model of number theory. The problem here is that there are an uncountable number of formulas in the language of $T' = Th\langle A, a\rangle_{a \in A}$. But every formula in this language corresponds to a formula in the language of $Th(A)$ in which some of the free variables have been replaced by constants. There only a countable number of formulas in the language of $Th(A)$. Enumerate these formulas: $\varphi_o, \varphi_1, \cdots$

Consider $$\varphi_o = \varphi_o(v_o, \cdots, v_k, v_{k+1}, \cdots, v_{k+n}).$$

Replace the free variables $v_o, \cdots, v_k$ by constants $a_o, \cdots, a_k$. It can be shown that there is a definable cofinal sequence in Ⓐ such that for any choice of $a_1, \cdots, a_k$, properly ordered n-tuples of the sequence satisfy $\varphi(a_o, \cdots, a_k, v_{k+1}, \cdots, v_{k+n})$ or $\sim\varphi(a_o, \cdots, a_k, v_{k+1}, \cdots, v_{k+n})$ provided the n-tuples are sufficiently large. This statement is proved by the formal analog of finite induction holding within the model. Again we add the that if the formula obtained is of the form $t(v_o, \cdots, v_n) < a$ where t is a Skolem term which contains the constants $a_o, \cdots, a_k$, we must find an a_1 for which $t(v_o, \cdots, v_n) = a_1$ holds.

It is easy to see how to continue this process using Ramsey's Theorem as a formal theorem of number theory and the formal analog of finite induction indicated above.

8. Big Cardinals

Consider a linearly ordered set X and define $X^{(\omega)}$ to be the subsets of X or order type ω. results of Section 4 do not generalize to $X^{(\omega)}$. A set $Y \subset X$ is said to be homogenous for a partition of $X^{(\omega)}$ into μ classes if $Y^{(\omega)}$ lies entirely within one class. There are arbitrarily large sets X, $X^{(\omega)}$ $C_0 \cup C_1$, but with no infinite subset Y with $Y^{(\omega)} \subset C_i$, $i = 0, 1$. Extending the notation of section 2:

THEOREM 8.1. For all K, not $K \to (\omega)_2^{\omega}$. We define a weaker notion. Let X be a linearly ordered set and define $X^{<\omega}$ to be all the (properly ordered) finite subsets of X. By a partition of $X^{<\omega}$ into μ classes we shall always mean that for each n, $X^{(n)}$ is partitioned into μ classes. A subset Y of X is homogenous for a partiton of $X^{<\omega}$ into μ classes if for all n, $Y^{(n)}$ lies in one partition class (which may depend on n). We write $K \to (\lambda)_{\mu}^{<\omega}$ if for every partition of $K^{<\omega}$ into at most μ classes, there is a subset Y of K, $|Y| = \lambda$, homogenous for the partition. A partition f of $X^{<\omega}$ into μ classes $f: X^{<\omega} \to \mu$ is equivalent to a countable set of functions $f_n: X^{(n)} \to \mu$.

Assume that X is linearly ordered and that $f: X^{(n)} \to \mu$. For $m > n$, f induces a map $f_m': X^{(m)} \to \mu$ by setting $f_m'(x_1, \cdots, x_m) = f(x_1, \cdots, x_n)$ when $x_1 < \cdots < x_m$.

Let $f_i: K^{<\omega} \to \mu$, $i \in I$, be a countable collection of maps. Each map f_i is associated with a set of maps $f_{i,n}$ i $K^{(n)} \to \mu$. Since $|\omega^2| = \omega$, we can map pairs of integers (i,n) onto the integers in a one-one fashion. If (i,n) maps into m under this mapping, we associate with the map $f_{i,n}$ the map $(f_i)_m'$ induced by $f_{i,n}$. The countable set of maps arising in this fashion yields a single mapping $f: K^{<\omega} \to \mu$. On the other hand a countable set of maps $f_i: K^{<\omega} \to \mu$ is equivalent to one map of $K^{<\omega}$ into μ^{ω}. Hence,

THEOREM 8.2. $K \to (\lambda)_{\mu}^{<\omega}$ implies $K \to (\lambda)_{\mu^{\omega}}^{<\omega}$.

COROLLARY 8.3. Not $\omega \to (\omega)_2^{<\omega}$.

Indeed we shall see that whenever $K \to (\lambda)_\mu^{<\omega}$, then K is very large. We consider some properties of large cardinals.

A cardinal is strongly inaccessible (I) if it is regular and uncountable and $\lambda < K$ implies $2^\lambda < K$. The existence of strongly inaccessible cardinals can not be proved from the usual axioms of set theory. A cardinal K has property C, E_2, E_1, E_o if $K \to (K)_2^2$, $K \to (K)_2^{<\omega}$, $K \to (\omega_1)_2^{<\omega}$ or $K \to (\omega)_2^{<\omega}$ respectively. A cardinal is <u>measureable</u> (M) if there is a K-additive ($K > \omega$) two valued non-trivial measure on K. This means there is a function on the power set, $P(K)$, of K into $\{0,1\}$ with $m(K) = 1$, $m(x) = 0$ if $|x| = 1$ and for any collection C of disjoint subsets of K with $|C| < K$, $m(U(x)) = \Sigma m(x)$, $(x \in C)$. Thus the union of fewer than K sets of measure 0 has measure 0 and the intersection of fewer than K sets of measure 1 has measure 1.

THEOREM 8.3. E_2 implies E_1 implies E_o. E_2 implies C implies I.

<u>Proof</u>: All are obvious except C implies I. That follows immediately from Corollary 4.6 and Theore 4.11.

If a cardinal has property E_1 or E_o, then every cardinal greater than it does. Hence neither E_1 nor E_o can imply I. However,

THEOREM 8.4. Suppose $\lambda_1 \geq \lambda_2 \geq \omega$ and K_1 and K_2 are the least cardinals such that $K_1 \to (\lambda_1)_2^{<\omega}$ and $K_2 \to (\lambda_2)_2^{<\omega}$. Then (i) $K_1 > K_2$ and (ii) K_1 and K_2 are both strongly inaccessible.

<u>Proof</u>: (i) Clearly $K_1 \geq K_2$. Suppose $K_1 = K_2 = K$. Let $(X, <)$ be a linearly ordered set of type K and let $x \in X$. The initial segment X_x of elements $< x$ has power $< K$. Hence for each x there is a function

$f_x : X_x^{<\omega} \to \{0,1\}$ which has no homogenous subset Y of power $\geq \lambda_2$. Define $g : X^{<\omega} \to \{0,1\}$ by $g(x_1,\cdots,x_n) = f_{x_n}(x_1,\cdots,x_{n-1})$ where $x_1 < \cdots < x_n$. Suppose $Y \subset X$ is homogenous for g. By the construction of g, it follows that for every $y \in Y$ the set $\{x; x \in X \text{ and } x < y\}$ must have power $> \lambda_2$. Therefore Y has power $\leq \lambda_2 < \lambda_1$. This contradicts $K_1 \to (\lambda_1)^{<\omega}$. (ii) Assume that K is the least cardinal satisfying $K \to (\lambda)_2^{<\omega}$. We must show that K is regular and that $\mu < K$ implies $2^\mu < K$. First, suppose $\mathrm{cf}\ K < K$. Then K can be partitioned, $K = \cup X_i$, $i \in I$, with $|I| = \mathrm{cf}\ K$ and $|X_i| < K$ for all i. By hypothesis, there are partitions f_i of $X_i^{<\omega}$ and g of $I^{<\omega}$ which do not possess homogenous sets of power λ. We define a partiton f of $K^{<\omega}$:

$y_1 \in K^{<\omega}$, $y_2 \in K^{<\omega}$ are in the same partition class if

(a) $y_1 \subset X_{i_1}$, $y_2 \subset X_{i_2}$ and $f_{i_1}(y_1) = f_{i_2}(y_2)$ or

(b) the sets $i_1 = \{i; X_i \cap y_1 \neq \emptyset\}$ and $i_2 = \{i; X_i \cap y_2 \neq \emptyset\}$ be in the same g partition classes and $|i_1|$ and $|i_2| > 1$.

Thus any set homogenous for f must either lie entirely in one class X_i and be homogenous for f_i or have at most one element in each X_i and the set $\{i; Y \cap X_i \neq \emptyset\}$ be homogenous for g. In either case $|Y| < \lambda$.

Next we show that if not $\mu \to (\lambda)_2^{<\omega}$, then not $2^\mu \to (\lambda)_2^{<\omega}$. Recall that Theorem 8.2 asserts that $2^\mu \to (\lambda)_2^{<\omega}$ implies $2^\mu \to (\lambda)_{2^\omega}^{<\omega}$. We shall actually prove not $\mu \to (\lambda)_2^{<\omega}$ implies not $2^\mu \to (\lambda)_\omega^{<\omega}$ which will suffice.

Let f be any partition of $\mu^{<\omega}$. We shall associate with f a partition f^* of the finite sequences of μ, that is, of $\bigcup_{n\in\omega} \mu^n$. The partition class of a finite sequence will depend not only on the f partition class determined by the set of distinct elements appearing in it, but also on the order in which these elements are arranged. Thus the partition f^* of $\bigcup_{n\in\omega} \mu^n$ is a refinement of the partition f considered as operating on $\bigcup_{n\epsilon\omega} \mu^n$. Sequences consisting of identical elements but differing in the arrangement of these elements will belong to distinct f^* partition classes.

As 2^μ may be represented as all sequences of length μ of 0's and 1's, we can lexicographically order 2^μ. Choose a partition f of $\mu^{<\omega}$ into two classes which does not possess an homogenous set of power λ. Define a partition f' of $(2^\mu)^{<\omega}$ as follows: for any finite set $t_o < \cdots < t_n$ let $f\ (t_o, \cdots, t_n) = f^*(x_o, \cdots, x_{n-1})$ where x_i, an element of μ, is the first co-ordinate of t_i satisfying $t_i(x_i) \neq t_{i+1}(x_i)$.

Let t_o and t_1 be elements of 2^μ with $t_o < t_1$. Define $X(t_o, t_1)$ to be the first co-ordinate where the sequence representing t_o differs from the sequence representing t_1. Let Y be a subset of 2^μ homogenous for f: We prove that the subset X of μ defined as

$$Z = \{X(t_o, t_1);\ t_o < t_1,\ t_o,\ t_1 \text{ elements of } Y\}$$

which is obviously homogenous for f, is in one-one correspondence with Y.

Consider a properly ordered triple $t_o < t_1 < t_2$ where the t_i are elements of Y. Note that $X(t_o, t_1) \neq X(t_1, t_2)$. For at the first co-ordinate where a properly ordered pair differs, the change must be from a 0 to a 1 and such a change can occur only once at a given co-ordinate. Since Y is homogenous for f: for all properly ordered triples $t_o < t_1 < t_2$ of Y, either $X(t_o, t_1) < X(t_1, t_2)$ or $X(t_1, t_2) < X(t_o, t_1)$; assume $X(t_o, t_1) < X(t_1, t_2)$. (A similar argument to the one we present with the roles of t_o and t_2 reversed holds for the opposite ordering.) I assert that $X(t_o, t_1)$ depends only on t_o, i.e. $X(t_o, t_1) = X(t_o, t_2)$. For $X(t_o, t_1) < X(t_1, t_2)$ which implies that for all co-ordinates less than $X(t_o, t_1)$, the sequences t_o and t_2 agree. Hence we have established a one-one correspondence between X and Y. Therfore any set Y homogenous for f' must have power $< \lambda$.

We conclude this section by showing:

THEOREM 8.5. M implies E_2.

Proof: Suppose m is a K-additive measure on K. Let f be a partition of $K^{<\omega}$, i.e., let $f_n: K^{(n)} \to (I, II)$. (We use I, II to avoid confusion with the values of the measure.) The key to the proof is the fact that any $g: K^{(n+1)} \to (I, II)$ induces a map $g': K^{(n)} \to (I, II)$. Simply set $g'(x_o, \cdots, x_{n-1}) = I$ if $\{x; g(x_o, \cdots, x_{n-1}, x) = I\}$ has measure 1 and similarly for $g'(x_o, \cdots, x_{n-1}) = II$. Define by induction for each $m \leq n$, $f_{n,m}: K^{(m)} \to (I, II)$, by setting $f_{n,n} = f_n$ and $f_{n,m-1} = f'_{n,m}$. When $m = 1$, the $f_{n,1}$ partition $K = K^{(1)}$ into two classes, exactly one of which, the preferred class, has measure 1. A set Y is preferred homogenous if each $f_{n,1}$ maps Y into the preferred class and for every $y_1 < \cdots < y_m$ in Y, $f_{n,m}(y_1, \cdots, y_m) = f_{n,1}(y_1)$. Since preferred homogeneity is a property of finite character, there is a maximal preferred homogenous set Y. Suppose it had power $< K$. Then its complement has measure 1. Further, for each $m < n$ and $y_1, \cdots, y_m$ in Y, there is a set Z of measure 1 such that $f_{n,m+1}(y_1, \cdots, y_m, z)$ has the preferred value for all $z \in Z$. The intersection of all such z has measure 1 (since Y has power $< K$) and hence is non-empty. Adjoin any element of this intersection to Y to obtain a proper extension of Y which is preferred homogenous, contradicting the maximality of Y. Hence $|Y| = K$ and the theorem is proved.

9. Some Non-elementary Languages

Suppose L is an elementary langauge and K a cardinal. We form the language L_K by adding variables v_α, $(\alpha < K)$, and two new rules of formation for formulas:

(iv) if $\{\psi_\alpha; \alpha < \beta < K\}$ are formulas, then $\bigvee_{\alpha<\beta} \psi_\alpha$ and $\bigwedge_{\alpha<\beta} \psi_\alpha$ are formulas.

(v) if $\beta < K$ and ψ is a formula, then $(\exists v_\alpha | \alpha < \beta)\psi$ is a formula.

In this last rule (v), $(\exists v_\alpha | \alpha < \beta)$ is to be interpreted as a sequence of quantifiers. Notice that we can express an infinite sequence of universal quantifiers by $\sim(\exists v_\alpha | \alpha < \beta) \sim \psi$. If $K = \omega$, L_K is just the elementary language L. If $K = \omega_1$, the formulas are countable long. However, L_{ω_1}

has more formulas than $\hat{L}$ (Section 3) since $\hat{L}$ has (iv) but not (v). In particular for $K > \omega_1$, there is a sentence $\sim(\exists v_\alpha | \alpha < \omega)(v_0 > v_1 \wedge v_1 > v_2 \wedge \cdots)$ which says that an ordering is a well ordering. This sentence can not be expressed in $\hat{L}$. If one allows infinitary functions, one may define Skolem functions for L_K and the results of Section 2 apply.

A cardinal is weakly compact if a set of power K of sentences in L_K has a model whenever all its subsets of smaller power have models.

THEOREM 9.2. (i) ω is weakly compact. (ii) K^+ is not weakly compact. (iii) $\mathrm{cf}\, K < K$ implies K is not weakly compact. (iv) The smallest inaccessible cardinal is not weakly compact.

<u>Proof</u>: (i) Is simply the compactness theorem for elementary languages.

(ii) We form a theory as follows:

Take individual constants b and $C_\alpha (\alpha < K^+)$, a binary symbol $<$, a ternary relation symbol R, and axioms asserting:

(1) $<$ is a well ordering

(2) $(v)(v < C_\alpha \rightarrow \bigvee_{\beta<\alpha} (v = C_\beta))$ for each $\alpha < K^+$.

(3) for each x, $R(x,y,z)$ is a 1-1 correspondence between the elements $< x$ and a set of elements $< C_K$.

(4) $b \neq C_\alpha$ (for each $\alpha < K^+$).

This set of axioms can not have a model since the elements $< K^+$ can not be mapped one-one into the K elements $< C_K$. But every subset of sentences of power K does have a model in which b is the first ordinal greater than all the C_α's mentioned.

(iii) The proof is similar to (ii) with the axiom (3) replaced by $(v_0)(\exists v_1)(v_1 > v_0 \wedge \bigvee_{\alpha\in I}(v_1 = C_\alpha))$ where I is some cofinal subset of K.

(iv) Similar to (ii). Axiom (3) is replaced by a formula which asserts for every x the correspondence between y and $\{z; R(x,y,z)\}$ is one-one on some cofinal subset of x and further

$(\exists u)(w < x \wedge (y)(z)(R(x,y,z) \to z < \mu))$, that is, $\mathrm{cf}\ x < 2^u$ for some $u < x$.

The condition that K is regular, uncountable, and not of the form K^+ is called <u>weak inaccessibility</u>. Assuming the generalized continuum hypothesis, it is equivalent to <u>strong</u> <u>inaccessiblity</u>.

COROLLARY 9.3. Assuming the generalized continuum hypothesis, all uncountable weakly compact cardinals are strongly inaccessible.

THEOREM 9.4. (i) If $K \to (K)^2_2$, then K is weakly compact. (ii) If K is weakly compact and strongly inaccessible, then $K \to (K)^2_2$.

<u>Proof</u>: If $K = \omega$, trivial. Assume $K > \omega$.

(i) Suppose $K \to (K)^2_2$ and T a theory in L_K with K sentences such that each subtheory of lower power has a model. As mentioned earlier, L_K may be supplied with Skolem functions. Assume that T is a Skolem theory. Let $\{S_\beta; \beta < K\}$ be a well ordering of the sentences of L_K and let $T_\alpha = T \cap \{S_\beta; \beta < \alpha\}$. By hypothesis, each T_α has a model. Let P consist of all well ordered sequences $p_{Ⓐ}$ of o's and 1's of the following form: Ⓐ is a model of T_α, $p_{Ⓐ}$ is of length α, and the βth element of p is 1 if and only if Ⓐ $\models S_\beta$. The set P is a binary ramification system and by hypothesis has a linearly ordered subset of power K. This subset determines a binary sequence p_∞ of length K with every sequence p of the linearly ordered subset an initial segment of p_∞. In turn, p_∞ will determine a theory $T_\infty = \{S_\beta; \beta\text{th co-ordinate of } p_\infty = 1\}$. By the construction of P, $T \subseteq T_\infty$. Define an equivalence relation on the constant terms of L_K by $t_1 \approx t_2$ if and only if $t_1 = t_2$ is a sentence of T_∞. Let A be the set of equivalence classes. Define a relation structure Ⓐ with universe A as follows: for each n-ary symbol R in L_K, let the interpretation of R in Ⓐ be $R^A = \{\hat{t}_1, \cdots, \hat{t}_k, R(t_1, \cdots, t_k) \in T_\infty\}$ where $\hat{t}_i$ is the equivalence class of t_i. That Ⓐ satisfies all the sentences of T_∞ is proved by induction on the length of the sentences of L_K.

(ii) Suppose K is strongly inaccessible but not $K \to (K)^2_2$. By Theorem 4.1, there is a binary ramification system P of power K all of whose linearly ordered subsets have power $< K$. Since K is strongly inaccessible, there are at most K maximal linearly ordered subsets. Construct a theory T as follows: for each $p \in P$, there is a symbol $\underline{p}$, a symbol $\prec$, an individual constant b, and axioms stating,

1. $\prec$ is a partial ordering which is a binary ramification system,
2. $\underline{p_1} \prec \underline{p_2}$ for each $p_1 \prec p_2$ in P,
3. $\underline{p}$ has only one immediate successor (for each $p \in P$ with only one immediate successor),
4. $\sim(\exists v) \bigwedge_{i \in I} v \succ p_i$ (for every maximal linearly ordered set $\{p_i\}_{i \in I}$ in P),
5. $(v)(v \prec \underline{p} \to v \to \bigwedge_{i \in I} v = \underline{p_i})$ where $\{p_i\}_{i \in I}$ is the set of element $\prec p$ in P.
6. $b \neq \underline{p}$ for each $p \in P$.

T has no model but every subset of power $\prec K$ does. Hence K can not be weakly compact.

COROLLARY 9.5. The first cardinal with property C is greater than the first inaccessible ordinal.

Let θ_M, θ_C, θ_I, θ_2, θ_1, θ_o be the least cardinals having properties M, C, I, E_2, E_1, E_0 respectively. From results already established, we have $\theta_M \geq \theta_2 \geq \theta_C > \theta_I$ and $\theta_2 > \theta_1 > \theta_0 \geq \theta_I$. To completely order these cardinals, we could show (but shall not) that $\theta_M > \theta_2$ and $\theta_0 > \theta_C$: Hence $\theta_M > \theta_2 > \theta_1 > \theta_0 > \theta_C > \theta_I$.

10. Omitting Types of Sequences

Let T be a countable set of sentences in an elemantary language L and Σ a countable set of formulas. A model Ⓐ of T omits Σ if there is no sequence $a_0, a_1, a_2, \cdots$ in A which satisfies all the formulas of Σ. (The difference between the work of this section and the work of Section 5 is that here Σ is not restricted to formulas with some fixed finite number of free variables.)

THEOREM 10.1. Suppose $K \to (\omega_1)_2^{<\omega}$ and T has a model of power K omitting Σ. Then T has an E.-M. set Ψ with the property that for every well ordered set X, $M(\Psi, X)$ omits Σ.

Proof: Let $\mathfrak{A}$ be a Skolem model of T of power K omitting Σ. Impose a well ordering on A and partition $A^{<\omega}$ into 2^{ω} classes such that sequences in the same class satisfy the same set of formulas. By assumption $K \to (\omega_1)_2^{<\omega}$ and by Theorem 8.2 $(K \to (\omega_1)_{2^\omega}^{<\omega}$. So there is a sequence Y of length ω_1 homogenous for this partition. Let Ψ be the set of formulas satisfied by properly ordered subsequences of Y and let X be any well ordered set. I assert that Σ is omitted in $M(\Psi, X)$. For suppose not. Then there is a sequence $b_0, b_1, \cdots$ in $M(\Psi, X)$ which satisfies Σ. Each b_n is represented by a term $t_n(x_0, \cdots, x_k)$ with $x_0 < x_1, \cdots < x_n$ in X. The entire sequence $b_0, b_1, \cdots$ can be represented using only a countable subset $X_0 \subset X$. The set X_0 being well ordered and countable, can be mapped order isomorphically onto a subset $Y_0 \subseteq Y$. Letting $a_n = t_n(y_0, \cdots, y_k)$ with $y_0 < \cdots < y_k$ corresponding to $x_0 < \cdots < x_k$, we obtain a sequence $a_0, a_1, \cdots$ in which satisfies Σ, contradicting the hypothesis of the theorem.

COROLLARY 10.2. Let T have a well ordered model of power K with $K \to (\omega_1)_2^{<\omega}$. Then there is an E.-M. set Ψ such that for every well ordered set X, $M(\Psi, X)$ is well ordered.

Proof: A model is well ordered if it omits the type of sequence $\Sigma = \{"v_n > v_{n+1}"; n \in \omega\}$.

COROLLARY 10.3. Same result for partly ordered models in which the linearly ordered part is well ordered of power K where $K \to (\omega_1)_2^{<\omega}$.

11. The Constructible Universe

In this section we assume some familiarity with the notions of axiomatic set theory. In particular, familiarity with the function $F(\alpha)$ which enumerates the sets constructible in the sense of Gödel and the following weak form of the Gödel isomorphism theorem is assumed.

THEOREM 11.1. Suppose Ⓐ is a well founded model of Zermelo-Frankel set theory, i.e., one in which the ordinals are well ordered when viewed from the outside. Then the necessarily unique order preserving map of the ordinals in the model Ⓐ to an initial segment $\{\alpha; \alpha < \gamma\}$ of ordinals induces an isomorphism (with respect to ϵ) of the constructible universe of the model to $\{F(\alpha); \alpha < \gamma\}$.

The existence of large cardinals has implications for the construtible universe.

THEOREM 11.2. Suppose there is a K such that

$$K \to (\omega_1)_2^{<\omega}.$$

Then

(i) any infinite constructible set Y has at most $|Y|$ subsets and hence

(ii) not every set is constructible (in fact not every subset of ω is constructible);

(iii) there is a countable ordinal γ such that $\langle\{F(\alpha), \alpha < \gamma\}, \epsilon\rangle$ is an elementary subsystem of the constructible universe and hence

(iv) not only are there only a countable number of definable sets in the constructible universe but every definable set is itself countable.

Proof: Let K be the least cardinal for which $K \to (\omega_1)_2^{<\omega}$ holds. By Theorem 8.4, K is strongly inaccessible. But then $\langle\{F(\alpha), \alpha < K\}, \epsilon\rangle$ is a model of Z.-F. plus the axiom of constructibility. According to Section 10, we can find an E.-M. set Ψ for which $M(\Psi, X)$, X any well ordered set, is a well founded model of Z.-F. plus axiom of constructibility. From the Gödel Isomorphism Theorem it follows that, given any ordinal γ, $\langle\{F(\alpha), \alpha < \gamma\}, \epsilon\rangle$ can be isomorphically embedded in $M(\Psi, X)$ provided that X is large enough.

We prove (i). Choose γ so large that Y and all its constructible subsets are included in $\langle\{F(\alpha), \alpha < \gamma\}, \epsilon\rangle$. Consider $\langle\{F(\alpha), \alpha < \gamma\}, \epsilon\rangle$ embedded isomorphically in $M(\Psi, X)$. Then Y and all its subsets are included in a subsystem generated by a subset W of X with $|W| = |Y|$. The elements of a subset of Y satisfy a certain set of formulas with parameters from Y and, according to the axiom of extensionality, elements of different subsets satisfy different sets of formulas. Thus different subsets of Y correspond to different types with respect to Y. The proof now proceeds as the proof of Corollary 5.6. The number of types with respect to Y is bounded by the number of types with respect to W. Since W is well ordered and $M(\Psi, X)$ has only a countable number of terms, the number of types with respect to W is at most $|W|$.

The proof of (iii) depends on the existence of a special sort of E.-M. set. Ψ is called remarkable if for every Skolem term t it contains the following formulas:

(a) $t(v_o, \ldots, v_{n-1}) < v_n$

(b) $t(v_o, \ldots, v_{m+n}) \leq v_m \to t(v_o, \ldots, v_{m+n}) = t(v_o, \ldots, v_m, v_{m+n+1} \ldots v_{m+2n})$

(c) $v_n < t(v_o, \ldots, v_{n-1}, v_{n+1}, \ldots, v_m) \to v_{n+1} < t(v_o, \ldots, v_{n-1}, v_{n+1}, \ldots, v_m)$.

The proof of the existence and uniqueness of a remarkable E.-M. set for Z.-F. plus the axiom of constructibility will be omitted. The assertion that Ψ is a remarkable E.-M. set for Z.-F. plus the axiom of constructibility is equivalent to the fact that for uncountable cardinals K, the ordinals of

$M(\Psi,K)$ have order type K and under the elementary monomorphism of $M(\Psi,K)$ into $M(\Psi,\mu)$, $\mu > K$, the ordinals of $M(\Psi,K)$ correspond to the first K ordinals of $M(\Psi,\mu)$. (For a more complete discussion of remarkable E.-M. sets, see [28].)

Condition (b) guarantees that for limit ordinals β, not only is $M(\Psi,\beta) \prec M(\Psi,\gamma)$, $\gamma > \beta$, but the ordinals of $M(\Psi,\beta)$ actually are an initial segment of the ordinals of $M(\Psi,\gamma)$. That is, $M(\Psi,\gamma)$ is an end extension of $M(\Psi,\beta)$.

Next we prove that for uncountable cardinals K, $M(\Psi,K)$ is actually an elementary subsystem of the constructible universe L. Note that since the ordinals of $M(\Psi,K)$ have order type K, $M(\Psi,K)$ is isomorphic to $<\{F(\alpha),\ \alpha < K\},\ \epsilon>$. Given any formula Φ of Z.-F., the reflection principle guarantees the existence of an uncountable cardinal $\mu \geq K$ such that Φ is absolute with respect to $<\{F(\alpha),\ \alpha < \mu\},\ \epsilon>$ or equivalently with respect to $M(\Psi,\mu)$. But since $M(\Psi K) \prec M(\Psi,\mu)$, Φ is absolute with respect to $M(\Psi,K)$.

In particular, we have that $M(\Psi,\omega_1)$ is an elementary subsystem of L. Also, $M(\Psi,\omega_1)$ is an end extension of its elementary subsystem $M(\Psi,\omega)$. Recall that $M(\Psi,\omega_1)$ is isomorphic to $<\{F(\alpha),\ \alpha < \omega_1\},\ \epsilon>$. But since $M(\Psi,\omega_1)$ is an end extension of $M(\Psi,\omega)$, the isomorphism of $M(\Psi,\omega_1)$ onto $<\{F(\alpha),\ \alpha < \omega_1\},\ \epsilon>$ must carry the ordinals of $M(\Psi,\omega)$ onto an initial segment of the ordinals of $<\{F(\alpha),\ \alpha < \omega_1\},\ \epsilon>$. Hence $M(\Psi,\omega)$ is isomorphic to $<\{F(\alpha),\ \alpha \quad \gamma\},\ \epsilon>$ for some countable ordinal γ. It follows immediately that $<\{F(\alpha),\ \alpha \quad \gamma\},\ \epsilon>$ is a countable elementary subsystem of the constructible universe L. The elements of each set of the model $<\{F(\alpha),\ \alpha < \gamma\},\ \epsilon>$ are also in the model. Since every definable set of L is definable in any of its elementary subsystems and since every set of $<\{F(\alpha),\ \alpha < \gamma\},\ \epsilon>$ is countable, we have that every definable set of L is countable.

12. Sources

No attempt has been made to give a complete bibliography of model theory. (A rather complete list (through 1963) is available in the Theory of Models, North Holland Publishing Company, Amsterdam, 1965.) We list here only the main sources we have used in preparing these lectures. Special mention must be made of Jack Silver's doctoral dissertation [28] which we have drawn on very extensively. It is a very fine paper, both in its results and its exposition.

Sources for some particular parts follow:

The notion of elementary extension is due to Tarski and Vaught [30]. Theorem 1.8 was first proved by Keisler by a different method.

Theorem 2.1 has been stated by various people in various forms. See, for example, [31]. For reference to infinitely long formulas see [15] or [29].

Theorem 3.2 was proved, using ultraproducts, by Chang and Keisler [5]. Theorem 3.3 is due to Vaught [25]. The proof given here as well as Cor. 3.4 and 3.5 are from Keisler [18]. Theorem 3.8 is by Chang [3]. Fuhrken [11], [12] is the source for the relationships between K-like models, Q_K-models, and two cardinal theorems as well as theorems 3.6 and 3.7.

For partition theorems, section 4 and 8, one should see [8], [9] and [28].

The basic source for section 5 is [7] but see also [6] and [23].

Theorem 6.1 is due to Morley [24] and 6.5 to Vaught [32]. Vaught [33] is a good source for both section 3 and section 6.

For end extensions of models of number theory see [22], [13] and [17]. For those of set theory see [20].

Results of section 8 are in [8], [14] and [28].

The exposition of section 9 is patterned on that of [28] though many of the results are originally due to Hanf [15].

Sections 10 and 11 are adapted from the doctoral dissertation of Jack Silver [28].

These notes are based on a series of lectures given at The University, Leeds in the Summer of 1967. I am grateful to my wife, Vivienne, who attended these lectures and wrote this paper.

1. C. C. Chang, On the formula there exist an x such that f(x) for all f∈F, Notices, Amer. Math. Soc. 70 (1964) p. 587.

2. __________, A simple proof of the Rabin-Keisler theorem, Bull. Amer. Math. Soc. 71 (1965) pp. 642-643.

3. __________, A note on the two cardinal theorem, Proc. AMS, 11 (1964) pp. 1148-1155.

4. __________, Ultraproducts and other methods of constructing models to appear in Sets, Models, and Recursion Theory, editor John Crossley, North Holland.

5. __________, and H. J. Keisler, Applications of ultraproducts of pairs of cardinals to the theory of models, Pacific J. Math. 12 (1962) pp. 835-845.

6. A. Ehrenfeucht, On theories categorical in power, Fund. Math. 44 (1957) pp. 241-248.

7. __________, and A. Mostowski, Models of axiomatic theories admitting automorphisms, Fund. Math. 43 (1956) pp. 50-68.

8. P. Erdos, A. Hajnel, and R. Rado, Partition relations for cardinal numbers, Acta Mathematica Academiae Scientiarum Hungaricae, 16 (1965) pp. 93-196.

9. P. Erdos and R. Rado, A partition calculus in set theory, Bull. Amer. Math. Soc. 62 (1956) pp. 195-228.

10. G. Fuhrken, Bemerkung zu einer Arbeit E. Engelers, Z. Math. Logik Grundlagen Math., 8 (1962) pp. 91-93.

11. __________, Languages with added quantifier "there exist at least $_\alpha$", Proc. 1963 Berkeley Symposium on Theory of Models, North Holland Publ. Co., 1965 pp. 121-131.

12. __________, Skolem type normal forms for first order languages with generalized quantifiers, Fund. Math. 64 (1964) pp. 291-302.

13. Haim Gaifman, Uniform extension operators for models and their applications, Technical Report No. 21, U. S. Office of Naval Research, Information Systems Branch.

14. W. Hanf, Incompactness in languages with infinitely long expressions, Fund. Math. 53 (1964) pp. 309-324.

15. Carol Karp, Languages with expressions of infinite length, Amsterdam, 1964, xx + 183 pp.

16. H. J. Keisler, A survey of ultraproducts, Proc. of the 1964 Interm. Congress for Logic, Methodology and the Philosophy of Science, North Holland Publ. Co., 1965, pp. 112-126.

17. __________, Ultraproducts of finite sets, Jour. of Symbolic Logic, 32 (1967) pp. 47-57.

18. __________, Some model theoretic results for ω-logic, Israel Jour. of Math., 4 (1966) pp. 249-261.

19. __________, Models with orderings, to appear in Proc. of the 1967 Intern. Congress for Logic, Methodology and the Philosophy of Science.

20. __________, and M. Morley, Elementary extensions of models of set theory, to appear Israel Jour. of Math.

21. __________, and J. Silver, Well-founded extensions of models of set theory, Notices Amer. Math. Soc., 14 (1967) pp. 256-257 (abstract).

22. R. MacDowell and E. Specker, Modelle der Arithmetik, in Infinitistic Methods, Proc. Symp. on the Foundations of Mathematics, Warsaw, September, 1959, Pergamon Press, 1961, pp. 257-263.

23. M. Morley, Categoricity in power, Trans. Amer. Math. Soc. 114 (1965) pp. 514-538.

24. __________, Omitting classes of elements, in The Theory of Models, Proc. of the 1963 Intern. Symp., North-Holland Publ. Co., 1965, pp. 265-274.

25. __________, and V. Morley, The Hanf number for K-logic, Notices Amer. Math. Soc. 14 (1967) p. 556 (abstract).

26. __________, and R. L. Vaught, Homogenous universal models, Math. Scand. 11 (1962) pp. 37-57.

27. F. Rowbottom, Large cardinals and small constructible sets, Doctoral Dissertation, University of Wisconsin, 1964.

28. Jack Silver, Some applications of model theory in set theory, Doctoral Dissertation, University of California, Berkeley, 1966.

29. Dana Scott, Logic with denumerably long formulas and finite strings of

quantifiers, in The Theory of Models, Proc. of the 1963 Intern. Symp., North Holland Publ. Co., 1965, pp. 329-341.

30. A. Tarski and R. L. Vaught, Arithmetical extensions of relational systems, Comp. Math. 13 (1957) pp. 81-102.

31. R. L. Vaught, Denumerable models of complete theories, Proc. Symp. Foundations of Math., Infinitistic Methods, Pergamon Press, 1961, pp. 303-321.

32. __________, A Löwenheim-Skolem theorem for cardinals far apart, in The Theory of Models, Proc. of the Intern. Symp., North Holland Publ. Co., 1963, pp. 390-401.

33. __________, The Löwenheim-Skolem Theorem, in Proc. of the 1964 Intern. Congress for Logic, Methodology and the Philosophy of Science, North Holland Publ. Co., 1965, pp. 81-89.

KLASSEN REKURSIVER FUNKTIONEN

von

D. Rödding

§ 1: Vergleiche zwischen verschiedenen Definitionsschemata

1.) Einführung

Die im folgenden dargestellten Methoden und Resultate betreffen Klassen rekursiver Funktionen über dem Bereich der natürlichen Zahlen. Zur Charakterisierung dieser Klassen werden im wesentlichen zwei Ansätze betrachtet:

a.) Funktionenklassen werden definiert mit Hilfe gewisser "Anfangsfunktionen" und gewisser Definitionsschemata zur Erzeugung (evtl.) neuer Funktionen aus bereits gegebenen. Im einfachsten Falle stellt sich eine solche Funktionenklasse dar als die kleinste Klasse, welche die Anfangsfunktionen enthält und abgeschlossen ist gegenüber Anwendungen der Definitionsschemata. Die Inklusionsverhältnisse zwischen so erzeugten Funktionenklassen geben (bei gleichen Anfangsfunktionen) Auskunft über die "Reichweite" der verwendeten Definitionsschemata.

b.) Man kann idealisierte Maschinen (z.B. TURING-Maschinen) zur Berechnung rekursiver Funktionen zugrundelegen und jeder Funktion f und jeder Maschine M, die f berechnet, eine "Schrittzahlfunktion" s_f zuordnen: durch $s_{f,M}(x_o,\dots,x_n)$ wird die Anzahl der Rechenschritte beschrieben, die für die Berechnung des Funktionswertes $f(x_o,\dots,x_n)$ aus den Argumenten $x_o,\dots,x_n$ erforderlich sind. Man kann dann in einer Klasse $\mathfrak{K}$ diejenigen f zusammenfassen, für die $s_{f,M}$ (bei geeignetem M) von einer vorgegebenen Wachstumsordnung ist; diese Wachstumsordnung ist ein Mass für die (maximale) Kompliziertheit der Funktionen in $\mathfrak{K}$.

Zwischen den in (a) und (b) angedeuteten Charakterisierungsmöglichkeiten bestehen mannigfaltige Zusammenhänge; sie sind Gegenstand einer Reihe neuerer Arbeiten und sollen hier unter einem einheitlichen Gesichtspunkt dargestellt werden. Auf die im Literaturanhang zusammengestellten Originalarbeiten wird dabei nicht im einzelnen hingewiesen. Vorausgesetzt werden die Grundlagen der Theorie der rekursiven, der primitiv rekursiven und der elementaren Funktionen, wie sie z.B. in den Lehrbüchern des Literaturanhangs entwickelt werden.

2.) Symbolik:

a.) N sei der Bereich der natürlichen Zahlen 0,1,2,...

b.) Kleine lateinische Buchstaben (evtl.indiziert) und Sonderzeichen dienen zur Bezeichnung der Elemente von N und der Funktionen über N.

c.) Kleine deutsche Buchstaben $\mathfrak{x},\mathfrak{y},\ldots$ bezeichnen Tupel $x_o,\ldots,x_n$ (jeweils fester Länge) von natürlichen Zahlen.

d.) Zur Unterscheidung der Funktion f und des Funktionswertes $f(x_o,\ldots,x_n)$ für Argumente $x_o,\ldots,x_n$ findet die λ-Symbolik Verwendung: $\lambda x f(x,y)$ bezeichnet für jedes y diejenige Funktion g, deren Funktionswert g(x) für jedes Argument x gerade f(x,y) ist. $\lambda xy f(x,y)$ bezeichnet die Funktion f.

e.) Grosse deutsche Buchstaben bezeichnen Klassen von Funktionen über N. Insbesondere sei

$\mathfrak{R}$ die Klasse der rekursiven,

$\mathfrak{P}$ " " " primitiv rekursiven

und $\mathfrak{E}$ " " " elementaren

Funktionen über N.

f.) Die Logiksymbole $\neg, \wedge, \vee, \rightarrow, \leftrightarrow, \bigwedge, \bigvee$ (nicht, und, oder, wenn-so, genau dann-wenn, für alle, es gibt) werden zur Mitteilung (also nicht im Rahmen einer formalen Sprache) verwendet.

3.) Definition:

$u_n^i := \lambda x_o \ldots x_n x_i \quad (i \leq n)$

$c_n^i := \lambda x_o \ldots x_n i$

4.) Definition: $\mathfrak{A}$ sei eine beliebige Klasse von Funktionen über N; dann sei $A(\mathfrak{A})$ die kleinste Funktionenklasse, welche

a.) $\mathfrak{A}$ umfasst,

b.) alle Funktionen u_n^i, c_n^i enthält und

c.) mit den Funktionen $f, g_o, \ldots, g_r$ auch stets die Funktion $\lambda \mathfrak{x} f(g_o(\mathfrak{x}), \ldots, g_r(\mathfrak{x}))$ enthält.

5.) Definition: Zu beliebigem $\mathfrak{A}$ sei $A_\Sigma^o(\mathfrak{A}) := A(\mathfrak{A})$ und $A_\Sigma^{n+1}(\mathfrak{A})$ die kleinste Funktionenklasse, welche

a.) $A_\Sigma^n(\mathfrak{A})$ umfasst,

b.) mit jeder Funktion f aus $A_\Sigma^n(\mathfrak{A})$ auch die Funktion $\lambda \mathfrak{x} y \sum_{i \leq y} f(\mathfrak{x}, i)$ enthält und

c.) mit den Funktionen $f, g_o, \ldots, g_r$ auch stets die Funktion

$$\lambda \mathfrak{x} f(g_o(\mathfrak{x}), \ldots, g_r(\mathfrak{x}))$$

enthält.

Ferner sei $A_\Sigma(\mathfrak{A}) := \bigcup_{n \in N} A_\Sigma^n(\mathfrak{A})$

6.) Definitionen:

a.) Die Funktionenklassen $A_\Sigma^n(\mathfrak{A})$, $A_\Sigma(\mathfrak{A})$ sind erklärt unter Bezug auf das Definitionsschema der Summation, durch welches jeder

Funktion f die Funktion $\lambda \mathfrak{x} y \sum_{i \leq y} f(\mathfrak{x},i)$ zugeordnet wird. In entsprechender Weise werden zum Schema der Produktbildung, durch welches jeder Funktion f die Funktion $\lambda \mathfrak{x} y \prod_{i \leq y} f(\mathfrak{x},i)$ zugeordnet ist, die Klassen $A^n_{\Pi}(\mathfrak{A})$, $A_{\Pi}(\mathfrak{A})$ erklärt.

b.) Das Definitionsschema "PR" der primitiven Rekursion, durch welches je zwei Funktionen g,h (geeigneter Stellenzahl) diejenige Funktion f zugeordnet wird, die durch die Gleichungen

$$f(\mathfrak{x},0)=g(\mathfrak{x})$$
$$f(\mathfrak{x},y+1)=h(\mathfrak{x},y,f(\mathfrak{x},y))$$

eindeutig festgelegt ist, gibt in analoger Weise Anlass zur Einführung der Klassen $A^n_{PR}(\mathfrak{A})$, $A_{PR}(\mathfrak{A})$.

c.) Auch das Schema "BR" der beschränkten primitiven Rekursion, durch welches je drei Funktionen f_o,g,h (geeigneter Stellenzahl) die durch die Bedingungen

$$f(\mathfrak{x},0)=g(\mathfrak{x})$$
$$f(\mathfrak{x},y+1)=h(\mathfrak{x},y,f(\mathfrak{x},y),$$
$$f(\mathfrak{x},y)\leq f_o(\mathfrak{x},y)$$

eindeutig definierte Funktion f zugeordnet wird (falls ein solches f existiert), bestimmt in der gleichen Weise wie vorher Klassen $A^n_{BR}(\mathfrak{A})$, $A_{BR}(\mathfrak{A})$.

7.) Im Rahmen der in [6] soeben eingeführten Terminologie stellen sich die Funktionenklassen $\mathfrak{P}$ und $\mathfrak{E}$ (vgl.[2]) wie folgt dar:

$$\mathfrak{P} = A_{PR}(\{\lambda x(x+1)\}),$$
$$\mathfrak{E} = A_{BR}(\{\lambda xy(x+y),\ \lambda x 2^x\}).$$

<u>Anmerkung:</u> Die Klasse $\mathfrak{E}$ ist ursprünglich anders definiert worden,

und zwar mit Hilfe der Definitionsschemata der Summation und Produktbildung, und mit Anfangsfunktionen, die über N die "Elementaroperationen" der Addition, Multiplikation, Subtraktion und Division darstellen. Jedoch führen diese beiden Beschreibungsweisen zu koextensionalen Klassen. Aus der hier gewählten Darstellung ergeben sich unmittelbar einige Abgeschlossenheitseigenschaften von $\mathfrak{E}$, die weiter unten (vgl.[9]) zusammengestellt sind.

8.) <u>Definitionen:</u>

a.) $p_0, p_1, p_2, \ldots$ sei die Folge der Primzahlen 2,3,5,...

b.) Es sei $\langle x_0,\ldots,x_n\rangle := \prod_{i\le n} p_i^{x_i}$, und für $x = \langle x_0,\ldots,x_n\rangle$ und $i\le n$ sei $(x)_i := x_i$.

c.) Im folgenden sollen Prädikate über N mit grossen lateinischen Buchstaben bezeichnet werden. Wie in [2:(d)] soll auch hier die λ-Symbolik verwendet werden. Zu einem Prädikat P sei die "charakteristische Funktion" χ_P wie folgt definiert

$$\chi_P(\mathfrak{x}) := \begin{cases} 1, & \text{falls } P(\mathfrak{x}) \\ 0 & \text{sonst} \end{cases}$$

("$P(\mathfrak{x})$" drückt hier aus, dass P auf das Argumentetupel $\mathfrak{x}$ zutrifft). Die Redeweise "$P\in\mathfrak{A}$" soll bedeuten, dass χ_P in $\mathfrak{A}$ liegt.

d.) Es sei

$$\mu_{i\le y}\, i \; P(\mathfrak{x},i)$$

das kleinste i mit $i\le y$ und $P(\mathfrak{x},i)$, falls ein solches existiert, und 0 sonst.

9.) Im folgenden sollen einige Abgeschlossenheitseigenschaften der Funktionenklasse $\mathcal{E}$ zusammengestellt werden:

a.) $f\in\mathcal{E} \rightarrow \lambda\mathfrak{x}y \sum_{i\leq y} f(\mathfrak{x},i,y),\ \lambda\mathfrak{x}y \prod_{i\leq y} f(\mathfrak{x},i,y) \in\mathcal{E}$

b.) $P\in\mathcal{E} \rightarrow \lambda\mathfrak{x}y\ \mu i_{i\leq y}\ P(\mathfrak{x},i,y)\in\mathcal{E}$

c.) $\lambda x p_x,\ \lambda\mathfrak{x}\langle\mathfrak{x}\rangle,\ \lambda xy(x)_y\ \in\mathcal{E}$

d.) $P\in\mathcal{E} \rightarrow \lambda\mathfrak{x}\neg P(\mathfrak{x})\in\mathcal{E}$

e.) $P\in\mathcal{E}\wedge Q\in\mathcal{E} \rightarrow \lambda\mathfrak{x}(P(\mathfrak{x})\wedge Q(\mathfrak{x}))\in\mathcal{E}$

f.) $P\in\mathcal{E} \rightarrow \lambda\mathfrak{x}y \bigwedge_{i\leq y} P(\mathfrak{x},i,y)\in\mathcal{E}$

g.) $P\in\mathcal{E}\wedge \bigwedge_{\rho\leq r} f_\rho\in\mathcal{E} \rightarrow \lambda\mathfrak{x}P(f_o(\mathfrak{x}),\ldots,f_r(\mathfrak{x}))\in\mathcal{E}$

h.) <u>Vor.</u>:

$$f(\mathfrak{x}) = \begin{cases} f_o(\mathfrak{x}), & \text{falls } P_o(\mathfrak{x}) \\ \ldots\ldots\ldots & \\ f_r(\mathfrak{x}), & \text{falls } P_r(\mathfrak{x}) \end{cases}$$

$f_o,\ldots,f_r\in\mathcal{E};\ P_o,\ldots,P_r\in\mathcal{E}$

<u>Beh.</u>: $f\in\mathcal{E}$

10.) Der folgende Satz vermittelt eine Auskunft über die Leistungsfähigkeit der Definitionsschemata der Summation, Produktbildung und beschränkten primitiven Rekursion, wenn sie jeweils auf Anfangsfunktionen angewendet werden, unter denen mindestens die elementaren Funktionen vertreten sind:

<u>Satz:</u> Für $\mathfrak{A}\supseteq\mathcal{E}$ gilt:

$$A_{BR}(\mathfrak{A}) = A_{\Pi}(\mathfrak{A}) = A_{\Sigma}(\mathfrak{A}) = A^1_{\Pi}(\mathfrak{A}) = A^1_{\Sigma}(\mathfrak{A})$$

<u>**Beweis:**</u> Es genügt zu zeigen:

a.) $A_{BR}(\mathfrak{A}) \subseteq A_{\Pi}(\mathfrak{A})$,

b.) $A_{\Pi}(\mathfrak{A}) \subseteq A^1_{\Pi}(\mathfrak{A})$,

c.) $A_{\Pi}(\mathfrak{A}) \subseteq A^1_{\Sigma}(\mathfrak{A})$,

d.) $A_{\Sigma}(\mathfrak{A}) \subseteq A_{BR}(\mathfrak{A})$.

<u>**Beweis von (a):**</u> Mit Rücksicht auf die Definition der Klassen $A_{BR}(\mathfrak{A})$, $A_{\Pi}(\mathfrak{A})$ genügt es, zu zeigen:

Falls $f(\mathfrak{x},0) = g_1(\mathfrak{x})$,

$f(\mathfrak{x},y+1)=g_2(\mathfrak{x},y,f(\mathfrak{x},y))$,

$f(\mathfrak{x},y)\leq g_3(\mathfrak{x},y)$,

$g_1,g_2,g_3 \in A_{\Pi}(\mathfrak{A})$,

so gilt $f\in A_{\Pi}(\mathfrak{A})$.

Sei $h(\mathfrak{x},y):= \prod_{i\leq y} p_i^{f(\mathfrak{x},i)}$, so gilt:

$h(\mathfrak{x},y)=z$

$$\leftrightarrow z = \prod_{i\leq y} p_i^{(z)_i} \wedge (z)_o=g_1(\mathfrak{x}) \wedge \bigwedge_{i\leq y} (z)_{i+1}=g_2(\mathfrak{x},i,(z)_i).$$

Es folgt

$$\lambda\mathfrak{x}yz(h(\mathfrak{x},y)=z) \in A_{\Pi}(\mathfrak{A}).$$

Sei $q(x):= \mu y_{y\leq x}\ x=p_y$,

$H(\mathfrak{x},y,z):\leftrightarrow h(\mathfrak{x},y)=z$,

$g_4(\mathfrak{x},y):= \prod_{i\leq y} p_i^{g_3(\mathfrak{x},i)}$

dann folgt:

$$f(\mathfrak{x},y)=(q(\prod_{i\leq g_4(\mathfrak{x},y)} p_i^{\chi_H(\mathfrak{x},y,i)}))_y.$$

Daher liegt f in $A_{\Pi}(\mathfrak{A})$.

Damit ist (a) gezeigt.

<u>**Beweis von (b):**</u>

Sei (zu beliebigem f)

$$f^*(x_0,\ldots,x_n) := \prod_{i_0 \leq x_0} \cdots \prod_{i_n \leq x_n} p_{\langle i_0,\ldots,i_n\rangle}^{f(i_0,\ldots,i_n)}$$

Dann genügt es, zu zeigen:

$$(b_1):\quad f \in \mathfrak{A} \rightarrow f^* \in A^1_{\Pi}(\mathfrak{A}),$$

$$(b_2):\quad g^*,\ h^*_0,\ldots,h^*_r \in A^1_{\Pi}(\mathfrak{A}) \rightarrow (\lambda\mathfrak{x}\, g(h_0(\mathfrak{x}),\ldots,h_r(\mathfrak{x})))^* \in A^1_{\Pi}(\mathfrak{A}),$$

$$(b_3):\quad g^* \in A^1_{\Pi}(\mathfrak{A}) \succ (\lambda\mathfrak{x}y \prod_{i \leq y} g(\mathfrak{x},i))^* \in A^1_{\Pi}(\mathfrak{A}),$$

denn aus $(b_1),(b_2),(b_3)$ folgt sofort

$$f \in A_{\Pi}(\mathfrak{A}) \rightarrow f^* \in A^1_{\Pi}(\mathfrak{A})$$

und daraus wegen

$$f(\mathfrak{x}) = (f^*(\mathfrak{x}))_{\langle\mathfrak{x}\rangle}$$

die Behauptung (b).

Ad (b_1):

Sei (für $\mathfrak{x} = (x_0,\ldots,x_n)$)

$$q_n(\mathfrak{x},i) := \begin{cases} p_i, & \text{falls } i = \langle (i)_0,\ldots,(i)_n\rangle \wedge (i)_0 \leq x_0 \wedge \ldots \wedge (i)_n \leq x_n \\ 1 & \text{sonst} \end{cases}$$

Dann gilt:

$$f^*(\mathfrak{x}) = \prod_{i \leq \langle\mathfrak{x}\rangle} q_n(\mathfrak{x},i)^{f((i)_0,\ldots,(i)_n)},$$

also (da $q_n \in \mathfrak{E} \subseteq \mathfrak{A}$): $f^* \in A^1_{\Pi}(\mathfrak{A})$, q.e.d.

Ad (b_2):

Sei (für $\mathfrak{x} = (x_0,\ldots,x_n)$, $\mathfrak{z} = (z_0,\ldots,z_r)$)

$$Sb_{n,r}(\mathfrak{x},y,\mathfrak{z}) := \prod_{i_0 \leq x_0} \cdots \prod_{i_n \leq x_n} p_{\langle i_0,\ldots,i_n\rangle}^{(y)_{\langle (z_0)_{\langle i_0,\ldots,i_n\rangle},\ldots,(z_r)_{\langle i_0,\ldots,i_n\rangle}\rangle}}$$

und

$$f(\mathfrak{x}) = g(h_o(\mathfrak{x}),\ldots,h_r(\mathfrak{x}))$$

Dann gilt:

$$f^*(\mathfrak{x}) = Sb_{n,r}(\mathfrak{x},g^*(h_o^*(\mathfrak{x}),\ldots,h_r^*(\mathfrak{x})),\ h_o^*(\mathfrak{x}),\ldots,h_r^*(\mathfrak{x})),$$

also (da $Sb_{n,r}\in\mathfrak{E}\subseteq\mathfrak{A}$): $f^*\in A^1_{\Pi}(\mathfrak{A})$, q.e.d.

Ad (b_3):

Sei (für $\mathfrak{x}=(x_o,\ldots,x_n)$)

$$Pr_n(\mathfrak{x},y,z) :=\prod_{i_o\leq x_o}\ldots\prod_{i_n\leq x_n}\ \prod_{i_{n+1}\leq y} p_{\langle i_o,\ldots,i_{n+1}\rangle}^{\prod_{j\leq i_{n+1}}(z)_{\langle i_o,\ldots,i_n,j\rangle}};$$

und

$$f(\mathfrak{x},y) = \prod_{i\leq y} g(\mathfrak{x},i)$$

dann gilt:

$$f^*(\mathfrak{x},y) = Pr_n(\mathfrak{x},y,g^*(\mathfrak{x},y))$$

also (da $Pr_n\in\mathfrak{E}\subseteq\mathfrak{A}$): $f^*\in A^1_{\Pi}(\mathfrak{A})$, q.e.d.

<u>Beweis von (c)</u>:

Zu beliebigem f sei

$$f^+(x_o,\ldots,x_n) := \sum_{i_o\leq x_o}\ldots\sum_{i_n\leq x_n} 2^{\langle i_o,\ldots,i_n\ ,f(i_o,\ldots,i_n)\rangle}$$

Dann existiert zu beliebigem n eine elementare Funktion r_n, welche für jede (n+1)-stellige Funktion f die folgende Umformung leistet (*): $r_n(\mathfrak{x},f^+(\mathfrak{x})) = f^*(\mathfrak{x})$.

Sei

$$s_n(\mathfrak{x},i) := \begin{cases} 1, \text{ falls } i=\langle(i)_o,\ldots,(i)_n\rangle\wedge(i)_o\leq x_o\wedge\ldots\wedge(i)_n\leq x_n \\ 0 \text{ sonst}\end{cases}$$

Dann gilt stets:

$$f^{+}(\mathfrak{x}) = \sum_{i \leq \langle \mathfrak{x} \rangle} 2^{\langle (i)_0, \ldots, (i)_n, f((i)_0, \ldots, (i)_n) \rangle} \cdot s_n(\mathfrak{x}, i) ,$$

und wegen $s_n \in \mathfrak{E} \subseteq \mathfrak{A}$ liegt für jede Funktion f aus $\mathfrak{A}$ die Funktion f^{+} in $A^1_{\Sigma}(\mathfrak{A})$. Wegen (*) (und $r_n \in \mathfrak{E}$) folgt dann auch $f^{*} \in A^1_{\Sigma}(\mathfrak{A})$.

Nun lässt sich der Beweis von (b_2), (b_3) übertragen, und es ergibt sich (unter Verwendung der dort eingeführten elementaren Funktionen $Sb_{n,r}$ und Pr_n):

$$f^{*}, g_0^{*}, \ldots, g_r^{*} \in A^1_{\Sigma}(\mathfrak{A}) \rightarrow (\lambda \mathfrak{x} f(g_0(\mathfrak{x}), \ldots, g_r(\mathfrak{x})))^{*} \in A^1_{\Sigma}(\mathfrak{A}),$$

$$f^{*} \in A^1_{\Sigma}(\mathfrak{A}) \rightarrow (\lambda \mathfrak{x} y \prod_{i \leq y} f(\mathfrak{x}, i))^{*} \in A^1_{\Sigma}(\mathfrak{A}),$$

also

$$f \in A_{\Pi}(\mathfrak{A}) \rightarrow f^{*} \in A^1_{\Sigma}(\mathfrak{A})$$

und daraus (wie vorher, weil

$f(\mathfrak{x}) = (f^{*}(\mathfrak{x}))_{\langle \mathfrak{x} \rangle}$) die behauptete Inklusion (c).

<u>Beweis von (d)</u>:

Sei $g(\mathfrak{x}, y) = \sum_{i \leq y} (1 \dot{-} f(\mathfrak{x}, i))$.

Dann lässt sich g aus f mit einer beschränkten primitiven Rekursion definieren:

$$g(\mathfrak{x}, 0) = 1 \dot{-} f(\mathfrak{x}, 0),$$

$$g(\mathfrak{x}, y+1) = g(\mathfrak{x}, y) + (1 \dot{-} f(\mathfrak{x}, y+1)),$$

$$g(\mathfrak{x}, y) \leq y+1 .$$

Daher liegt in $A_{BR}(\mathfrak{A})$ mit f auch stets die Funktion $\lambda \mathfrak{x} y \sum_{i \leq y} (1 \dot{-} f(\mathfrak{x}, i))$.

Ferner gilt:

$$\operatorname*{Max}_{i \leq y} f(\mathfrak{x}, i) = f(\mathfrak{x}, \sum_{x \leq y} (1 \dot{-} \sum_{i \leq x} (1 \dot{-} \sum_{j \leq y} (1 \dot{-} (1 \dot{-} (f(\mathfrak{x}, j) \dot{-} f(\mathfrak{x}, i))))))).$$

Daher liegt in $A_{BR}(\mathfrak{A})$ mit f auch stets die Funktion $\lambda \mathfrak{x} y \operatorname*{Max}_{i \leq y} f(\mathfrak{x}, i)$.

Sei schliesslich $g(\mathfrak{x},y) = \sum_{i\leq y} f(\mathfrak{x},i)$; dann gilt

$$g(\mathfrak{x},0) = f(\mathfrak{x},0),$$

$$g(\mathfrak{x},y+1) = g(\mathfrak{x},y) + f(\mathfrak{x},y+1),$$

$$g(\mathfrak{x},y) \leq (y+1)\cdot \operatorname*{Max}_{i\leq y} f(\mathfrak{x},i),$$

so dass mit f auch stets g in $A_{BR}(\mathfrak{A})$ liegt. Damit ist (d) bewiesen.

11.) Der soeben bewiesene Satz zeigt, dass sich (in Anwesenheit der elementaren Anfangsfunktionen) das Schema der beschränkten primitiven Rekursion fast vollständig eliminieren lässt. Für eine Hierarchie von Funktionenklassen, welche ganz $\mathfrak{P}$ ausschöpft und mit $\mathfrak{E}$ beginnt, soll ein ähnlicher Sachverhalt durch den folgenden Satz nachgewiesen werden; zugleich wird eine weitere Charakterisierung der Klassen dieser Hierarchie (mit Hilfe der Anzahl "hintereinanderliegender" primitiver Rekursionen, die bei der Definition ihrer Funktionen benötigt werden) angegeben:

<u>Definition:</u>

$H_o(x,y) := 2^{x+y+1}$

$H_{n+1}(0,y) := H_n(y,0)$

$H_{n+1}(x+1,y) := H_{n+1}(x,H_{n+1}(x,y))$

<u>Satz:</u> Für $n\geq 3$ gilt:

$$A_{PR}^{n+2}(\{\lambda x(x+1)\})$$

$$= A_{BR}(\{\lambda x(x+1),\ H_n\})$$

$$= A_S(\mathfrak{E}\cup\{H_n\})$$

Der Beweis reduziert sich auf die folgenden Behauptungen:

a.) $A_{PR}^{n+2}(\{\lambda x(x+1)\}) \subseteq A_{BR}(\{\lambda x(x+1), H_n\})$

b.) $A_{BR}(\{\lambda x(x+1), H_n\}) \subseteq A_S(\mathfrak{E} \cup \{H_n\})$

c.) $\mathfrak{E} \subseteq A_{PR}^5(\{\lambda x(x+1)\})$

d.) Für $n \geq 3$ gilt:

$A_S(\mathfrak{E} \cup \{H_n\}) \subseteq A_{PR}^{n+2}(\{\lambda x(x+1)\})$

12.) Beweis von [11a.)]:

Es genügt, zu zeigen:

$$(a)^* \quad \bigwedge_{f \in A_{PR}^{n+1}(\{\lambda x(x+1)\})} \bigvee_K \bigwedge_{\mathfrak{x}} f(\mathfrak{x}) \leq H_n(K, \Sigma\mathfrak{x})$$

Hierbei ist $\Sigma\mathfrak{x} := \sum_{\rho \leq r} x_\rho$ für $\mathfrak{x} = (x_0, \ldots, x_r))$

Zunächst werden zwei Hilfssätze vorausgesetzt:

Hilfssatz 1: Vor.: $f(\mathfrak{x}) = g(h_0(\mathfrak{x}), \ldots, h_r(\mathfrak{x}))$,

$\bigwedge_{\mathfrak{y}} g(\mathfrak{y}) \leq H_n(K, \Sigma\mathfrak{y})$,

$\bigwedge_{\mathfrak{x}} h_\rho(\mathfrak{x}) \leq H_n(K_\rho, \Sigma\mathfrak{x})$ für $\rho \leq r$,

Beh.: $\bigvee_{K^*} \bigwedge_{\mathfrak{x}} f(\mathfrak{x}) \leq H_n(K^*, \Sigma\mathfrak{x})$

Hilfssatz 2: Vor.: $f(\mathfrak{x}, 0) = g_1(\mathfrak{x})$,

$f(\mathfrak{x}, y+1) = g_2(\mathfrak{x}, y, f(\mathfrak{x}, y))$,

$\bigwedge_{\mathfrak{x}} g_1(\mathfrak{x}) \leq H_n(K_1, \Sigma\mathfrak{x})$,

$\bigwedge_{\mathfrak{x}} \bigwedge_y \bigwedge_z g_2(\mathfrak{x}, y, z) \leq H_n(K_2, \Sigma\mathfrak{x} + y + z)$.

Beh.:

$\bigvee_{K^*} \bigwedge_{\mathfrak{x}} \bigwedge_y f(\mathfrak{x}, y) \leq H_{n+1}(K^*, \Sigma\mathfrak{x} + y)$.

Aus diesen Hilfssaätzen ergibt sich leicht $(a)^*$ durch eine Induktion nach n. Der Induktionsbeginn ist trivial, da alle in $A^1_{PR}(\{\lambda x(x+1)\})$ liegenden Funktionen höchstens linear wachsen. Im Induktionsschritt wird verwendet, dass jede Funktion aus $A_{PR}^{n+2}(\{\lambda x(x+1)\})$ durch endlich viele Einsetzungen aus Funktionen hervorgeht, die unmittelbar durch eine primitive Rekursion aus Funktionen aus $A_{PR}^{n+1}(\{\lambda x(x+1)\})$ entstanden sind. Man wendet daher zunächst auf diese Funktionen die Induktionsvoraussetzung und Hilfssatz 2 und dann noch endlich oft Hilfssatz 1 an. Damit ist $(a)^*$ auf die Hilfssätze 1 und 2 zurückgeführt; zu ihrem Beweis werden einige Eigenschaften der Funktionen H_n benötigt:

(a_1): $H_n(x,y) > x+y$

(a_2): $H_n(x+1,y) > H_n(x,y)$

(a_3): $H_n(x,y+1) > H_n(x,y)$

(a_4): $H_n(x+1,y) \geq H_n(x,y+1)$

(a_5): $H_{n+1}(x,y) \geq H_n(x,y)$

(a_6): $2\,H_n(x,y) \leq H_n(x+1,y)$

(a_7): $\bigwedge\limits_{K_0} \dots \bigwedge\limits_{K_r} \bigvee\limits_{K} \bigwedge\limits_{y} \sum\limits_{\rho\leq r} H_n(K_\rho,y) \leq H_n(K,y)$

<u>Ad</u>(a_1): Der Beweis wird durch Induktion über n geführt. Der Induktionsbeginn ist trivial. Im Induktionsschritt wird die Behauptung $\bigwedge\limits_{xy} H_{n+1}(x,y) > x+y$ ihrerseits durch eine Induktion nach x gezeigt: der Beginn dieser Induktion ist ebenfalls trivial, da nach Voraussetzung der Induktion über n bereits $H_n(y,0) > y$ gilt. Im Induktionsschritt ergibt sich nun die folgende Abschätzung

$$H_{n+1}(x+1,y) = H_{n+1}(x,H_{n+1}(x,y)) > x+H_{n+1}(x,y) > 2x+y,$$

wobei die beiden letzten Ungleichungen aus der Voraussetzung der Induktion über x resultieren. Es folgt

$$H_{n+1}(x+1,y) > x+1+y,$$

und damit ist die Induktion über x (also auch die Induktion über n) beendet, und (a_1) ist nachgewiesen.

Ad(a_2): Beweis durch Induktion über n, unter Verwendung von (a).

Ad(a_3): Beweis durch Induktion über n, im Induktionsschritt durch eine Induktion über x, unter Verwendung von (a_2) für den Beginn dieser Induktion.

Ad(a_4): Beweis durch Induktion über n, unter Verwendung von (a_3) im Induktionsschritt.

Ad(a_5): Beweis durch Induktion über x: Aus (a_4) ergibt sich zunächst $H_n(y,0)\geq H_n(0,y)$ und damit der Induktionsbeginn. Im Induktionsschritt erhält man unter Verwendung von (a_3):

$H_{n+1}(x+1,y)=H_{n+1}(x,H_{n+1}(x,y))\geq H_n(x,H_n(x,y))=H_n(x+1,y)$, q.e.d.

Ad(a_6): $2\,H_n(x,y)\leq H_n(0,H_n(x,y))\leq H_n(x+1,y)$.

Ad(a_7): trivial aus (a_6).

Beweis von Hilfssatz 1:

Für passendes K^* gilt:

$$
\begin{aligned}
& g(h_o(\mathfrak{x}),\ldots,h_r(\mathfrak{x})) \\
\underset{\text{n.Vor.}}{\leq}\; & H_n(K,\ \sum_{\rho\leq r} H_n(K_\rho,\Sigma\mathfrak{x})) \\
\underset{\text{vgl.}(a_7)}{\leq}\; & H_n(K^*,\ H_n(K^*,\Sigma\mathfrak{x})) \\
=\; & H_n(K^*+1,\Sigma\mathfrak{x}),\quad \text{q e.d.}
\end{aligned}
$$

Beweis von Hilfssatz 2:

Gezeigt wird zunächst

$$f(\mathfrak{x},y)+\Sigma\mathfrak{x}+y\leq H_n(K_1+K_2+y+1,\ \Sigma\mathfrak{x}+y)$$

durch Induktion über y:

Der Induktionsbeginn ist trivial.

Im Induktionsschritt ergibt sich die folgende Abschätzung:

$$f(\mathfrak{g},y+1)+\Sigma\mathfrak{g}+y+1$$

$$\leq \quad H_n(K_2,f(\mathfrak{g},y)+\Sigma\mathfrak{g}+y)+\Sigma\mathfrak{g}+y+1$$

n.Vor.über g_2

$$\leq \quad H_n(K_2,H_n(K_1+K_2+y+1,\Sigma\mathfrak{g}+y))+\Sigma\mathfrak{g}+y+1$$

n.Ind.-Vor.

$$\leq \quad H_n(K_2+1,\ H_n(K_1+K_2+y+1,\Sigma\mathfrak{g}+y))$$

vgl.(a_6)

$$\leq \quad H_n(K_1+K_2+y+2,\Sigma\mathfrak{g}+y+1), \qquad \text{q.e.d.}$$

Daraus folgt mit $K_3 := K_1+K_2+1$

$$\begin{aligned} f(\mathfrak{g},y) &\leq H_n(K_1+K_2+y+1,\Sigma\mathfrak{g}+y) \\ &\leq H_n(2\circ(\Sigma\mathfrak{g}+y+K_3),0) \\ &\leq H_{n+1}(0,H_{n+1}(0,\Sigma\mathfrak{g}+y+K_3)) \\ &\leq H_{n+1}(K_3+1,\Sigma\mathfrak{g}+y), \qquad \text{q.e.d.} \end{aligned}$$

13.) Beweis von [11 b.)]:

Zu zeigen ist

(a) $A_{BR}(\{\lambda x(x+1),\ H_n\}) \subseteq A_S(\mathfrak{E}\cup\{H_n\})$

Offenbar genügt es, zu zeigen

(b) $\bigwedge_{f\in A_{BR}(\{\lambda x(x+1),H_n\})} \bigvee_K \bigvee_{\dot f\in\mathfrak{E}} \bigwedge_{\mathfrak{g}} \bigwedge_y (y\geq H_{n+1}(K,\Sigma\mathfrak{g})\rightarrow f(\mathfrak{g})=\dot f(\mathfrak{g},y))$,

denn mit (b) ergibt sich (a) wie folgt: Zu beliebigem f aus $A_{BR}(\{\lambda x(x+1),\ H_n\})$ sei eine Konstante K und eine elementare Funktion $\dot f$ gemäss (b) gegeben, dann gilt insbesondere

$$f(\mathfrak{g}) = \dot f(\mathfrak{g},H_{n+1}(K,\Sigma\mathfrak{g}))$$

und

$$\lambda\mathfrak{x}H_{n+1}(K,\Sigma\mathfrak{x}) \in A_S(\mathcal{E}\cup\{H_n\}),$$

also

$$f \in A_S(\mathcal{E}\cup\{H_n\}), \quad \text{q.e.d.}$$

Sei $\mathfrak{A} := \{ f : \bigvee_K \bigvee_{\dot{f}\in\mathcal{E}} \bigwedge_{\mathfrak{x}} \bigwedge_y (y \geq H_{n+1}(K,\Sigma\mathfrak{x}) \rightarrow f(\mathfrak{x}) = \dot{f}(\mathfrak{x},y))\}$

Es genügt, zu zeigen:

(b_1): $H_n \in \mathfrak{A}$

(b_2): Mit $f, g_o, \ldots, g_r$ liegt auch $\lambda\mathfrak{x}f(g_o(\mathfrak{x}),\ldots,g_r(\mathfrak{x}))$ in $\mathfrak{A}$

(b_3): Falls f durch eine beschränkte Rekursion aus Funktionen g_1, g_2, g_3 aus $\mathfrak{A}$ definiert ist, so liegt auch f in $\mathfrak{A}$.

<u>Ad(b_1):</u>

Sei $F_n(0,y) := y$,

$F_n(x+1,y) := H_n(F_n(x,y),0)$.

Durch Induktion nach x zeigt man leicht:

$(*)$: $F_n(x,F_n(y,z)) = F_n(x+y,z)$,

und daraus ergibt sich (ebenfalls durch eine Induktion nach x):

$(**)$ $H_{n+1}(x,y) = F_n(2^x,y)$.

Es soll nun gezeigt werden, dass für beliebiges n die Funktionen $\lambda xyz(z\dot{-}H_n(x,y))$ und $\lambda xyz(z\dot{-}F_n(x,y))$ in $\mathcal{E}$ liegen;

dazu genügt es, zu beweisen:

(b_{11}): $\lambda xyz(z\dot{-}H_o(x,y)) \in \mathcal{E}$.

(b_{12}): Mit $\lambda xyz(z\dot{-}H_n(x,y))$ liegt auch $\lambda xyz(z\dot{-}F_n(x,y))$ in $\mathcal{E}$.

(b_{13}): Mit $\lambda xyz(z\dot{-}F_n(x,y))$ liegt auch $\lambda xyz(z\dot{-}H_{n+1}(x,y))$ in $\mathcal{E}$.

(b_{11}) und (b_{13}) ergeben sich unmittelbar aus der Definition von H_o bzw. $(**)$.

Ad(b_{12}):

Nach (a_1) aus [12] gilt

$$H_n(x,0) > x,$$

also $F_n(x+1,y) > F_n(x,y)$. Daraus folgt:

$$z \dot{-} F_n(0,y) = z \dot{-} y,$$

$$z \dot{-} F_n(x+1,y) = z \dot{-} H_n(z \dot{-} (z \dot{-} F_n(x,y)),0),$$

$$z \dot{-} F_n(x,y) \leq z.$$

Also ist $\lambda xyz(z \dot{-} F_n(x,y))$ durch eine beschränkte primitive Rekursion aus $\lambda xyz(z \dot{-} H_n(x,y))$ definierbar, q.e.d.

Sei $P_n(x,y,z) \leftrightarrow H_n(x,y)=z$. Wegen

$$P_n(x,y,z) \leftrightarrow z \dot{-} H_n(x,y)=0 \wedge (z+1) \dot{-} H_n(x,y) \neq 0$$

liegt P_n in $\mathfrak{E}$, also auch die Funktion $\dot{H}_n$ mit

$$\dot{H}_n(x,y,z) = \mu_{i \leq z} i \; P_n(x,y,i).$$

Für $z \geq H_{n+1}(0,x+y)$ gilt

$$z \geq H_{n+1}(0,x+y) = H_n(x+y,0) \geq H_n(x,y)$$

und daher

$$H_n(x,y) = \dot{H}_n(x,y,z), \quad \text{q.e.d.}$$

Ad(b_2):

Sei $f(\mathfrak{y}) = \dot{f}(\mathfrak{y},z)$ für $z \geq H_n(K,\Sigma\mathfrak{y})$,

$g_o(\mathfrak{x}) = \dot{g}_o(\mathfrak{x},y)$ für $y \geq H_n(K_o,\Sigma\mathfrak{x})$,

...

$g_r(\mathfrak{x}) = \dot{g}_r(\mathfrak{x},y)$ für $y \geq H_n(K_r,\Sigma\mathfrak{x})$,

$\dot{f},\dot{g}_o,\ldots,\dot{g}_r \in \mathfrak{E}$.

Nach (a_7) in [12] gibt es ein K^* mit

$$H_n(K, \sum_{\rho \leq r} H_n(K_\rho,\Sigma\mathfrak{x})) \leq H_n(K^*,\Sigma\mathfrak{x}).$$

Für $y \geq H_n(K^*, \Sigma\mathfrak{g})$ gilt daher:

$$f(g_o(\mathfrak{g}),\ldots,g_r(\mathfrak{g}))$$
$$= \dot{f}(\dot{g}_o(\mathfrak{g},y),\ldots,\dot{g}_r(\mathfrak{g},y),y), \quad \text{q.e.d.}$$

Ad(b_3):

Sei
$$f(\mathfrak{g})=g_1(\mathfrak{g}),$$
$$f(\mathfrak{g},y+1)=g_2(\mathfrak{g},y,f(\mathfrak{g},y)),$$
$$f(\mathfrak{g},y)\leq g_3(\mathfrak{g},y),$$
$$g_1(\mathfrak{g})=\dot{g}_1(\mathfrak{g},y) \quad \text{für } y\geq H_n(K_1,\Sigma\mathfrak{g}),$$
$$g_2(\mathfrak{g},y,z)=\dot{g}_2(\mathfrak{g},y,z,u) \quad \text{für } u\geq H_n(K_2,\Sigma\mathfrak{g}+y+z),$$
$$g_3(\mathfrak{g},y)=\dot{g}_3(\mathfrak{g},y,z) \quad \text{für } z\geq H_n(K_3,\Sigma\mathfrak{g}+y),$$
$$\dot{g}_1,\dot{g}_2,\dot{g}_3 \in \mathfrak{E},$$

dann ist zu zeigen, dass für eine geeignete Konstante K und eine geeignete Funktion $\dot{f}$ aus $\mathfrak{E}$ gilt:

$$f(\mathfrak{g},y)=\dot{f}(\mathfrak{g},y,u) \quad \text{für } u\geq H_n(K,\Sigma\mathfrak{g}+y).$$

Zum Beweis sei zunächst

$$F(\mathfrak{g},y,z,u)$$
$$\leftrightarrow z = \prod_{i\leq y} p_i^{(z)_i} \wedge (z)_o=\dot{g}_1(\mathfrak{g},u) \wedge \bigwedge_{i<y} (z)_{i+1}=\dot{g}_2(\mathfrak{g},i,(z)_i,u)$$

Dann gilt $F \in \mathfrak{E}$.

Ferner existiert (wegen $n>0$ und $\dot{g}_3 \in \mathfrak{E}$) eine Konstante K_4 mit

$$\dot{g}_3(\mathfrak{g},y,H_n(K_3,\Sigma\mathfrak{g}+y))\leq H_n(K_4,\Sigma\mathfrak{g}+y).$$

Zu K_1, K_2 und K_4 existiert ein K mit:

$$K_1 \leq K$$
$$H_n(K_2,\Sigma\mathfrak{g}+y+H_n(K_4,\Sigma\mathfrak{g}+y))\leq H_n(K,\Sigma\mathfrak{g}+y)$$
$$\prod_{i\leq y} p_i^{H_n(K_4,\Sigma\mathfrak{g}+i)} \leq H_n(K,\Sigma\mathfrak{g})$$

Für beliebiges $u \geq H_n(K,\Sigma\mathfrak{x}+y)$ ergeben sich nun die beiden folgenden Beziehungen:

$$(b_1): \quad F(\mathfrak{x},y,z,u) \leftrightarrow z = \prod_{i \leq y} p_i^{f(\mathfrak{x},i)}$$

$$(b_2): \quad F(\mathfrak{x},y,z,u) \rightarrow z \leq H_n(K,\Sigma\mathfrak{x}+y)$$

Mit

$$\dot{f}(\mathfrak{x},y,u) := (\mu_{z \leq u} z \; F(\mathfrak{x},y,z,u))_y$$

folgt dann $\dot{f} \in \mathfrak{E}$ und

$$f(\mathfrak{x},y) = \dot{f}(\mathfrak{x},y,u) \text{ für } u \geq H_n(K,\Sigma\mathfrak{x}+y), \qquad \text{q.e.d.}$$

14.) <u>Beweis von [11c.)]</u>:

a.) Sei $\vartheta_o(x) := x$

$$\vartheta_{n+1}(x) := 2^{\vartheta_n(x)}$$

$$\vartheta_n(\mathfrak{x}) := \vartheta_n(\Sigma\mathfrak{x}) \qquad \text{(vgl. 12 zu } \Sigma\mathfrak{x})$$

Dann gilt der folgende

<u>Hilfssatz 1</u>:

$$\bigwedge_{f \in \mathfrak{E}} \bigvee_{K} \bigwedge_{\mathfrak{x}} f(\mathfrak{x}) \leq \vartheta_K(\mathfrak{x}).$$

(Diese Behauptung ergibt sich leicht durch eine Induktion über den Aufbau der Funktionenklasse $\mathfrak{E}$ gemäss [7].)

b.) Es sei

$$[x:y] := \mu_{z \leq x} z \; y \cdot (z+1) > x,$$

$$[\sqrt{x}] := \mu_{z \leq x} z \; (z+1)^2 > x,$$

$$\mathfrak{A}_1 := A(\{\lambda xy(x+y), \lambda xy(x \dot{-} y), \lambda xy(x \cdot y), \lambda x 2^x, \lambda x[\sqrt{x}], \lambda xy[x:y]\}),$$

$$r(x,y) := x \dot{-} [x:y]\cdot y,$$

$$\beta(x,y,z) := r(x,1+(z+1)\cdot y),$$

$$\sigma_2(x,y) := (x+y)^2+x,$$

$$\sigma_{21}(x) := x \dot{-} ([\sqrt{x}])^2,$$

$$\sigma_{22}(x) := [\sqrt{x}] \dot{-} \sigma_{21}(x),$$

$$\gamma(x,y) := \beta(\sigma_{21}(x), \sigma_{22}(x),y).$$

Dann gilt:

$$\vartheta_n, r,\beta, \sigma_2, \sigma_{21}, \sigma_{22},\gamma \in \mathfrak{A}_1 \subseteq \mathfrak{E},$$

$$[\sqrt{\sigma_2(x,y)}] = x+y$$

und daher

$$\sigma_{21}(\sigma_2(x,y))=x, \ \sigma_{22}(\sigma_2(x,y))=y.$$

c.) Sei ferner $a_o,\ldots,a_n$ eine beliebige Folge natürlicher Zahlen. Sei S das Maximum der Zahlen $n,a_o,\ldots,a_n$. Dann lässt sich das System

$$r(c,d_o) = a_o$$

$$\vdots$$

$$r(c,d_n) = a_n$$

simultaner Kongruenzen lösen durch Zahlen c, $d_i=1+(i+1)d$, derart dass c,d den Abschätzungen

$$d \leq S!,$$

$$c \leq \prod_{i\leq n} (1+(i+1)d)$$

genügen. Für solche c,d gilt also

$$\bigwedge_{i\leq n} \beta(c,d,i) = a_i \ .$$

d.) $\mathfrak{A}_2$ sei die kleinste Prädikatenklasse, welche

d_1.) zu jeder Funktion f aus $\mathfrak{A}_1$ das Prädikat $\lambda\mathfrak{x}(f(\mathfrak{x})=0)$ enthält,

d_2.) mit dem Prädikat P und zu jeder beliebigen Funktion f aus $\mathfrak{A}_1$ auch stets die Prädikate

$$\lambda\mathfrak{x} \bigvee_{y\leq f(\mathfrak{x})} P(\mathfrak{x},y),\ \lambda\mathfrak{x} \bigwedge_{y\leq f(\mathfrak{x})} P(\mathfrak{x},y)$$

enthält.

Dann gilt:

d_3.) $\mathfrak{A}_2$ enthält mit P und Q auch die Prädikate $\lambda\mathfrak{x}(P(\mathfrak{x})\wedge Q(\mathfrak{x}))$ und $\lambda\mathfrak{x}(P(\mathfrak{x})\vee Q(\mathfrak{x}))$.

e.) Sei $\mathfrak{A}_3 := \{f : \lambda\mathfrak{x}y(f(\mathfrak{x})=y)\in \mathfrak{A}_2\}$.

<u>Hilfssatz 2:</u> $\mathfrak{E} \subseteq \mathfrak{A}_3$

<u>Beweis:</u> Es genügt, zu zeigen:

(e_1): Mit $f,g_0,\ldots,g_r$ liegt auch $\lambda\mathfrak{x}f(g_0(\mathfrak{x}),\ldots,g_r(\mathfrak{x}))$ in $\mathfrak{A}_3$, falls die Funktionen $g_0,\ldots,g_r$ elementar sind.

(e_2): Falls f durch beschränkte primitive Rekursion aus Funktionen g_1,g_2,g_3 aus $\mathfrak{A}_3$ definiert ist und g_3 elementar ist, so liegt auch f in $\mathfrak{A}_3$.

<u>Ad(e_1):</u> Nach (b) gibt es für die elementaren Funktionen $g_0,\ldots,g_r$ Konstanten $K_0,\ldots,K_r$, so dass gilt:

$$g_0(\mathfrak{x})\leq \vartheta_{K_0}(\mathfrak{x}),\ldots,g_r(\mathfrak{x})\leq\vartheta_{K_r}(\mathfrak{x}).$$

Daraus ergibt sich die folgende Äquivalenz

$$f(g_0(\mathfrak{x}),\ldots,g_r(\mathfrak{x}))=y$$

$$\leftrightarrow \bigvee_{y_0\leq\vartheta_{K_0}(\mathfrak{x})} \cdots \bigvee_{y_r\leq\vartheta_{K_r}(\mathfrak{x})} (g_0(\mathfrak{x})=y_0 \wedge \ldots\wedge g_r(\mathfrak{x})=y_r\wedge f(y_0,\ldots,y_r)=y\ ),$$

aus der (mit (d_3)) sofort (e_2) folgt.

<u>Ad(e_2):</u>

Es sei
$$f(\mathfrak{x},0) = g_1(\mathfrak{x})$$
$$f(\mathfrak{x},y+1) = g_2(\mathfrak{x},y,f(\mathfrak{x},y)),$$
$$f(\mathfrak{x},y) \leq g_3(\mathfrak{x},y),$$
$$g_3 \in \mathcal{E}.$$

Nach (b) existiert ein K mit $g_3(\mathfrak{x},y) \leq \vartheta_K(\mathfrak{x},y)$. Die Folge $a_0,\ldots,a_y$ sei nun durch die Gleichungen
$$a_i := f(\mathfrak{x},i) \quad \text{(für } i \leq y\text{)}$$
bestimmt. Nach (c) gibt es Konstanten s,c,d,K_0 mit
$$s = \mathrm{Max}(y,a_0,\ldots,a_y) \leq \vartheta_K(\mathfrak{x},y),$$
$$d \leq s! \leq 2^{2^s},$$
$$c \leq \prod_{i \leq y} (1+(i+1)d) \leq \vartheta_{K_0}(y,d),$$
$$\bigwedge_{i \leq y} \beta(c,d,i) = a_i,$$
d.h. es gibt Konstanten K_1, K_2 mit
$$c \leq \vartheta_{K_1}(\mathfrak{x},y),$$
$$d \leq \vartheta_{K_2}(\mathfrak{x},y),$$
$$\bigwedge_{i \leq y} \beta(c,d,i) = a_i \, .$$

Aus dieser Überlegung ergibt sich die Gültigkeit der folgenden Äquivalenz
$$f(\mathfrak{x},y) = z$$
$$\leftrightarrow \bigvee_{c \leq \vartheta_{K_2}(\mathfrak{x},y)} \; \bigvee_{d \leq \vartheta_{K_2}(\mathfrak{x},y)} (\beta(c,d,0) = g_1(\mathfrak{x})$$

$$\wedge \bigwedge_{i \leq y} (i=y \vee \beta(c,d,i+1) = g_2(\mathfrak{x},i,\beta(c,d,i))) \wedge \beta(c,d,y)=z),$$

aus welcher (mit (d_3)) sofort $f \in \mathcal{A}_3$ folgt.

f.) Hilfssatz 3:

Zu jedem Prädikat P aus $\mathcal{A}_2$ gibt es Funktionen f_1, f_2, f_3 aus $\mathcal{A}_1$, so dass gilt:

$$P(\mathfrak{x}) \leftrightarrow \bigvee_{a \leq f_1(\mathfrak{x})} \bigwedge_{b \leq f_2(\mathfrak{x},a)} f_3(\mathfrak{x},a,b)=0$$

Beweis: Die Behauptung ergibt sich sofort aus der Definition der Klasse $\mathcal{A}$ durch endlichfache Anwendung der beiden folgenden Lemmata, durch welche es ermöglicht wird, jeweils zwei gleichartige Quantoren zu einem zusammenzufassen und verschiedenartige Quantoren (in einer Richtung) zu vertauschen:

Lemma 1:

Falls $f_2(\mathfrak{x},y) \leq \vartheta_K(\mathfrak{x},y)$, so gilt

$$\bigvee_{a \leq f_1(\mathfrak{x})} \bigvee_{b \leq f_2(\mathfrak{x},a)} P(\mathfrak{x},a,b) \leftrightarrow \bigvee_{c \leq \sigma_2(f_1(\mathfrak{x}),\vartheta_K(\mathfrak{x},f_1(\mathfrak{x})))} (\sigma_{21}(c) \leq f_1(\mathfrak{x})$$

$$\wedge\ \sigma_{22}(c) \leq f_2(\mathfrak{x},\sigma_{21}(c)) \wedge P(\mathfrak{x},\sigma_{21}(c),\sigma_{22}(c)))$$

und

$$\bigwedge_{a \leq f_1(\mathfrak{x})} \bigwedge_{b \leq f_2(\mathfrak{x},a)} P(\mathfrak{x},a,b)$$

$$\leftrightarrow \bigwedge_{c \leq \sigma_2(f_1(\mathfrak{x}),\ \vartheta_K(\mathfrak{x},f_1(\mathfrak{x})))} (\sigma_{21}(c) > f_1(\mathfrak{x})$$

$$\vee\ \sigma_{22}(c) > f_2(\mathfrak{x},\sigma_{21}(c)) \vee P(\mathfrak{x},\sigma_{21}(c),\ \sigma_{22}(c)))\ .$$

Lemma 2:

Falls $f_1(\mathfrak{x}) \leq \vartheta_{K_1}(\mathfrak{x})$, $f_2(\mathfrak{x},y) \leq \vartheta_{K_2}(\mathfrak{x},y)$, so gibt es ein K mit:

$$\bigwedge_{a\leq f_1(\mathfrak{x})} \bigvee_{b\leq f_2(\mathfrak{x},a)} P(\mathfrak{x},a,b)$$

$$\leftrightarrow \bigvee_{e\leq \vartheta_K(\mathfrak{x})} \bigwedge_{a\leq f_1(\mathfrak{x})} (\gamma(e,a)\leq f_2(\mathfrak{x},a)\wedge P(\mathfrak{x},a,\gamma(e,a))) \; .$$

Die beiden in Lemma 1 behaupteten Äquivalenzen ergeben sich unmittelbar aus der Voraussetzung $f_2(\mathfrak{x},y)\leq\vartheta_K(\mathfrak{x},y)$. Der Beweis von Lemma 2 beruht auf einer Anwendung von (c):

Sei vorausgesetzt

$$\bigwedge_{a\leq f_1(\mathfrak{x})} \bigvee_{b\leq f_2(\mathfrak{x},a)} P(\mathfrak{x},a,b),$$

$$f_1(\mathfrak{x})\leq\vartheta_{K_1}(\mathfrak{x}),$$

$$f_2(\mathfrak{x})\leq\vartheta_{K_2}(\mathfrak{x},y) \; .$$

Sei $n := f_1(\mathfrak{x})$. Dann existiert eine Folge $b_0,\ldots,b_n$ mit

$$\bigwedge_{y\leq n} P(\mathfrak{x},y,b_y),$$

$$\bigwedge_{y\leq n} b_y \leq f_2(\mathfrak{x},y)$$

Zu dieser Folge seien die Zahlen c,d gemäss (c) bestimmt; es folgt (für passendes K_3):

$$d\leq(\vartheta_{K_2}(\mathfrak{x},y))! \leq \vartheta_{K_2+2}(\mathfrak{x},y)$$

$$c\leq \prod_{i\leq y} (1+(i+1)d)\leq\vartheta_{K_3}(y,d),$$

$$\bigwedge_{y\leq n} \beta(c,d,y) = b_y.$$

Es gibt ein K mit

$$\sigma_2(\vartheta_{K_3}(y,\vartheta_{K_2+2}(\mathfrak{x},y)),\ \vartheta_{K_2+2}(\mathfrak{x},y))\leq\vartheta_K(\mathfrak{x},y).$$

Mit $e := \sigma_2(c,d)$ ergibt sich daher

$$e\leq\vartheta_K(\mathfrak{x},y),$$

$$\bigwedge_{y\leq n} \gamma(e,y) = b_y,$$

d.h. $$\bigwedge_{y\leq n} P(\mathfrak{x},y,\gamma(e,y)).$$

Insgesamt gilt also

$$\bigvee_{e\leq\vartheta_K(\mathfrak{x},y)} \bigwedge_{a\leq f_1(\mathfrak{x})} (\gamma(e,a)\leq f_2(\mathfrak{x},a)\wedge P(\mathfrak{x},a,\gamma(e,a))),$$

so dass die Richtung " $\rightarrow$ " der Äquivalenz aus Lemma 2 bewiesen ist. Die umgekehrte Richtung "$\leftarrow$" ist trivialerweise richtig.

Damit ist der Beweis von Hilfssatz 3 abgeschlossen .

g.) Es sei nun f eine beliebige elementare Funktion. Nach (e),(f) gibt es Funktionen f_1,f_2,f_3 aus $\mathfrak{E}_1$ mit

$$(g_1):\ f(\mathfrak{x})=y \leftrightarrow \bigvee_{a\leq f_1(\mathfrak{x},y)} \bigwedge_{b\leq f_2(\mathfrak{x},y,a)} f_3(\mathfrak{x},y,a,b) = 0.$$

Alle Funktionen aus $\mathfrak{E}_1$ lassen sich mit höchstens drei primitiven Rekursionen definieren. Daher liegen die Funktionen f_1,f_2,f_3 in $A^3_{PR}(\{\lambda x(x+1)\})$.

Es sei $$G(\mathfrak{x},y,a):\leftrightarrow \bigwedge_{b\leq f_2(\mathfrak{x},y,a)} f_3(\mathfrak{x},y,a,b)=0.$$

χ_G liegt also in $A^4_{PR}(\{\lambda x(x+1)\})$.

Zu f existiert nach Hilfssatz 1 eine Konstante K mit

$$f(\mathfrak{x})\leq\vartheta_K(\mathfrak{x}).$$

Zu f_1 existiert ebenfalls ein K_1 mit

$$f_1(\mathfrak{x},y)\leq\vartheta_{K_1}(\mathfrak{x},y).$$

Es gibt daher ein K_2 mit

$$\sigma_2(\vartheta_K(\mathfrak{x}),\vartheta_{K_1}(\mathfrak{x},\vartheta_K(\mathfrak{x})))\leq\vartheta_{K_2}(\mathfrak{x}).$$

Sei

$$F(\mathfrak{x},z) \;:\leftrightarrow \sigma_{22}(z)\leq f_1(\mathfrak{x},\sigma_{21}(z)) \wedge G(\mathfrak{x},\sigma_{21}(z),\ \sigma_{22}(z)) \,.$$

Zum Nachweis von $f\in A^5_{PR}(\{\lambda x(x+1)\})$ genügt es, zu zeigen, dass f aus F durch <u>eine</u> Anwendung des beschränkten λ-Operators gewonnen werden kann: Nach (g_1) gibt es zu jedem $\mathfrak{x}$ ein z mit $z\leq\vartheta_{K_2}(\mathfrak{x})$ und $F(\mathfrak{x},z)$. Falls umgekehrt $z\leq\vartheta_{K_2}(\mathfrak{x})$ und $F(\mathfrak{x},z)$, so kann $z=\sigma_2(y,a)$ (für geeignete y,a) angenommen werden. Es folgt $a\leq f_1(\mathfrak{x},y)$, $G(\mathfrak{x},y,a)$, nach (g_1) also $f(\mathfrak{x})=y$. Aus diesen Überlegungen ergibt sich für f die Darstellung

$$f(\mathfrak{x})=\sigma_{21}(\ \mu z_{z\leq\vartheta_{K_2}(\mathfrak{x})}\ F(\mathfrak{x},z)),$$

aus der sofort $f\in A^5_{PR}(\{\lambda x(x+1)\})$ folgt.

15.) <u>Abschluss des Beweises von [11]:</u>

Zu zeigen ist nur noch [11 d.)]; dazu genügt der Nachweis von

(a): $H_n \in A^{n+2}_{PR}(\{\lambda x(x+1)\})$

durch Induktion über n. Der Induktionsbeginn ist trivial, da die Funktion $\lambda x 2^x$ mit zwei primitiven Rekursionen (und Einsetzungen) aus den Funktionen u^i_n, c^i_n und $\lambda x(x+1)$ definiert werden kann. Falls $H_n \in A^{n+2}_{PR}(\{\lambda x(x+1)\})$, so liegt F_n (vgl. den Beweis von (b_1) in [13]) in $A^{n+3}_{PR}(\{\lambda x(x+1)\})$, und H_{n+1} (vgl.(**) in [13])liegt in $A^{n+3}_{PR}(\{\lambda x(x+1)\})$, q.e.d.

§ 2: Aufzählung und Einsetzung.

16.) Einführung:

Gegeben seien eine Funktionenklasse $\mathfrak{A}$ und eine zweistellige Funktion E, derart dass die Funktionen $\lambda x E(a,x)$ genau die einstelligen Funktionen von $\mathfrak{A}$ durchlaufen, wenn a die natürlichen Zahlen durchläuft; man wird dann E eine "Aufzählungsfunktion" für die einstelligen Funktionen von $\mathfrak{A}$ nennen. Falls $\mathfrak{A}$ nur aus einstelligen Funktionen besteht, wird man E als Aufzählungsfunktion von $\mathfrak{A}$ bezeichnen; nur derartige Funktionenklassen $\mathfrak{A}$ sollen hier im Zusammenhang mit Aufzählungen betrachtet werden. - Für die folgenden Überlegungen wird der Begriff einer einstelligen Aufzählungsfunktion E von $\mathfrak{A}$ benötigt; unter Rückgriff auf die Tatsache, dass $\lambda xy\langle x,y\rangle$ (vgl.[8b.)]) eine eineindeutige Abbildung der Paare natürlicher Zahlen auf natürliche Zahlen vermittelt, wird eine einstellige Funktion E Aufzählungsfunktion von $\mathfrak{A}$ heissen, falls

$$\mathfrak{A} = \{\lambda x E(\langle a,x\rangle) : a \in N\} .$$

Unter gewissen naheliegenden Voraussetzungen über $\mathfrak{A}$ liegt E nicht in $\mathfrak{A}$: Wenn nämlich $\mathfrak{A}$

a.) die Funktionen λxx, $\lambda x(x+1)$ enthält und

b.) mit f,g stets auch $\lambda x\langle f(x),g(x)\rangle$, $\lambda x f(g(x))$ enthält, so führt die Annahme, E liegt in $\mathfrak{A}$, auf den Widerspruch

$$\bigvee_a \bigwedge_x E(\langle x,x\rangle)+1 = E(\langle a,x\rangle).$$

Zu einer einstelligen Funktion φ sei $\mathfrak{A}^\varphi$ die kleinste Klasse einstelliger Funktionen, welche φ und gewisse andere Anfangsfunktionen enthält und abgeschlossen ist gegen Anwendungen gewisser Definitionsschemata. Zu den Anfangsfunktionen sollen insbesondere λxx und $\lambda x(x+1)$ gehören; $\mathfrak{A}^{\varphi}$ soll zumindest abgeschlossen sein gegen die Definitionsmöglichkeiten, die sich aus dem Übergang von f,g zu $\lambda x\langle f(x),g(x)\rangle$ und $\lambda xf(g(x))$ ergeben. Jeder Funktion φ sei eine Aufzählungsfunktion E^{φ} der Klasse $\mathfrak{A}^{\varphi}$ zugeordnet. Also liegt E^{φ} nicht in $\mathfrak{A}^{\varphi}$. Falls $\lambda x0$ zu den Anfangsfunktionen (ausser φ) der Klassen $\mathfrak{A}^{\varphi}$ gehört, gilt darüber hinaus $\mathfrak{A}^{\varphi} \subseteq \mathfrak{A}^{E^{\varphi}}$ (also $\mathfrak{A}^{\varphi} \subset \mathfrak{A}^{E^{\varphi}}$), denn zu beliebigem a liegt die Funktion $\lambda xE^{\varphi}(\langle a,x\rangle)$ in $\mathfrak{A}^{E^{\varphi}}$. Der Übergang von $\mathfrak{A}^{\varphi}$ zu $\mathfrak{A}^{E^{\varphi}}$ bedeutet daher eine (echte) Vergrösserung (im Sinne der Inklusion) von $\mathfrak{A}^{\varphi}$. Dieses Prinzip der Vergrösserung einer Funktionenklasse unter Benutzung einer Aufzählungsfunktion soll im folgenden systematisch angewendet werden; dabei werden nur die beiden schon angegebenen Definitionsschemata der Einsetzung zugrundegelegt.

17.) <u>Definitionen</u>:

a.) φ sei eine beliebige einstellige Funktion. Dann sei $\mathfrak{A}^{\varphi}$ die Klasse einstelliger Funktionen, welche

a_1.) die Funktionen φ, λxK (für beliebiges K), $\lambda x(x)_0$, $\lambda x(x)_1$, λxx und $\lambda x(x+1)$ enthält und

a_2.) mit f und g auch stets $\lambda x\langle f(x),g(x)\rangle$ und $\lambda xf(g(x))$ enthält.

b.) Eine Aufzählungsfunktion E^{φ} von $\mathfrak{A}^{\varphi}$ sei wie folgt erklärt:

$$E^{\varphi}(\langle\langle 0\rangle,x\rangle) := \varphi(x)$$
$$E^{\varphi}(\langle\langle 1,K\rangle,x\rangle) := K$$
$$E^{\varphi}(\langle\langle 2\rangle,x\rangle) := (x)_0$$
$$E^{\varphi}(\langle\langle 3\rangle,x\rangle) := (x)_1$$
$$E^{\varphi}(\langle\langle 4\rangle,x\rangle) := x$$
$$E^{\varphi}(\langle\langle 5\rangle,x\rangle) := x+1$$

$E^{\varphi}(\langle\langle 6,\langle a,b\rangle\rangle,x\rangle) := \langle E^{\varphi}(\langle a,x\rangle),\ E^{\varphi}(\langle b,x\rangle)\rangle$

$E^{\varphi}(\langle\langle 7,\langle a,b\rangle\ ,x\rangle) := E^{\varphi}(\langle a,E^{\varphi}(\langle b,x\rangle)\rangle)$

$E^{\varphi}(y) := 0$ für Argumente y, die durch die ersten 8 Zeilen der Definition nicht erfasst sind.

c.) Die Funktionenfolge E_n sei (bei zunächst noch unspezifiziertem E_o) definiert durch:

$$E_{n+1} := E^{E_n}$$

d.) Es sei $\mathcal{A}_n := \mathcal{A}^{E_n}$

e.) Die beiden Folgen E_n, $\mathcal{A}_n$ sollen erweitert werden für "transfinite Indizes" durch die folgenden Festsetzungen:

$$E_\omega(x) := E_{((x)_o)_o}(\langle((x)_o)_1,\ (x)_1\rangle),$$

$$E_{\omega+n+1}(x) := E^{E_{\omega+n}}(x)$$

$$\mathcal{A}_{\omega+n} := \mathcal{A}^{E_{\omega+n}}\ .$$

(Die Indizes n,ω+n werden hier als Ordinalzahlen aufgefasst, ω als Ordnungszahl von N.)

f.) $[a]_n := \lambda x E_n(\langle a,x\rangle)$

g.) $\mathfrak{P}^1$ sei die Klasse der einstelligen primitiv rekursiven Funktionen.

18.) Für die in [17] eingeführte Hierarchie von Funktionenklassen $\mathcal{A}_n$, $\mathcal{A}_{\omega+n}$ gilt (nach [16]) offenbar

$$\mathcal{A}_n \sqsubset \mathcal{A}_{n+1},$$

$$\mathcal{A}_{\omega+n} \sqsubset \mathcal{A}_{\omega+n+1}.$$

Ferner gilt: $\bigcup_n \mathcal{A}_n \sqsubset \mathcal{A}_\omega$. Denn erstens gibt es zu beliebigem n und f aus $\mathfrak{A}_n$ ein a mit $f=[a]_{n+1}=\lambda x E_\omega(\langle\langle n+1,a\rangle,x\rangle)\in \mathfrak{A}_\omega$, also gilt $\bigcup_n \mathfrak{A}_n \subseteq \mathfrak{A}_\omega$; zweitens liegt E_ω in keiner der Klassen $\mathfrak{A}_n$, denn aus $E_\omega \in \mathfrak{A}_n$ würde auch $\lambda x(E_\omega(\langle\langle n+1,x\rangle,x\rangle)+1)\in \mathfrak{A}_n$ folgen, woraus sich für geeignetes a der Widerspruch

$$\lambda x E_{n+1}(\langle x,x\rangle)+1 = [a]_{n+1}$$

ergibt. Es soll nun gezeigt werden, dass für ein geeignet gewähltes E_o aus $\mathfrak{P}^1$ die Hierarchie der $\mathfrak{A}_n$ gegen $\mathfrak{P}^1$ "konvergiert": $\bigcup_n \mathfrak{A}_n = \mathfrak{P}^1$:

19.) <u>Definition</u>:

$\zeta(\langle 0,y\rangle) := (y)_1$

$\zeta(\langle x+1,y\rangle) := \langle 7,\langle (y)_o, \zeta(\langle x,y\rangle)\rangle\rangle$

$\zeta(z) := 0$ für Argumente z, die durch die ersten beiden Zeilen der Definition nicht erfasst werden.

<u>Satz</u>:

<u>Vor.</u>: $E_o = \zeta$

<u>Beh.</u>: $\bigcup_n \mathfrak{A}_n = \mathfrak{P}^1$

<u>Bew.</u>: Sei $E_o = \zeta$:

a.) <u>Beh.1</u>: $\bigcup_n \mathfrak{A}_n \subseteq \mathfrak{P}^1$

<u>Bew.1</u>: Sei $E_n^* := \lambda xy E_n(\langle x,y\rangle)$; dann liegt E_o^* in $\mathfrak{P}$, und aus der Definition von E_{n+1} aus E_n lässt sich eine Definition von E_{n+1}^* aus E_n^* gewinnen, bei der eine Wertverlaufsrekursion mit Einschachtelungen an der Parameterstelle verwendet wird. Daher liegt mit E_n^* auch E_{n+1}^* in $\mathfrak{P}$. Also sind alle E_n^* primitiv-rekursiv; daher liegen alle E_n in $\mathfrak{P}^1$. Es folgt $\mathfrak{A}_n \subseteq \mathfrak{P}^1$ für beliebiges n. Damit ist Beh.1 gezeigt.

b.) Für beliebige einstellige Funktionen f sei

$$f^{o}(x) := x,$$

$$f^{n+1}(x) := f(f^{n}(x)).$$

Sei $\mathfrak{P}_1$ die kleinste Klasse einstelliger Funktionen, welche

a_1.) die Funktionen $\lambda x 0, \lambda x (x)_o, \lambda x (x)_1, \lambda x x, \lambda x (x+1)$ enthält und

a_2.) mit f,g auch die Funktionen $\lambda x \langle f(x), g(x)\rangle$, $\lambda x f(g(x))$ und $\lambda x f^{(x)_o}(g((x)_1))$ enthält.

<u>Beh.2:</u> $\mathfrak{P}_1 = \mathfrak{P}^1$

<u>Bew.2:</u> Die Inklusion " $\subseteq$ " ist trivial. Zum Nachweis der umgekehrten Inklusion " $\supseteq$ " soll zunächst eine Modifikation der Funktionen $\lambda\mathfrak{g}\langle\mathfrak{g}\rangle$ eingeführt werden:

Sei $\langle x\rangle^{+} := x$,

$$\langle\mathfrak{g}, y\rangle^{+} := \langle\langle\mathfrak{g}\rangle^{+}, y\rangle,$$

dann gilt:

$$\lambda\mathfrak{g}\langle\mathfrak{g}\rangle^{+} \in \mathfrak{P}$$

Für beliebiges n und i mit $i \leq n$ gibt es in $\mathfrak{P}_1$ eine Funktion $f_{n,i}$ mit

$$f_{n,i}(\langle x_o, \ldots, x_n\rangle^{+}) = x_i;$$

sei $(x)^{+}_{n,i} := f_{n,i}(x)$.

<u>Beh.2.1:</u> $\bigwedge_{f\in\mathfrak{P}} \bigvee_{f^{*}\in\mathfrak{P}_1} \bigwedge_{\mathfrak{g}} f(\mathfrak{g}) = f^{*}(\langle\mathfrak{g}\rangle^{+})$.

<u>Bew.2.1:</u> Es genügt, zu zeigen:

b_1.) Falls $f = \lambda\mathfrak{g}(g_o(\mathfrak{g}), \ldots, g_r(\mathfrak{g}))$,

$$g(\mathfrak{y}) = g^{*}(\langle\mathfrak{y}\rangle^{+}),\ g_o(\mathfrak{g}) = g_o^{*}(\langle\mathfrak{g}\rangle^{+}), \ldots, g_r(\mathfrak{g}) = g_r^{*}(\langle\mathfrak{g}\rangle^{+}),$$

$$g^{*}, g_o^{*}, \ldots, g_r^{*} \in \mathfrak{P}_1,$$

so gibt es eine Funktion f^* aus $\mathfrak{P}_1$ mit

$$f(\mathfrak{x}) = f^*(\langle\mathfrak{x}\rangle^+).$$

b_2.) Falls $f(\mathfrak{x},0)=g_1(\mathfrak{x})$

$$f(\mathfrak{x},y+1)=g_2(\mathfrak{x},y,f(\mathfrak{x},y))$$

$$g_1(\mathfrak{x})=g_1^*(\langle\mathfrak{x}\rangle^+),$$

$$g_2(\mathfrak{x},y,z)=g_2^*(\langle\mathfrak{x},y,z\rangle^+)$$

$$g_1^*, g_2^* \in \mathfrak{P}_1,$$

so gibt es eine Funktion f^* aus $\mathfrak{P}_1$ mit

$$f(\mathfrak{x},y)=f^*(\langle\mathfrak{x},y\rangle^+).$$

<u>Ad(b_1):</u>

Die Funktion

$$f^* := \lambda x g^*(\langle g_o^*(x),\ldots,g_r^*(x)\rangle^+)$$

leistet das Verlangte.

<u>Ad(b_2):</u>

Die Voraussetzungen über f, g_1, g_2, g_1^*, g_2^* seien erfüllt. Sei

$$h_1 := \lambda x\langle (x)^+_{n,o},\ldots,(x)^+_{n,n}, 0, g_1^*(x)\rangle^+,$$

$$h_2 := \lambda x\langle (x)^+_{n+2,o},\ldots,(x)^+_{n+2,n},(x)^+_{n+2,n+1}+1, g_2^*(x)\rangle^+,$$

$$f' := \lambda x h_2^{(x)_1}(h_1((x)_o)),$$

$$f^* := \lambda x(f'(x))^+_{n+2,n+2} \quad .$$

Offenbar liegen mit g_1^*, g_2^* auch die Funktionen h_1, h_2 in $\mathfrak{P}_1$, und wegen

$$f'(x) = \lambda x h_2^{(x)_o}(h_1((x)_1))\ (\langle (x)_1,(x)_o\rangle)$$

liegen auch f' und damit f^* in $\mathfrak{P}_1$. Dass f^* das Verlangte leistet, ergibt sich aus der folgenden

Beh.2.1.1.:

$$f'(\langle \mathfrak{x},y\rangle^{+}) = \langle \mathfrak{x},y,f(\mathfrak{x},y)\rangle^{+}$$

Bew.2.1.1.: Induktion nach y:

Für y=0 gilt:

$$\begin{aligned} f'(\langle \mathfrak{x},0\rangle^{+}) &= h_2^{0}(h_1(\langle \mathfrak{x}\rangle^{+})) \\ &= h_1(\langle \mathfrak{x}\rangle^{+}) \\ &= \langle \mathfrak{x},0,g_1^{*}(\langle \mathfrak{x}\rangle^{+})\rangle^{+} \\ &= \langle \mathfrak{x},0,g_1(\mathfrak{x})\rangle^{+} \\ &= \langle \mathfrak{x},0,f(\mathfrak{x},0)\rangle^{+}. \end{aligned}$$

Im Induktionsschritt ergibt sich:

$$\begin{aligned} f'(\langle \mathfrak{x},y+1\rangle^{+}) &= h_2^{y+1}(h_1(\langle \mathfrak{x}\rangle^{+})) \\ &= h_2(f'(\langle \mathfrak{x},y\rangle^{+})) \\ &= h_2(\langle \mathfrak{x},y,f(\mathfrak{x},y)\rangle^{+}) \end{aligned}$$

n. Ind.-Vor.

$$\begin{aligned} &= \langle \mathfrak{x},y+1,g_2^{*}(\langle \mathfrak{x},y,f(\mathfrak{x},y)\rangle^{+})\rangle^{+} \\ &= \langle \mathfrak{x},y+1,g_2(\mathfrak{x},y,f(\mathfrak{x},y))\rangle^{+} \\ &= \langle \mathfrak{x},y+1,f(\mathfrak{x},y+1)\rangle^{+}, \quad \text{q.e.d.} \end{aligned}$$

Damit ist (b_2), also Beh.2 nachgewiesen.

Zu f aus $\mathfrak{P}^1$ existiert also ein f^* aus $\mathfrak{P}_1$ mit

$$f(x) = f^{*}(\langle x\rangle^{+}).$$

Wegen $\langle x\rangle^{+}=x$ bedeutet das $\mathfrak{P}^1 \subseteq \mathfrak{P}_1$.

c,) Es ist also nur noch zu zeigen, dass $\mathfrak{P}_1 \subseteq \bigcup_n \mathfrak{A}_n$. Im Hinblick auf die Abgeschlossenheitseigenschaften der $\mathfrak{A}_n$ genügt dann der Nachweis der folgenden

<u>Beh.3</u>: Für $f,g \in \mathcal{E}_n$ liegt $\lambda x f^{(x)_0}(g((x)_1))$ in $\mathcal{E}_{n+1}$.

<u>Bew.3</u>: Zunächst soll die folgende Identität bewiesen werden:

<u>Beh.3.1</u>: $[\zeta(\langle x,\langle a,b\rangle\rangle)]_{n+1}(y)=[a]_{n+1}^{x}([b]_{n+1}(y))$

<u>Bew.3.1</u>: Induktion nach x:

Für $x=0$ gilt nach Definition von ζ(1.Zeile):

$$[\zeta(\langle 0,\langle a,b\rangle\rangle)]_{n+1}(y) = [b]_{n+1}(y)=[a]_{n+1}^{0}([b]_{n+1}(y)).$$

Im Induktionsschritt ergibt sich aus der 2.Zeile der Definition von ζ, der 8.Zeile der Definition von E_{n+1} und der Induktionsvoraussetzung:

$$\begin{aligned}
&[\zeta(\langle x+1,\ \langle a,b\rangle\rangle)]_{n+1}(y)\\
&\quad = [\langle 7,\langle a,\zeta(\langle x,\langle a,b\rangle\rangle)\rangle\rangle]_{n+1}(y)\\
&\quad = [a]_{n+1}([\zeta(\langle x,\langle a,b\rangle\rangle)]_{n+1}(y))\\
&\quad = [a]_{n+1}([a]_{n+1}^{x}([b]_{n+1}(y)))\\
&\quad = [a]_{n+1}^{x+1}([b]_{n+1}(y)), \quad \text{q.e.d.}
\end{aligned}$$

Seien nun f und g Funktionen aus $\mathcal{E}_n$. Sei $f=[a]_{n+1}$, $g=[b]_{n+1}$. Dann liegt in $\mathcal{E}_{n+1}$ mit ζ auch die Funktion $\lambda x[\zeta(\langle (x)_0,\langle a,b\rangle\rangle)]_{n+1}((x)_1)$, und das ist (nach Beh.3.1.) gerade $\lambda x f^{(x)_0}(g((x)_1))$, q.e.d.

20.) In [19] ist gezeigt worden, dass für passend gewähltes primitiv-rekursives E_0 die Hierarchie der $\mathcal{E}_n$ schon "unterhalb $\mathcal{E}_\omega$" alle primitiv-rekursiven Funktionen umfasst. Die dabei zugrundegelegte Funktion E_0 (d.h.: ζ, vgl.[19]) ist recht kompliziert. Für ein "einfacheres" E_0 soll nun gezeigt werden, dass jedenfalls "oberhalb $\mathcal{E}_\omega$" alle Funktionen aus $\mathcal{P}^1$ erfasst werden:

Satz:

Vor.: $E_o = \lambda x((x)_1)_1$

Beh.: $\mathfrak{P}^1 \subseteq \bigcup_n \mathfrak{A}_{\omega+n}$.

Bew.: Aus [19] wird die Definition der Funktionenklasse $\mathfrak{P}_1$ übernommen. Es gilt also $\mathfrak{P}^1 = \mathfrak{P}_1$, und daher genügt es, wie in [19: c.)] zu zeigen:

Beh.1: Für $f, g \in \mathfrak{A}_{\omega+n}$ liegt $\lambda x f^{(x)_o}(g((x)_1))$ in $\mathfrak{A}_{\omega+n+1}$.

Bew.1:

In dem Beweis der entsprechenden Beh.3 aus [19: c.)] wird die dortige Voraussetzung $E_o = \zeta$ nur insoweit ausgewertet, als sichergestellt ist, dass in $\mathfrak{A}_o$ eine Funktion η (nämlich E_o) mit

(a.) $\eta(\langle 0,y\rangle) = (y)_1$,

(b.) $\eta(\langle x+1,y\rangle) = \langle 7,\langle (y)_o,\eta(\langle x,y\rangle)\rangle\rangle$

liegt. Es würde zum Nachweis von Beh.1 also genügen, zu zeigen:

Beh.1.1: In $\mathfrak{A}_\omega$ liegt eine Funktion η, welche die Bedingungen (a), (b) erfüllt.

Bew.1.1: Diesem Beweis wird zugrundegelegt das folgende

Rekursionstheorem: Zu jedem a und jeder Funktionenfolge φ_n mit $\varphi_n = [a]_{n+1}$ gibt es ein b mit $[b]_{n+1} = \lambda x \varphi_n(\langle b,x\rangle)$.

Der Beweis des Rekursionstheorems wird in [21] geführt. Es wird hier angewendet auf die Funktionenfolge

$\varphi_n := \lambda x\langle 7,\langle ((x)_1)_o, E_n(x)\rangle\rangle$;

mit $a := \langle 6,\langle\langle 1,7\rangle,\langle 6,\langle\langle 7,\langle\langle 2\rangle,\langle 3\rangle\rangle\rangle,\langle 0\rangle\rangle\rangle\rangle\rangle$

gilt in der Tat $[a]_{n+1} = \varphi_n$. Also gibt es ein b mit

$$[b]_{n+1}(x) = \varphi_n(\langle b,x\rangle),$$

d.h.

$$E_{n+1}(\langle b,x\rangle) = \langle 7,\langle (x)_o, E_n(\langle b,x\rangle)\rangle\rangle.$$

Mit Rücksicht auf die Gleichung

$$E_o(\langle b,x\rangle) = (x)_1$$

ergibt sich aus

$$\eta := \lambda x E_{(x)_o}(\langle b,(x)_1\rangle),$$

dass die Funktion η den Forderungen (a),(b) genügt. Nach Definition von E_ω gilt

$$\eta = \lambda x E_\omega(\langle\langle (x)_o,b\rangle, (x)_1\rangle).$$

η liegt also in $\mathcal{A}_\omega$, und nur das war noch zu zeigen.

21) <u>Rekursionstheorem</u>:

<u>Vor.</u>: $\bigwedge\limits_n \bigwedge\limits_x [a]_{n+1}(x) = \varphi_n(x)$

<u>Beh.</u>: Für passendes b gilt

$$\bigwedge_n \bigwedge_x [b]_{n+1}(x) = \varphi_n(\langle b,x\rangle)$$

<u>Bew.</u>:

Sei $S(x) := \langle 7,\langle (x)_o, \langle 6,\langle\langle 1,(x)_1\rangle, \langle 4\rangle\rangle\rangle\rangle\rangle$;

dann gibt es ein a_o mit $[a_o]_{n+1} = S$ für alle n.

Zu a und a_o wird nun ein b_o konstruiert, welches für alle n die Bedingung

$$[b_o]_{n+1} = \lambda x \varphi_n(\langle S(\langle (x)_o,(x)_o\rangle), (x)_1\rangle)$$

erfüllt.

Sei

$$b_o := \langle 7,\langle a,\langle 6,\langle\langle 7,\langle a_o,\langle 6,\langle\langle 2\rangle,\langle 2\rangle\rangle\rangle\rangle\rangle,\langle 3\rangle\rangle\rangle\rangle\rangle \ .$$

Dieses b_o leistet das Verlangte.

Sei $b := S(\langle b_o,b_o\rangle)$. Dann gilt in der Tat

$$\begin{aligned} [b]_{n+1}(x) &= [S(\langle b_o,b_o\rangle)]_{n+1}(x) \\ &= [b_o]_{n+1}(\langle b_o,x\rangle) \\ &= \varphi_n(\langle S(\langle b_o,b_o\rangle),x\rangle) \\ &= \varphi_n(\langle b,x\rangle), \quad \text{q.e.d.} \end{aligned}$$

§ 3: Rechenzeit als Kompliziertheitsmass für rekursive Funktionen

22.) Einführung:

a.) Für die Berechnung rekursiver Funktionen auf idealisierten Maschinen wird im allgemeinen das Konzept der TURING-Maschine zugrundegelegt. Im folgenden soll jedoch ein anderes Maschinenkonzept betrachtet werden, nämlich das einer "Registermaschine": Eine Registermaschine (RM) besteht aus einem Speicher und einem Programm zur schrittweisen Bearbeitung dieses Speichers. Ein Speicher besteht aus unendlich vielen Registern $R_o, R_1, \ldots$. Jedes dieser Register enthält eine natürliche Zahl, so dass der Inhalt des Speichers einer RM beschrieben wird durch eine Zahlenfolge $x_o, x_1, \ldots$, in welcher x_i den Inhalt des Registers R_i darstellt.- Das Programm einer RM besteht aus einer endlichen Folge von Instruktionen. Jede Instruktion ist ein 5-Tupel $(i, \alpha_i, \omega_i, \varphi_i^=, \varphi_i^+)$: i ist die Nummer der Instruktion im Programm; α_i ist die Adresse eines Registers; ω_i bezeichnet eine Operation an dem Register mit der Adresse α_i (oder den Stop der RM); $\varphi_i^=$ ist die Nummer derjenigen Instruktion, welche im Programmablauf als nächste aufgerufen wird, falls das Register mit der Adresse α_i vor Ausführung von ω_i die Zahl 0 enthält; φ_i^+ ist (entsprechend) die Nummer der nächsten Instruktion, falls dieses Register eine von 0 verschiedene Zahl enthält. - Der Ablauf eines Programms beginnt mit der Ausführung der Instruktion $(0, \alpha_o, \omega_o, \varphi_o^=, \varphi_o^+)$ und vollzieht sich danach schrittweise und determiniert durch das Programm, evtl. abbrechend. Es soll vorausgesetzt werden, dass die Instruktionen fortlaufend von 0 bis i_o numeriert sind, und dass die Instruktion mit der Nummer 1 als einzige die Operation "Stop" für die RM enthält. Die Registeradressen α_i sollen natürliche Zahlen sein, derart dass α_i die Adresse des Registers R_{α_i} ist. Drei Operationen ω_i am Inhalt x_j $(j=\alpha_i)$ eines Registers R_j sind möglich: $x_j \Rightarrow x_j$, $x_j \Rightarrow x_j+1$ und $x_j \Rightarrow x_j \dot{-} 1$.

b.) Im folgenden werden alle Rechnungen auf RM nur an Speichern ausgeführt, in welchen höchstens endlich viele Register eine von 0 verschiedene Zahl enthalten. Dann lässt sich in jedem Schritt der Rechnung der Speicherinhalt durch eine Zahl $\langle x_o, \dots, x_r \rangle$ kodieren (für $\rho > r$ enthalte R_ρ also die Zahl 0). Jeder Registermaschine M lassen sich nun zwei partiell rekursive Funktionen r_M, s_M wie folgt zuordnen: Falls x das Kodifikat des Speicherinhaltes zu Beginn der Rechnung von M ist, so sind zwei Fälle möglich: Entweder stoppt M niemals:, dann seien $r_M(x)$ und $s_M(x)$ undefiniert. Falls aber M stoppt, so nach einer wohldefinierten Zahl von Schritten; diese sei $s_M(x)$. Dann ist der Anfangsinschrift x in eindeutiger Weise eine Endinschrift y des Speichers am Ende der Rechnung zugeordnet; diese sei $r_M(x)$.

c.) Falls M eine RM und f Funktion über N ist, so soll die Redeweise "M berechnet f" den folgenden Sachverhalt beschreiben:

$$\bigwedge_{\mathfrak{x}} r_M(\langle \mathfrak{x} \rangle) = \langle \mathfrak{x}, f(\mathfrak{x}) \rangle .$$

d.) Eine Funktion g heisst "Schrittzahlfunktion für f", wenn es eine Registermaschine M gibt, die f berechnet und für deren Schrittzahlfunktion s_M die Gleichung

$$\bigwedge_{\mathfrak{x}} g(\mathfrak{x}) = s_M(\langle \mathfrak{x} \rangle)$$

gilt. Beliebige Schrittzahlfunktionen für f sollen mit "s_f" bezeichnet werden.

e.) Dass das Konzept einer RM für eine adäquate Charakterisierung des Berechenbarkeitsbegriffs geeignet ist, ergibt sich aus der Tatsache, dass genau die rekursiven Funktionen auf Registermaschinen berechenbar (im Sinne von (c)) sind. Darauf soll hier nicht näher eingegangen werden. - Im folgenden soll die Rechenzeit, welche eine Funktion f auf einer sie berechnenden Maschine M benötigt, also die Wachstumsordnung der Funktion $\lambda\mathfrak{x}\, s_M(\langle\mathfrak{x}\rangle)$, als Mass für die Kompliziertheit von f angesetzt werden. Insbesondere werden Funktionen "polynomialer Kompliziertheit" betrachtet, d.h. Funktionen f mit einer Schrittzahlfunktion s_f, die für ein geeignetes Polynom Q der Abschätzung

$$\bigwedge_{\mathfrak{x}} s_f(\mathfrak{x}) \leq Q(\mathfrak{x})$$

genügt. Es ergibt sich, dass eine beliebige rekursive Funktion f aus jeder ihrer Schrittzahlfunktionen s_f mit einer Funktion g von höchstens polynomialer Kompliziertheit allein durch Einsetzungen gewonnen werden kann: $f \in A(\{s_f, g\})$. - Den Schluss dieses Paragraphen bilden Untersuchungen über eine Hierarchie von Funktionen wachsender Kompliziertheit, welche, ausgehend von den Funktionen polynomialer Kompliziertheit, gegen $\mathfrak{E}$ "konvergiert".

23.) Definition: $\mathfrak{T}_p$ sei die Klasse der Funktionen f, zu denen eine Schrittzahlfunktion s_f und ein Polynom Q mit $\bigwedge_{\mathfrak{x}} s_f(\mathfrak{x}) \leq Q(\mathfrak{x})$ existiert.

Satz: Es gibt in $\mathfrak{T}_p$ eine Funktion U mit der folgenden Eigenschaft: Zu jeder einstelligen rekursiven Funktion f und jeder ihrer Schrittzahlfunktionen s_f gibt es eine Konstante p und ein

Polynom Q, derart dass gilt:

$$\bigwedge_x \bigwedge_y (y \geq Q(x,s_f(x)) \rightarrow f(x)=U(p,x,y))$$

Der Beweis dieses Satzes beruht auf der Beschreibung der Berechnung einer Funktion f auf einer Maschine M mit Hilfe einer geeigneten "Konfigurationsfunktion" aus $\mathcal{T}_p$. Er wird in [24] bis [35] geführt. Zunächst wird die Definition von "$\langle \mathfrak{x} \rangle$" und "$(x)_i$" abgeändert, mit dem Ziel, $\lambda \mathfrak{x} \langle \mathfrak{x} \rangle \in \mathcal{T}_p$ und $\lambda x i (x)_i \in \mathcal{T}_p$ zu erreichen. Zu diesem Zweck müssen (in [24] bis [30]) einige Hilfsmittel über die Funktionenklasse $\mathcal{T}_p$ bereitgestellt werden. Das geschieht, indem aus der Klasse aller RM eine Teilklasse $\mathfrak{M}$ solcher RM ausgesondert wird, deren Struktur besonders übersichtlich ist und daher eine leichte Abschätzung der zugehörigen Schrittzahlfunktionen s_M ermöglicht. Unter Einschränkung auf Maschinen aus $\mathfrak{M}$ wird dann eine Funktionenklasse $\mathcal{T}_p^*$ ähnlich wie $\mathcal{T}_p$ definiert; nach Definition gilt $\mathcal{T}_p^* \subseteq \mathcal{T}_p$. (Die umgekehrte Inklusion ergibt sich in [36] als Folgerung aus dem oben formulierten Darstellungssatz.) Dass $\mathcal{T}_p^*$ gegen beschränkte primitive Rekursionen abgeschlossen ist, ist der Inhalt von [28]. Diese Tatsache wird dann bei der Konstruktion der oben erwähnten "Konfigurationsfunktion" (die den Abschluss des Beweises bildet) entscheidend ausgenutzt.

24.) <u>Abänderung der Definitionen</u> von $\langle \mathfrak{x} \rangle$ und $(x)_i$:

Im folgenden seien $\alpha_0, \alpha_1, \alpha_2$ drei beliebige Symbole.

a.) Jeder Zahl x wird zunächst eine endliche Folge $\underline{x}$ dieser Symbole zugeordnet durch die folgende <u>Definition</u>:

Falls $x=0$, so sei $\underline{x} := \Lambda$ (die leere Folge). Falls $x \neq 0$, so gibt es eine eindeutig bestimmte Zahlenfolge $i_0,\ldots,i_r$ mit

$x = \sum_{\rho \leq r} i_\rho 2^\rho$ und $i_\rho \in \{1,2\}$ für $\rho \leq r$. Dann sei $\underline{x} := \alpha_{i_r} \ldots \alpha_{i_o}$.

b.) Jeder endlichen Folge aus den Symbolen $\alpha_o, \alpha_1, \alpha_2$ wird eine natürliche Zahl φ zugeordnet durch die folgende

Definition $\overline{\alpha_{i_r} \ldots \alpha_{i_o}} := \sum_{\rho \leq r} i_\rho 3^\rho$

$\Lambda := 0$

c.) Def.:

$\langle x_o, \ldots, x_r \rangle := \overline{\underline{x_r}\, \alpha_o\, \underline{x_{r-1}}\, \alpha_o \ldots \underline{x_1}\, \alpha_o\, \underline{x_o}}$

Es gilt offenbar:

$\langle \mathfrak{x} \rangle = \langle \mathfrak{x}, 0 \rangle$

$\langle x_o, \ldots, x_r \rangle = \langle y_o, \ldots, y_r \rangle \rightarrow \bigwedge_{\rho \leq r} x_\rho = y_\rho$

$\bigwedge_x \bigvee_{\mathfrak{x}} x = \langle \mathfrak{x} \rangle$

Daher ist die folgende Definition sinnvoll

$(\langle x_o, \ldots, x_r \rangle)_\rho := x_\rho$ für $\rho \leq r$

25.) Definition einer Klasse $\mathfrak{M}$ einfacher Registermaschinen

a.) Zu jedem i gibt es eine Registermaschine "a_i" mit

$$(r_{a_i}(x))_j = \begin{cases} (x)_j + 1, & \text{falls } i = j \\ (x)_j & \text{sonst} \end{cases}$$

und

$$s_{a_i}(x) = 1.$$

$\mathfrak{M}$ enthalte alle diese Maschinen a_i.

b.) Zu jedem i gibt es eine Registermaschine "s_i" mit

$$(r_{s_i}(x))_j = \begin{cases} (x)_j \dot{-} 1, & \text{falls } i=j \\ (x)_j & \text{sonst} \end{cases}$$

und

$$s_{s_i}(x) = 1.$$

$\mathfrak{M}$ enthalte alle diese Maschinen s_i.

c.) Zu zwei Registermaschinen M_1, M_2 gibt es stets eine Registermaschine "M_1M_2" mit:

$$r_{M_1M_2}(x) = r_{M_2}(r_{M_1}(x))$$

und

$$s_{M_1M_2}(x) \leq s_{M_2}(r_{M_1}(x)) + s_{M_1}(x)$$

(das kann durch eine passende Zusammensetzung der Programme von M_1 und M_2 zum Programm von M_1M_2 stets erreicht werden).
$\mathfrak{M}$ enthalte mit M_1 und M_2 stets M_1M_2.

d.) Zu beliebigem i und einer beliebigen Registermaschine M gibt es stets eine Registermaschine $(M)_i$, deren Verhalten durch das folgende "Strukturdiagramm" angedeutet wird:

(*):

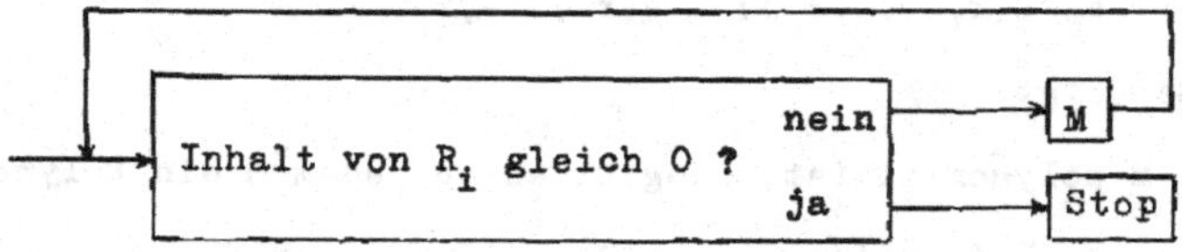

(d.h.: $(M)_i$ repetiert M solange, bis der Inhalt des Registers R_i gleich 0 ist, und stoppt dann; $(M)_i$ stoppt also evtl. niemals.) Für $(M)_i$ gilt, falls bis zum Stop die Schleife im Strukturdiagramm (*) gerade j-mal durchlaufen wird:

$$r_{(M)_i}(x) = r_M^j(x) \quad .$$

Ferner kann das Programm von $(M)_i$ aus dem von M in einer Weise konstruiert werden, die für $s_{(M)_i}$ die folgende Abschätzung erlaubt:

$$s_{(M)_i}(x) \leq 2 + \sum_{k<j} (s_M(r_M^k(x))+1).$$

$\mathfrak{M}$ enthalte zu beliebigem i mit M stets $(M)_i$.

e.) $\mathfrak{M}$ sei die kleinste Klasse von RM, welche den Bedingungen (a),(b),(c),(d) genügt.

__Bemerkung__: Die in (a) bis (d) auftretenden Gleichungen und Ungleichungen sind sinngemäss für partielle Funktionen zu verstehen.

26.) __Definition:__

M heisst "polynomial", wenn zu jedem n ein Polynom Q existiert, derart dass für beliebige $x_1,\dots,x_n$ die Abschätzung

$$s_M(\langle x_1,\dots,x_n\rangle) \leq Q(x_1,\dots,x_n)$$

gilt. Es folgt:

a.) Falls M polynomial ist, so gibt es zu jedem n ein Polynom Q mit

$$\bigwedge_{x_1}\dots\bigwedge_{x_n} r_M(\langle x_1,\dots,x_n\rangle) \leq Q(x_1,\dots,x_n)$$

b.) $\mathfrak{F}_P$ ist die Klasse der auf polynomialen RM berechenbaren Funktionen

c.) a_i, s_i, $(a_i s_j)_j$, $(a_{i_1} a_{i_2} s_j)_j$, $(s_i s_j)_j$ sind polynomiale Maschinen.

d.) Mit M_1, M_2 ist auch $M_1 M_2$ polynomial.

e.) Zu jeder polynomialen Maschine M gibt es ein Polynom Q mit $\bigwedge_x s_M(x) \leq Q(x)$. Denn falls n die grösste Adresse eines Registers ist, die im Programm von M unter den Adressen α_i vorkommt, so gibt es zu diesem n nach Definition ein Polynom Q_n mit

$$s_M(\langle x_o, \ldots, x_n \rangle) \leq Q_n(x_o, \ldots, x_n) .$$

Wegen $x_i \leq \langle x_o, \ldots, x_n \rangle$ für $x \leq n$ folgt

$$s_M(\langle x_o, \ldots, x_n \rangle) \leq Q(\langle x_o, \ldots, x_n \rangle)$$

für ein passendes Polynom Q. Da die Funktionen $\lambda \mathfrak{x} \langle \mathfrak{x} \rangle$ monoton wachsend in allen Argumenten sind, folgt für beliebiges x:

falls $x = \langle x_o, \ldots, x_n, x_{n+1}, \ldots, x_m \rangle$, so gilt:

$s_M(x) = s_M(\langle x_o, \ldots, x_n \rangle) \leq Q(\langle x_o, \ldots, x_n \rangle) \leq Q(x)$, q.e.d.

f.) Falls $\bigwedge_x s_M(x) \leq Q(x)$ (Q ein Polynom), so ist M polynomial. Denn aus der modifizierten Definition von $\langle \mathfrak{x} \rangle$ ergibt sich zunächst, dass zu jedem n ein Polynom Q_n^* mit

$$\langle x_o, \ldots, x_n \rangle \leq Q_n^*(x_o, \ldots, x_n)$$

existiert (vgl.auch [30 i.)]). Mit $Q_n := \lambda x_o \ldots x_n Q(Q^*(x_o, \ldots, x_n))$ folgt $s_M(\langle x_o, \ldots, x_n \rangle) \leq Q_n(x_o, \ldots, x_n)$, q.e.d.

27.) <u>Definition:</u>

$\mathfrak{F}_P^*$ sei die Klasse der Funktionen, die auf polynomialen RM aus $\mathfrak{M}$ berechenbar sind.

Ein Kriterium für die Zugehörigkeit einer Funktion zu $\mathcal{T}_P^*$ ergibt sich aus dem folgenden

<u>Lemma</u>:

<u>Vor.</u>: M berechnet f,

$M \in \mathfrak{M}$,

$s_M(\langle \mathfrak{x} \rangle) \leq Q(\mathfrak{x})$,

Q ein Polynom.

<u>Beh.</u>: $f \in \mathcal{T}_P^*$

<u>Bew.</u>:

M, f und Q seien nach Voraussetzung gegeben. Sei $\mathfrak{x} = \langle x_0,\ldots,x_n \rangle$. Sei $m > n$, und m sei grösser als jede Adresse eines Registers, die im Programm von M vorkommt. Dann gibt es eine polynomiale Maschine M_0 aus $\mathfrak{M}$ mit

$$r_{M_0}(\langle x_0,\ldots,x_m \rangle) = \langle x_0,\ldots,x_n \rangle .$$

Es folgt: Die Maschine M_0M liegt in $\mathfrak{M}$, ist polymial und berechnet f, q.e.d.

28.) <u>Satz</u>:

$A_{BR}(\mathcal{T}_P^*) \subseteq \mathcal{T}_P^*$

<u>Beweis</u>:

Es ist zu zeigen:

a.) Die Funktionen u_n^i, c_n^i sind mit polynomialen Maschinen aus $\mathfrak{M}$ berechenbar.

b.) Mit $g_0,\ldots,g_r$ liegt auch $\lambda \mathfrak{x} g_0(g_1(\mathfrak{x}),\ldots,g_r(\mathfrak{x}))$ in $\mathcal{T}_P^*$

c.) Falls

$$f(\mathfrak{x},0)=g_1(\mathfrak{x})$$

$$f(\mathfrak{x},y+1)=g_2(\mathfrak{x},y,f(\mathfrak{x},y))$$

$$f(\mathfrak{x},y) \leq g_3(\mathfrak{x},y)$$

und $g_1,g_2,g_3 \in \mathcal{T}^*_P$, so gilt: $f \in \mathcal{T}^*_P$.

(a) ergibt sich unmittelbar, unter Verwendung von [26c.)]

Ad(b): $M_o,\ldots,M_r$ seien polynomiale Maschinen aus $\mathfrak{M}$ zur Berechnung von $g_o,\ldots,g_r$. $g_1,\ldots,g_r$ seien k-stellig. t sei die grösste Adresse eines Registers, die im Programm einer der Maschinen $M_1,\ldots,M_r$ verwendet wird.

Durch Umindizierung der Register in M_o erhält man eine polynomiale Maschine M^*_o aus $\mathfrak{M}$ mit

$$r_{M^*_o}(\langle x_o,\ldots,x_t,y_1,\ldots,y_r\rangle) = \langle x_o,\ldots,x_t,y_1,\ldots,y_r,g_o(y_1,\ldots,y_r)\rangle.$$

Die Maschine

$$M_1(a_{t+1}\ s_{k+1}\ )_{k+1}\ldots M_r(a_{t+r}s_{k+1})_{k+1}\ M^*_o(s_{k+1}s_{t+1})_{t+1}\ \ldots$$

$$\ldots\ (s_{k+1}\ s_{t+r})_{t+r}\ (a_{k+1}s_{t+r+1})_{t+r+1}$$

liegt in $\mathfrak{M}$, berechnet $\lambda\mathfrak{x}g_o(g_1(\mathfrak{x}),\ldots,g_r(\mathfrak{x}))$ und ist polynomial.

Ad(c):

M_1 und M_2 seien polynomiale Maschinen aus $\mathfrak{M}$ zur Berechnung von g_1 und g_2. g_1 sei k+1-stellig. t sei die grösste Adresse eines Registers, die in M_2 verwendet wird. Ohne Einschränkung der Allgemeinheit kann $t>k+2$ angenommen werden. Zu M_1 gibt es eine polynomiale Maschine M^*_1 aus $\mathfrak{M}$ mit

$$r_{M^*_1}(\langle\mathfrak{x},y\rangle) = \langle\mathfrak{x},0,g_1(\mathfrak{x}),0,\ldots,0,y\rangle$$

(hier sei $\mathfrak{x}$ ein $(k+1)$-Tupel, und y stehe nach Ablauf der Rechnung von M_1^* in R_{t+1}).

Zu M_2 gibt es eine polynomiale Maschine M_2^* aus $\mathfrak{M}$ mit

$$r_{M_2^*}(\langle \mathfrak{x},a,b,y_{k+3},\ldots,y_t,y\rangle) = \langle \mathfrak{x},a+1,g_2(\mathfrak{x},a,b),\ y_{k+3},\ldots,y_t,\ y \dot{-} 1\ \rangle.$$

Für $j \leq y$ folgt

$$r^j_{M_2^*}(\langle \mathfrak{x},0,g_1(\mathfrak{x}),y_{k+3},\ldots,y_t,\ y\rangle) = \langle \mathfrak{x},j,f(\mathfrak{x},j),y_{k+3},\ldots,y_t,\ y \dot{-} j\ \rangle.$$

Die Maschine $M_1^*(M_2^*)_{t+1}$ liegt in $\mathfrak{M}$ und berechnet f. Die zugehörige Schrittzahlfunktion sei s'_f. Für s'_f ergibt sich die folgende Abschätzung:

$$\begin{aligned}
s'_f(\langle \mathfrak{x},y\rangle) &= s_{M_1^*(M_2^*)_{t+1}}(\langle \mathfrak{x},y\rangle) \\
&\leq s_{M_1^*}(\langle \mathfrak{x},y\rangle) + s_{(M_2^*)_{t+1}}(r_{M_1^*}(\langle \mathfrak{x},y\rangle)) \\
&= s_{M_1^*}(\langle \mathfrak{x},y\rangle) + s_{(M_2^*)_{t+1}}(\langle \mathfrak{x},0,g_1(\mathfrak{x}),0,\ldots,0,y\rangle) \\
&\leq s_{M_1^*}(\langle \mathfrak{x},y\rangle) \\
&\quad + 2 + \sum_{j<y} (s_{M_2^*}(r^j_{M_2^*}(\langle \mathfrak{x},0,g_1(\mathfrak{x}),0,\ldots,0,y\rangle))+1) \\
&\leq P(\mathfrak{x},y) + \sum_{j<y} (s_{M_2^*}(\langle \mathfrak{x},j,f(\mathfrak{x},j),0\ldots,0,y \dot{-} j\rangle)+1) \\
&\leq P(\mathfrak{x},y) + \sum_{j<y} Q(\mathfrak{x},y,j,f(\mathfrak{x},j))
\end{aligned}$$

für passende Polynome P,Q. Wegen $g_3 \in \mathcal{T}^*_P$ gibt es ein Polynom R mit $g_3(\mathfrak{x},y) \leq R(\mathfrak{x},y)$, also $f(\mathfrak{x},y) \leq R(\mathfrak{x},y)$. Daher gilt

$$s_f^!(\langle \mathfrak{x},y\rangle) \leq S(\mathfrak{x},y)$$

für ein passend gewähltes Polynom S. Nach [27] liegt f in $\mathcal{T}_P^*$, q.e.d.

29.) Definition:

$\mathcal{T} := A_{BR}(\{\lambda xy(x+1), \lambda xy(x\cdot y)\})$

a.) Aus dem soeben bewiesenen Satz ergibt sich sofort die Inklusion

$$\mathcal{T} \subseteq \mathcal{T}_P^* \subseteq \mathcal{T}_P ,$$

da die Funktionen $\lambda xy(x+y)$ und $\lambda xy(x.y)$ in $\mathcal{T}_P^*$ liegen und die Inklusion $\mathcal{T}_P^* \subseteq \mathcal{T}_P$ trivial ist.

b.) Der in [23] behauptete Darstellungssatz für einstellige rekursive Funktionen soll nun dadurch bewiesen werden, dass innerhalb $\mathcal{T}$ eine Funktion U mit dem im Satz verlangten Eigenschaften konstruiert wird.

c.) Wendet man dann den Satz speziell auf Funktionen aus $\mathcal{T}_P$ an, so ergibt sich leicht die Inklusion $\mathcal{T}_P \subseteq \mathcal{T}$, also insgesamt

$$\mathcal{T} = \mathcal{T}_P^* = \mathcal{T}_P .$$

Das wird in [36] gezeigt.

30.) Zunächst soll gezeigt werden, dass die Funktionen $\lambda \mathfrak{x}\langle \mathfrak{x}\rangle$ und $\lambda xy(x)_y$ in $\mathcal{T}$ liegen:

a.) $\lambda xy(x \dot{-} y)$, $\lambda xy[x:y] \in \mathcal{T}$.

b.) Sei $2_y^x := y \dot{-} (y \dot{-} 2^x)$; dann gilt

$\lambda xy 2_y^x \in \mathcal{T}$, denn 2_y^x entsteht aus Funktionen aus $\mathcal{T}$ durch eine beschränkte Rekursion:

$$2_y^0 = y \dot{-} (y \dot{-} 1) \,,$$

$$2_y^{x+1} = y \dot{-} (y \dot{-} 2_y^x \cdot 2)$$

$$= y \dot{-} (y \dot{-} 2 \cdot (y \dot{-} (y \dot{-} 2^x))),$$

$$2_y^x \leq y \,.$$

c.) Sei $\zeta_2(x,\rho)$ der Koeffizent i_ρ in der Darstellung

$$x = \sum_{\sigma \leq r} i_\sigma \cdot 2^\sigma \quad (i_\sigma \in \{0,1\} \text{ für } \sigma \leq r) \quad \text{von } x.$$

Dann gilt:

$$\zeta_2(x,\rho) = [x : 2_x^\rho] \dot{-} [x : 2_{2x}^{\rho+1}] \cdot 2$$

Daher liegt ζ_2 in $\mathcal{T}$.

d.) Sei $\zeta_2^*(x,\rho)$ der Koeffizient i_ρ in der Darstellung

$$x = \sum_{\sigma \leq r} i_\sigma \cdot 2^\sigma \quad (i_\sigma \in \{1,2\} \text{ f. } \sigma \leq r) \quad \text{von } x,$$

falls $x > 0$ und $\rho \leq r$; sei ferner $\zeta_2^*(x,\rho) = 0$ für $\rho > r$ und $\zeta_2^*(0,\rho) = 0$ für alle ρ.

Dann gilt:

$$\zeta_2^*(x,\rho) = \mu j_{j \leq 2} \bigvee_{i \leq x} \bigvee_{r \leq x} (x = \sum_{\sigma \leq r} (\zeta_2(i,\sigma)+1) \cdot 2_r^\sigma \wedge \rho \leq r$$

$$\wedge \; 1 + \zeta_2(i,\rho) = j).$$

Daher liegt ζ_2^* in $\mathcal{T}$.

e.) Sei $\zeta_3(x,\rho)$ der Koeffizient i_ρ in der Darstellung

$$x = \sum_{\sigma \leq r} i_\rho \cdot 3^\sigma \quad (i_\sigma \in \{0,1,2\} \text{ f. } \sigma \leq r) \quad \text{von } x.$$

Dann gilt

$$\zeta_3(x,\rho) = [x : 3_x^\rho] \dot{-} [x : 3_{3x}^{\rho+1}] \cdot 3$$

mit $\mathfrak{z}_y^x := y \dot- (y \dot- \mathfrak{z}^x)$ und (wie in (b)) $\lambda xy\mathfrak{z}_y^x \in \mathcal{T}$.

Daher liegt ζ_3 in $\mathcal{T}$.

f.) Sei $\alpha(x,\rho)$ die Anzahl der σ mit $\sigma<\rho$ und $\zeta_3(x,\sigma)=0$. Dann gilt

$$\alpha(x,y) = \sum_{\sigma<\rho} (1 \dot- \zeta_3(x,\sigma))$$

Daher liegt α in $\mathcal{T}$.

g.) Sei $i(x,\rho)$ der "erste Index links von der ρ-ten Null in der ternären Darstellung von x", d.h.

$$i(x,\rho) = \mu\sigma_{\sigma\leq x} \quad \alpha(x,\sigma)=\rho .$$

Dann gilt offenbar: $\lambda xyi(x,y) \in \mathcal{T}$.

h.) Nun folgt:

$$(x)_\rho = \sum_{i(x,\rho)\leq\sigma<i(x,\rho+1)} \zeta_3(x,i) \qquad \cdot 2_x^{\sigma \dot- i(x,\rho)}$$

Daher liegt $\lambda xy(x)_y$ in $\mathcal{T}$.

i.) Es gilt:

$$\langle x_o+y_o,\ldots,x_r+y_r\rangle$$
$$\leq (1+\langle x_o,\ldots,x_r\rangle)\cdot 3^{r+1} \prod_{\rho\leq r} (y_\rho+1)^2$$

j.) Sei

$$l(x) = \mu r_{r\leq x} \quad x = \sum_{\sigma<r} \zeta_2^*(x,\sigma) \cdot 2_x^\sigma$$

Es folgt: $l \in \mathcal{T}$

k.) Es gilt:

$$\langle x\rangle = \sum_{\sigma<l(x)} \zeta_2^*(x,\sigma)\cdot 3^\sigma_{(x+1)^2} ,$$

$$\langle x_0,\ldots,x_r\rangle = \langle x_1,\ldots,x_r\rangle \cdot 3^{l(x_0)+1}_{3\cdot(x_0+1)^2} + \sum_{\sigma<l(x_0)} \zeta_2^*(x_0,\sigma)\cdot 3^{\sigma}_{(x_0+1)^2} \; .$$

Daraus folgt: $\lambda g\langle g\rangle \in \mathfrak{r}$.

31.) Ein weiterer vorbereitender Schritt ist die Kodifikation der Programme von Registermaschinen: Sei M eine RM mit dem Programm

$$\begin{matrix} 0 & \alpha_0 & \omega_0 & \varphi_0^- & \varphi_0^+ \\ \vdots & \vdots & \vdots & \vdots & \vdots \\ N & \alpha_N & \omega_N & \varphi_N^- & \varphi_N^+ \end{matrix}$$

Dieses Programm soll dargestellt werden durch ein Zahlenschema. Dabei kann auf die erste Spalte verzichtet werden, da sie keine Information über das Programm beisteuert. Die Adressen $\alpha_0,\ldots,\alpha_N$ sind natürliche Zahlen (vgl.[22a.)]). Den Operationen $\omega_0,\ldots,\omega_N$ sollen in der folgenden Weise Zahlen zugeordnet werden: Falls $i \leq N$, $\alpha_i = j$ und x_j der Inhalt des Registers R_j, so deute

die	Zahl	0	die	Operation	$x_j \Rightarrow x_j+1$,
"	"	1	"	"	$x_j \Rightarrow x_j \dot{-} 1$,
"	"	2	"	"	$x_j \Rightarrow x_j$,
"	"	3	"	"	Stop an.

$\varphi_0^-,\ldots,\varphi_N^-$, $\varphi_0^+,\ldots,\varphi_N^+$ sind als Nummern von Instruktionen natürliche Zahlen. Schliesslich sei $r := \underset{n\leq N}{\mathrm{Max}}\, \alpha_n$. r heisse "Rang" des Programms. Das Programm soll nun durch die Zahl p in der folgenden Weise kodifiziert werden:

$$p := \langle N, 3^r, \langle\alpha_0,\ldots,\alpha_N\rangle, \langle\omega_0,\ldots,\omega_N\rangle, \langle\varphi_0^-,\ldots,\varphi_N^-\rangle, \langle\varphi_0^+,\ldots,\varphi_N^+\rangle\rangle \; .$$

32.) <u>Def.</u>:

$Pr(x) :\leftrightarrow x$ ist Kodifikation eines Programms.

<u>Beh.</u>:

$\chi_{Pr} \in \mathcal{T}$

<u>Bew.</u>:

$$
\begin{aligned}
Pr(x) \leftrightarrow x = \langle (x)_o, \ldots, (x)_5 \rangle \\
\wedge \bigvee_{y\leq x} (x)_1 = 3_x^{((x)_2)_y} \\
\wedge \bigwedge_{y\leq x} ((y > (x)_o \rightarrow ((x)_2)_y = \ldots = ((x)_5)_y = 0) \\
\wedge (x)_1 \geq 3_x^{((x)_2)_y} \\
\wedge ((x)_3)_y \leq 3 \\
\wedge ((x)_4)_y \leq (x)_o \\
\wedge ((x)_5)_y \leq (x)_o),
\end{aligned}
$$
q.e.d

33.) Falls p, wie in [32] angegeben, die Kodifikation eines Programms, i die Adresse einer Instruktion dieses Programms und $x_o, \ldots, x_n$ eine Beschriftung der Speicher der durch p dargestellten Maschine ist, so soll die Zahl $k := \langle p, i, x_o, \ldots, x_n \rangle$ eine "Konfiguration" heissen; ad hoc sei definiert:

$$N(p) := (p)_o,$$
$$r(p) := ((p)_1)_1,$$
$$p(k) := (k)_o,$$

$$\begin{aligned} i(k) &:= (k)_1, \\ \alpha(k) &:= ((p(k))_2)_{i(k)}, \\ \omega(k) &:= ((p(k))_3)_{i(k)}, \\ \varphi^{=}(k) &:= ((p(k))_4)_{i(k)}, \\ \varphi^{\neq}(k) &:= ((p(k))_5)_{i(k)}. \end{aligned}$$

Sei $L(x)$ das grösste i mit $(x)_i \neq 0$. Aus

$$L(x) = \mu_{i \leq x} \left((x)_i \neq 0 \wedge \bigwedge_{j \leq x} (j > i \rightarrow (x)_j = 0) \right)$$

ergibt sich, dass L in $\mathfrak{F}$ liegt. Offenbar liegen auch die soeben definierten Funktionen $\lambda x N(x), \ldots, \lambda x\ \varphi^{\neq}(x)$ sämtlich in $\mathfrak{F}$.

Setzt man

$$\mathrm{Konf.}(k) :\leftrightarrow \mathrm{Pr}(p(k)) \wedge i(k) \leq N(p(k)) \wedge L(k) \leq r(p(k)) + 2,$$

so wird durch das Prädikat "Konf" (aus $\mathfrak{F}$) die Eigenschaft einer Zahl beschrieben, Konfiguration zu sein.

Die Relation R_F soll zwischen Zahlen k_1, k_2 dann vorliegen, wenn beide Konfigurationen sind und k_2 die Konfiguration ist, die sich nach einem Rechenschritt der durch $p(k_1)$ dargestellten Maschine aus k_1 ergibt. ("Folgekonfiguration"):

$$R_F(k_1, k_2) :\leftrightarrow$$

$$\mathrm{Konf}(k_1) \wedge \mathrm{Konf}(k_2) \wedge \bigwedge_{j \leq r(p(k_1))+2} (j \neq 1 \wedge j \neq \alpha(k_1)+2 \rightarrow (k_1)_j = (k_2)_j)$$

$$\begin{array}{lllllllll}
\wedge\ ((\omega(k_1) & = 0 \wedge & (k_1)_{\alpha(k_1)+2} & = 0 \wedge & (k_2)_{\alpha(k_1)+2} & = & (k_1)_{\alpha(k_1)+2} & +1 \wedge & i(k_2) = \varphi^{=}(k_1)) \\
\quad \vee (\ '' & = 0 \wedge & '' & \neq 0 \wedge & '' & = & '' & +1 \wedge & '' = \varphi^{\neq}(k_1)) \\
\quad \vee (\ '' & = 1 \wedge & '' & = 0 \wedge & '' & = & '' & \dot{-}1 \wedge & '' = \varphi^{=}(k_1)) \\
\quad \vee (\ '' & = 1 \wedge & '' & \neq 0 \wedge & '' & = & '' & \dot{-}1 \wedge & '' = \varphi^{\neq}(k_1)) \\
\quad \vee (\ '' & = 2 \wedge & '' & = 0 \wedge & '' & = & '' & \wedge & '' = \varphi^{=}(k_1)) \\
\quad \vee (\ '' & = 2 \wedge & '' & \neq 0 \wedge & '' & = & '' & \wedge & '' = \varphi^{\neq}(k_1)) \\
\quad \vee (\ '' & = 3 \wedge & k_1 = k_2)) & & & & & &
\end{array}$$

Offenbar liegt R_F in $\mathcal{T}$. Ferner gilt (vgl.[30 i.)]):

$$R_F(k_1,k_2) \rightarrow k_2 \leq (1+k_1)\cdot 3^{\,r(p(k_1))+3}\cdot(1+N(p(k_1)))^2\cdot 4$$

$$\leq Q\,(k_1)$$

für ein geeignetes Polynom Q. Setzt man

$$F(k) := \mu x_{x\leq Q(k)}\; R_F(k,x)\;,$$

so liegt F in $\mathcal{T}$, und F(k) stellt für Konfigurationen k die Folgekonfiguration von k dar.

34.) Es soll nun eine dreistellige Funktion K eingeführt werden, für welche gilt: Ist p die Kodifikation eines Programms, so ist K(p,x,y) die Konfiguration der durch p dargestellten Maschine im y-ten Rechenschritt, falls zu Anfang der Rechnung der Speicher mit der Adresse 0 die Zahl x trug und alle andern Speicher leer waren. Die folgende Definition leistet das Verlangte:

$$K(p,x,o) := \begin{cases} \langle p,0,x\rangle\;, & \text{falls } Pr(p) \\ 0 & \text{sonst} \end{cases}$$

$$K(p,x,y+1) := F(K(p,x,y)).$$

Für K ergibt sich nun die folgende Abschätzung (vgl.[30]):

$$K(p,x,y) \leq \langle \underbrace{p,N(p),x+y,y\ldots,y}_{r(p)+1} \rangle$$

$$\leq 3^{r(p)+3}\cdot(p+1)^2\cdot(N(p)+1)^2\cdot(x+y+1)^2\cdot(y+1)^{2r(p)}$$

Die Funktion K liegt nicht in $\mathcal{T}$. Aus der Abschätzung ergibt sich aber, dass für jedes r_o die Funktion K_{r_o} mit

$$K_{r_o}(p,x,y) := \begin{cases} K(p,x,y) & \text{für } r(p)\leq r_o \\ 0 & \text{sonst} \end{cases}$$

in $\mathcal{T}$ liegt.

35.) Es sei nun eine einstellige Funktion f gegeben. M_o berechne f. Die zugehörige Schrittzahlfunktion sei s_f. p sei die Kodifikation des Programms von M_o gemäss [34]. Da F (aus [34]) in $\mathfrak{T}$ liegt, existiert eine polynomiale Maschine M_F aus $\mathfrak{M}$ mit

$$r_{M_F}(\langle x,p,i,k_1,k_2,z\rangle)$$

$$= \langle x,p,i+1,F(k_1),\ F(k_2),\ |F(k_1) - F(k_2)|\ \rangle.$$

Es folgt

$$r^y_{M_F}(\langle x,p,0,K(p,x,0),\ K(p,x,1),z\rangle)$$

$$= \langle x,p,y,K(p,x,y),K(p,x,y+1),|K(p,x,y)-K(p,x,y+1)|\ \rangle$$

und daraus

$$r_{(M_F)_5}(\langle\ x,p,0,K(p,x,0),K(p,x,1),z\rangle)$$

$$= \langle x,p,s_f(x),K(p,x,s_f(x)),K(p,x,s_f(x)+1),\ 0\ \rangle.$$

Ferner gilt (vgl.[25]):

$$s_{(M_F)_5}(u) \leq 2 + \sum_{j<s_f(x)} (s_{M_F}(r^j_{M_F}(u))+1)$$

für $u = \langle x,p,0,K(p,x,0),K(p,x,1),z\ \rangle$.

Daraus folgt

$$s_{(M_F)_5}(u)$$

$$\leq 2 + \sum_{j<s_f(x)} (s_{M_F}(\langle x,p,j,K(p,x,j),K(p,x,j+1),$$

$$|K(p,x,j)-K(p,x,j+1)|\rangle)+1)$$

$\leq Q_1(x,s_f(x))$ für ein geeignetes Polynom Q_1.

Es gibt polynomiale Maschinen M_1, M_2 aus $\mathfrak{M}$ mit

$$r_{M_1}(\langle x\rangle) = \langle x,p,0,K(p,x,0),K(p,x,1),|K(p,x,0)-K(p,x,1)|\ \rangle$$

und

$$r_{M_2}(\langle x,p,y,k_1,k_2,z\rangle) = \langle x,(k_1)_3\rangle .$$

Sei $M := M_1(M_F)_5M_2$. M berechnet ebenfalls die Funktion f, und zwar mit einer Schrittzahlfunktion $s_f^!$, für die sich aus der oben angegebenen Abschätzung für $s_{(M_F)_5}$ eine polynomiale Abschätzung $s_f^!(x)\leq Q(x,s_f(x))$ ergibt. Andererseits benötigt M nur eine (unabhängig von f) beschränkte Zahl von Registern. r_o sei die grösste im Programm p' von M auftretende Adresse eines Registers. r_o ist unabhängig von f. Aus $r(p')\leq r_o$ folgt

$$K(p',x,y) = K_{r_o}(p',x,y)$$

und daraus

$$f(x)=(K_{r_o}(p',x,s_f^!(x)))_3.$$

Sei $U := \lambda pxy(K_{r_o}(p,x,y))_3$; es folgt

$$\bigwedge_x \bigwedge_{y\geq Q(x,s_f(x))} f(x)=U(p',x,y) ,$$

und die Funktion U liegt in $\mathcal{T}$. Damit ist die in [29] angekündigte Konstruktion von U realisiert, der Darstellungssatz aus [23] also bewiesen.

36.) Aus dem soeben bewiesenen Darstellungssatz sollen noch zwei Folgerungen gezogen werden:

a.) Sei f eine rekursive Funktion (beliebiger Stellenzahl n+1). M_1 berechne f. Die zugehörige Schrittzahlfunktion sei s_f. Zu M_1 erhält man durch Umindizierung der Register eine Maschine M_2 mit

$$r_{M_2}(\langle y,0,\mathfrak{x}\rangle) = \langle y,f(\mathfrak{x}),\mathfrak{x}\rangle$$

und

$$s_{M_2}(\langle y,0,\mathfrak{x}\rangle)=s_f(\mathfrak{x}).$$

Wegen $\lambda xy(x)_y \in \mathcal{T}$ gibt es polynomiale Maschinen M_3, M_4 mit

$$r_{M_3}(\langle x\rangle) = \langle x,0,(x)_o,\dots,(x)_n\rangle$$

und

$$r_{M_4}(\langle x,y,\mathfrak{x}\rangle) = \langle x,y\rangle$$

Sei $M := M_3M_2M_4$

und

$$h := \lambda xf((x)_o,\dots,(x)_n).$$

Dann berechnet M die Funktion h, und für die zugehörige Schrittzahlfunktion s_h gilt die Abschätzung

$$s_h(x) \leq Q_1(x,s_f((x)_o,\dots,(x)_n))$$

für ein geeignetes Polynom Q_1.

Zu h existiert nun gemäss [35] für passende p,Q eine Darstellung

$$h(x) = U(p,x,Q(x,s_h(x)))$$

$$= U(p,x,Q(x,Q_1(x,s_f((x)_o,\dots,(x)_n)))).$$

Mit $g := \lambda\mathfrak{x}yU(p,\langle\mathfrak{x}\rangle,Q(\langle\mathfrak{x}\rangle,Q_1(\langle\mathfrak{x}\rangle,y))))$

folgt $f(\mathfrak{x}) = g(\mathfrak{x},s_f(\mathfrak{x}))$, $g \in \mathcal{T}$,

also $f \in A(g,s_f)$. Das wurde in [22] behauptet.

b.) Wendet man (a) auf Funktionen aus $\mathcal{T}_P$ an, so ergibt sich, weil diese Funktionen sämtlich durch Polynome majorisiert werden können, die Inklusion

$\mathcal{T}_P \subseteq \mathcal{T}$, die in [29] behauptet wurde. Es gilt also

$$\mathcal{T} = \mathcal{T}_P^* = \mathcal{T}_P \ .$$

37.) Es soll nun eine Hierarchie von Funktionenklassen $\mathcal{T}_n$ betrachtet werden, welche von $\mathcal{T}$ ausgehend gegen $\mathcal{E}$ "konvergiert", d.h. für welche gilt:

$$\mathcal{T} \subseteq \mathcal{T}_n \subset \mathcal{T}_{n+1}, \quad \bigcup_n \mathcal{T}_n = \mathcal{E} .$$

Dazu sei an die Definition der Funktionen ϑ_n in [14] erinnert.

<u>Definition</u>:

$\mathcal{T}_n$ sei die Klasse der Funktionen f, zu denen es eine Schrittzahlfunktion s_f und ein Polynom Q gibt, derart dass gilt:

$$\bigwedge_{\mathfrak{x}} s_f(\mathfrak{x}) \leq \vartheta_n(Q(\mathfrak{x})) .$$

Aus dieser Definition ergibt sich unmittelbar:

a.) $\mathcal{T}_o = \mathcal{T}$

b.) $\mathcal{T}_n \subseteq \mathcal{T}_{n+1}$

c.) Nach [36] gibt es zu f aus $\mathcal{T}_n$ Polynome P,Q und eine Zahl p, so dass für alle $\mathfrak{x}$ die Identität

$f(\mathfrak{x}) = U(p, \langle\mathfrak{x}\rangle, P(\vartheta_n(Q(\mathfrak{x}))))$ gilt. Zu P und Q existiert ferner ein Polynom R mit

$$P(\vartheta_n(Q(\mathfrak{x}))) \leq \vartheta_n(R(\mathfrak{x}))$$

Daher besitzt $\mathcal{T}_n$ die folgende Darstellung:

$$\mathcal{T}_n = \{f : \bigvee_p \bigvee_{Q \text{Pol.}} \bigwedge_{\mathfrak{x}} f(\mathfrak{x}) = U(p, \langle\mathfrak{x}\rangle, \vartheta_n(Q(\mathfrak{x})))\} .$$

d.) Da alle Funktionen f aus $\mathcal{E}$ eine elementare Schrittzahlfunktion s_f besitzen und umgekehrt aus $s_f \in \mathcal{E}$ stets $f \in \mathcal{E}$ folgt, gilt

$$\bigcup_n \mathcal{T}_n = \mathcal{E} ,$$

denn zu jeder Funktion $g \in \mathcal{E}$ gibt es ein n und ein Polynom Q mit

$$\bigwedge_{\mathfrak{x}} g(\mathfrak{x}) \leq \vartheta_n(Q(\mathfrak{x})) .$$

38.) Die Klassen $\mathcal{T}_n$ sind nicht mehr abgeschlossen gegen Einsetzungen und beschränkte Rekursionen, jedoch lassen sich für beide Definitionsschemata Schranken für die "Reichweite" angeben:

a.) Falls $g \in \mathcal{T}_n$.

$h_o \in \mathcal{T}_{i_o}, \ldots, h_r \in \mathcal{T}_{i_r}$,

$f = \lambda \mathfrak{x} g(h_o(\mathfrak{x}), \ldots, h_r(\mathfrak{x}))$,

so liegt f in $\mathcal{T}_{n+\mathrm{Max}(i_o, \ldots, i_r)}$.

Beweis: Aus Maschinen, welche die Funktionen $g, h_o, \ldots, h_r$ berechnen (mit Schrittzahlen, die majorisierbar sind durch $\vartheta_n, \vartheta_{i_o}, \ldots \vartheta_{i_r}$) lässt sich eine Maschine zur Berechnung von f konstruieren, deren Schrittzahl majorisierbar ist durch $n+\mathrm{Max}(i_o, \ldots, i_r)$.

b.) Falls

$f(\mathfrak{x}, 0) = g_1(\mathfrak{x})$,

$f(\mathfrak{x}, y+1) = g_2(\mathfrak{x}, y, f(\mathfrak{x}, y))$,

$f(\mathfrak{x}, y) \leq g_3(\mathfrak{x}, y)$,

$g_1, g_2 \in \mathcal{T}_i$,

$g_3 \in \mathcal{T}_n$,

so liegt f in $\mathcal{T}_{n+i}$.

Auch diese Behauptung lässt sich durch eine geeignete Konstruktion der Maschine, die f berechnet, zeigen.

39.) Beh.: $\mathcal{T}_n \subset \mathcal{T}_{n+1}$

Bew.: Zu beliebigem n sei

$$U^*_{n+1}(k,x) := U((k)_o, \langle x \rangle, \vartheta_{n+1}((k)_1 \cdot (x+1)))$$

Es genügt, die folgende Aussage zu beweisen:

<u>Beh.1</u>: Für beliebige einstellige Funktionen f aus $\mathcal{T}_n$ gilt

$$\bigvee_k \bigwedge_x f(x) = U^*_{n+1}(k,x).$$

<u>Bew.1</u>:

Sei f aus $\mathcal{T}_n$. p sei gemäss [37 c.)] gewählt.

Dann gilt

$$f(x)=U(p,\langle x\rangle, \vartheta_n(Q(x))).$$

Zu dem Polynom Q gibt es eine Zahl q mit

$$Q(x)\leq 2^{q(x+1)} .$$

Sei k := $\langle p,q \rangle$,

dann gilt in der Tat für beliebige x:

$$f(x) = U^*_{n+1}(k,x) ,$$

q.e.d.

Als <u>Korollar</u> ergibt sich, dass die Funktion

$$\lambda pxy\ (1 \dot{-} U(p,\langle x\rangle, \vartheta_{n+1}(y)))$$

nicht in $\mathcal{T}_n$ liegt.

40.) Die in [38] angegebenen Schranken für die Reichweite der Einsetzung und der beschränkten Rekursion sind in dem folgenden Sinne optimal:

a.) Die Funktion $\lambda x \vartheta_n(\vartheta_i(x))$ liegt in $\mathcal{T}_{n+i}$, aber nicht in $\mathcal{T}_{n+i-1}$.

b.) Falls $i>0$, $n\geq 0$, so gibt es Funktionen g_1, g_2, g_3 mit $g_1, g_2 \in \mathcal{T}_i$, $g_3 \in \mathcal{T}_n$, derart dass die aus ihnen durch beschränkte Rekursion definierte Funktion f nicht in $\mathcal{T}_{n+i-1}$ liegt:

Sei $h(y,z) := \vartheta_{Min(z,n)}(y)$,

$$f(p,x,y,z) := \langle h(y,z+1),\ 1 \dot{-} U(p,\langle x\rangle, \vartheta_i(h(y,z)))\rangle ,$$

$$g(z,u) := \begin{cases} 2^u & \text{, falls } z<n \\ u & \text{sonst} \end{cases}$$

$$g_1(p,x,y) := \langle h(y,1), 1 \dot{-} U(p, \langle x \rangle, \vartheta_i(y)) \rangle,$$

$$g_2(p,x,y,z,u) := \langle g(z,(u)_o), 1 \dot{-} U(p, \langle x \rangle, \vartheta_i((u)_o)) \rangle,$$

$$g_3(p,x,y,z) := \langle \vartheta_n(y), 1 \rangle.$$

Dann gilt:

$$g,\ \lambda y h(y,1) \in \mathcal{T}_1 \subseteq \mathcal{T}_i, \text{ und daher } g_1, g_2 \in \mathcal{T}_i,$$

$$g_3 \in \mathcal{T}_n,$$

$$f(p,x,y,0) = g_1(p,x,y),$$

$$f(p,x,y,z+1) = g_2(p,x,y,z,f(p,x,y,z)),$$

$$f(p,x,y,z) \leq g_3(p,x,y,z).$$

f ist also aus g_1, g_2, g_3 mit beschränkter Rekursion definiert.

Andererseits gilt

$$1 \dot{-} U(p, \langle x \rangle, \vartheta_{i+n}(y)) = (f(p,x,y,n))_1 ,$$

so dass (vgl. das Korollar in [39]) f nicht in $\mathcal{T}_{i+n-1}$ liegen kann, q.e.d.

Literatur

Die hier vorausgesetzten Grundlagen der Theorie der rekursiven, primitiv-rekursiven und elementaren Funktionen finden sich in den folgenden Lehrbüchern:

A. Grzegorczyk. Fonctions récursives. Paris, Louvain 1961

H. Hermes. Aufzählbarkeit, Entscheidbarkeit, Berechenbarkeit. Berlin, Göttingen, Heidelberg 1961

S.C. Kleene, Introduction to metamathematics. Amsterdam 1952

R. Péter. Rekursive Funktionen. Budapest 1951

Ein grosser Teil der dargestellten Methoden und Ergebnisse wurde den folgenden Arbeiten entnommen:

P. Axt, Enumeration and the Grzegorczyk hierarchy. Zeitschrift math. Logik und Grundlagen der Math., Bd.9 (1963), S.53-65

A. Grzegorczyk, Some classes of recursive functions. Rozprawy Matematyczne IV, Warschau 1953

W. Heinermann. Untersuchungen über die Rekursionszahlen rekursiver Funktionen. Dissertation, Münster 1961

S.C. Kleene, Extension of an effectively generated class of functions by enumeration. Coll. Math., Bd.VI (1958), S.67-78

M.L. Minsky, Recursive unsolvability of Post's problem of "Tag" and other topics in the theory of Turing machines. Annals of Math.,Bd.74 (1961),S.437-455

D.M. Ritchie, Complexity classification of primitive recursive functions by their machine programs. Notices of the Amer.Math.Soc.,Bd.12 (1965),S.343

R.W.Ritchie, Classes of predictably computable functions. Transactions of the Amer.Math.Soc.,Bd.106 (1963), S.139-173

D. Rödding, Darstellungen der (im Kalmar-Csillag'schen Sinne) elementaren Funktionen. Archiv für math.Logik u. Grundlagenforschung, Bd.7 (1962), S. 139-158

D. Rödding, Über Darstellungen der elementaren Funktionen II. Archiv für math. Logik und Grundlagenforschung, Bd.9 (1964), S.36-48

H. Schwichtenberg, Rekursionszahlen und die Grzegorczyk-Hierarchie. Erscheint im Archiv für math. Logik und Grundlagenforschung.

J.C.Shepherdson und H.E. Sturgis, Computability of recursive functions. Journal of the Ass. for Computing Machinery, Bd.10 (1963), S.217-255

HYPERARITHMETIC ULTRAFILTERS

by J. P. Cleave

§1. Introduction

This chapter is a contribution to a predicative treatment of model theory to be compared with the treatment of classical analysis by Weyl (18) and Grzegorczyk (5) (see also Mostowski (13), Feferman (1)). In view of the well-known identification of "predicative" with "hyperarithmetic" (Kreisel (9)), our work uses techniques well-known in the theory of recursive functions & the hyperarithmetic hierarchy.

A major role is played by ultrafilters in many model-theoretic constructions. Accordingly we commence with two constructions of hyperarithmetic ultrafilters. Applications of this work to ultraproducts and saturated models will appear elsewhere.

A modest familiarity with the notion of "hyperarithmetic" is assumed. Exact definitions can be found in Kleene (7), Mostowski (12). Briefly, a set $\underline{0}$ of ordinal notations is defined (Kleene (6)). $\underline{0}$ is a set of natural numbers and is the field of a well-founded partial ordering $\leq_0$. With each number a in $\underline{0}$ is associated a set H_a of natural numbers. An object (set, predicate, etc) is hyperarithmetical if it is recursive in H_a for some $a \in \underline{0}$.

Definition 1.

Let $a \in \underline{0}$. B (a). (B (a)) is the collection of sets recursive (partial recursive, resp.) in H_b for some $b \leq_0 a$.

It is well-known (Spector (17)) that H_a, H_b are recursive in each other if the ordinals $|a|$, $|b|$ denoted by a,b respectively are equal. Hence

$$|a| = |b| \rightarrow B(a) = B(b) \ \& \ B^{\dagger}(a) = B^{\dagger}(b). \tag{1}$$

Further, if $a \leq_0 b$ then H_a is recursive in H_b, but H_b is not recursive in H_a. Hence

$$|a| < |b| \rightarrow B(a) \subset B(b) \ \& \ B^{\dagger}(a) \subset B^{\dagger}(b) \tag{2}$$

From (1), (2) it follows that the collection $H = \{B\ (a); a \in \underline{0}\}$ is well ordered by inclusion, of order-type ω_1 (least non-constructive ordinal). H is the hierarchy of hyperarithmetic sets.

Writing "$0_o = 1$", "$(n+1)_o = 2^{n_o}$" we have $n_o \in \underline{0}$ and $|n_o| = n$. Then $B(0_o)$ = all recursive sets,

$\bigcup_n B(n_o)$ = all arithmetically definable sets.

It is quite easy to see that for each $a \in \underline{0}$, B(a) is a field of (closed under (finite) union, intersection and complementation), and so Boolean algebra.

Definition 2.

Let $\underline{B} = \langle B; 0, 1, \cup, \cap, \bar{\ } \rangle$ be a Boolean algebra with zero element 0. A collection U of elements of $\underline{B}$ is an ultrafilter in $\underline{B}$ if

(i) $0 \notin U$
(ii) $x, y \in U \rightarrow x \cap y \in U$
(iii) $x \in U \,\&\, y \in B \rightarrow x \cup y \in U$
(iv) $x \in U \equiv \bar{x} \notin U$

An ultrafilter U is a principal ultrafilter if for some x in B, $U = \{y;\ y \in B \,\&\, y \geq x\}$.

(Note. If $\underline{B}$ is a field sets of natural numbers containing all finite sets then an ultrafilter in $\underline{B}$ containing all cofinite sets is non-principal).

Our aim is to construct ultrafilters in the fields B (a), $a \in \underline{O}$. Two methods of construction are presented - a direct method using maximal sets, and an inductive procedure, based upon a theorem concerning paths in finitely branched trees, giving "graded" ultrafilters.

§ 2. Indexing Hyperarithmetic Objects

2.1 Ordinal notations.

$\underline{O}$ plays a major role in hyperarithmetic constructions. For many purposes it is useful to use subsets Z of $\underline{O}$ such that no two members of Z denote the same ordinal number. Following Feferman, Spector (2) we make

Definition 3.

Z is a path within $\underline{O}$ if

(1) $Z \subset \underline{O}$
(2) $c, d \in Z \longrightarrow c \leq_O d \vee d \leq_O c$
(3) $c \in Z \,\&\, d \leq_O c \longrightarrow d \in Z$.

Z is a path through $\underline{O}$ if the order type of Z is ω_1.

In the notation of (6), for $a \in \underline{O}$, $C(a) = \{b;\ b \leq_O a\}$ is a recursively enumerable path within $\underline{O}$. It is shown, in (2) (theorem 3.7), (6) that there exists a Π^1_1 path through $\underline{O}$.

Also note that $\underline{O}$ has a recursive seccessor function; if $a \in \underline{O}$ then $2^a \in \underline{O}$ and $|2^a| = |a| + 1$. "a* " will sometimes be used instead of "2^a". We also define:

$$\dot{a} = \begin{cases} (\min c < a)(a = 2^c) & \text{if } (Ec < a)(a = 2^c) \\ a & \text{otherwise} \end{cases}$$

$$\ddot{a} = \begin{cases} a^* & \text{if } (Ec < a)(a = 2^c) \\ a & \text{otherwise.} \end{cases}$$

Thus if $a \in \underline{O}$ then $\dot{a}, \ddot{a} \in \underline{O}$ and

1) $|a| = |b|+1 \rightarrow |\dot{a}| = |b|, |\ddot{a}| = |a|+1$

2) $|a|$ a limit ordinal $\rightarrow |\dot{a}| = |\ddot{a}| = |a|$.

2.2 Recursive function Theory.

To effect our constructions well-known methods of recursive function theory will be used; the following notation, mostly from (8) will be employed.

$\langle x_0 \cdots x_n \rangle$ denotes the Gödel-number of the sequence $(x_0 \cdots x_n)$ of natural numbers.

$$\langle x_0 \cdots x_n \rangle = \prod_{i \leq n} p_i^{x_i+1}$$

where p_i = i + 1.st prime number. Functions

$(a)_i$, lh (a) are as in (8) p230. The components of a sequence can be recovered from the Godel-number by the function

$\underline{i} \cdot w = (w)_i \dot{-} 1.$

Thus, if $w = \langle x_0 \ldots x_n \rangle$, $i \leq n$ then $\underline{i} \cdot w = x_i$. Also, as in (8) p340, " $z(x_0 \ldots x_n)$" is an abbreviation for "$U(\min y\, T_n(z, x_0 \ldots x_n, y))$". The symbols "$\simeq$", "$\cong$" denote complete equality between partial functions and complete equivalence between partially defined predicates, respectively.

2.3. Hyperarithmetic Objects

The fundamental objects we shall use are the partial hyperarithmetic functions. Such a function f can be defined by specifying two numbers e, a where $a \in \underline{O}$ and e is the Gödel-number of the recursive procedure for computing f from H_a. Thus for each partial hyperarithmetic function f, there exists a number e and $a \in \underline{O}$ such that

$f(x_1 \ldots x_n) \simeq U(\min y\, T_n^{Ha}(e, x_1 \ldots x_n, y))$.

and we say that $\langle e,a \rangle$ is an <u>index</u> (or an <u>a-index</u> if we wish to specify the ordinal concerned) of f: letting $w = \langle e,a \rangle$ we shall use "$w : (x_1 \ldots x_n)$" as an abbreviation of "$U(\min y\, T_n^{Ha}(e, x_1 \ldots x_n, y))$".

Hyperarithmetic sets can be defined as the sets of zeros of partial hyperarithmetic functions: if $w = \langle e,a \rangle$, $a \in \underline{O}$ we write

$M(w) = \{x ;\ w : (x) = o\}$.

Hyperarithmetic relations can be defined by means of their graphs and indexed accordingly.

Thus if R is an n-ary relation, then w is an index of R if

$$M(w) = \{\langle x_1 \ldots x_n\rangle ;\ (x_1 \cdots x_n) \in R\}.$$

Constructions involving these basic objects are defined in terms of the indices of the objects involved: Thus a set U of sets is hyperarithmetic if there exists an index w such that

$$U = \{M(u);\quad u \in M(w)\}.$$

In this sense we can speak of hyperarithmetic ultrafilters of B (a).

Paths in $\underline{O}$ will be used for indexing our constructions: if Z is a path then

$$\text{Ind}(Z) = \{\langle e,a\rangle :\ a \in Z,\quad e = 0,1,2 \ldots\}$$

Where $Z = C(a)$, $a \in \underline{O}$ we write "Ind (a)" instead of "Ind (C(a))". Thus

$$\text{Ind}(a) = \{\langle e,b\rangle ;\ b \leq_O a,\ e = 0,1,2 \ldots\}.$$

We remark that Ind(a) is recursively enumerable.

An object is said to be a-(partial) recursive if it is (partial) recursive in H_a: $|a|$ is regarded as a measure of the complexity of the object.

A result needed later is

<u>Lemma 1.</u>

Let $b \in \underline{O}$. Let $v(x)$ be a b-recursive function and suppose that $(n)(v(n) \in \text{Ind}(b))$. Then the predicate P(x), where

$$P(x) \equiv (Ey)(\langle v(x)\rangle : (y) = 0)$$

is b-recursive.

§3. Ultrafilters and Maximal Sets.

A connection between maximal simple sets (Friedberg (3)) and ultrafilters was first observed by Tennenbaum (16). A construction of a maximal set is given in (3). Exploiting the analogy "recursively enumerable" $\sim \Pi_1^1$ and "recursive $\sim$ hyperarithmetic", a Π_1^1 maximal set is constructed in (10). A weaker notion than "maximal" is "r-maximal". Lachlan (11) has constructed an r-maximal set which is not maximal (Also (15)). Presumably the techniques of meta-recursion theory allow one to construct a Π_1^1-r-maximal set which is not Π_1^1-maximal.

Definitions of maximal and r-maximal, relativised to an arbitrary level of the hyperarithmetic hierarchy are given in

<u>Definition 4.</u>

Let $a \in \underline{O}$. A set M is <u>a-maximal</u> (a-r-maximal, resp.) if

(i) M is recursively enumerable in H_a.

(ii) M is infinite, and

(iii) for all sets R recursively enumerable in H_a (recursive in H_a resp). either $R \cap \overline{M}$ is finite or $\overline{R} \cap \overline{M}$ is finite.

Tennenbaum's result relativised is

Theorem 2.

Let $a \in \underline{O}$. M an a-maximal set. Then
$U(M) = \{ R \in B(a^*);\ R \cup M \text{ cofinite} \}$
is a non-principal ultrafilter in $B(a^*)$.

This theorem can be derived from the stronger

Theorem 3.

Let $a \in \underline{O}$, M recursively enumerable in H_a. Then
$U(M) = \{ R \in B(a^*);\ R \cup M \text{ cofinite} \}$
is a non-principal ultrafilter in B (a^*) if and only if M is a-r-maximal.

Proof (A) Suppose M is a-r-maximal. Then $\overline{M}$ is infinite. Hence $\phi \cup M$ is not cofinite.
Hence

$$\phi \notin U(M) \tag{3}$$

$U(M)$ is non-empty since N, the set of natural numbers is obviously in $U(M)$. Let $R, S \in U(M)$. Then $R, S \in B(a^*)$ and $R \cup M$, $S \cup M$ cofinite. Hence $R \cap S \in B(a^*)$ and $(R \cup M) \cap (S \cup M)$ cofinite. But $(R \cup M) \cap (S \cup M) = (R \cap S) \cup M$. Hence

$$R, S \in U(M) \longrightarrow R \cap S \in U(M) \tag{4}$$

Let $R \in U(M)$, $S \in B(a^*)$. Then $R \cup S \in B(a^*)$ and $R \cup M$ cofinite. Hence $(R \cup S) \cup M$ cofinite. Hence $R \cup S \in U(M)$.
Thus

$$R \in U(M),\ S \in B(a^*) \longrightarrow R \cup S \in U(M) \tag{5}$$

Now let $R \in B(a^*)$. Suppose $R \in U(M)$. Then $R \cup M$ is cofinite. Hence since $\overline{M}$ is infinite, $\overline{M} \cap (R \cup M) = \overline{M \cup R}$ is infinite. Thus $M \cup \overline{R}$ is not cofinite so that $\overline{R} \notin U(M)$.

Conversely suppose $\overline{R} \notin U(M)$. Then $R \cap \overline{M}$ is infinite. Now $R \in B(a^*)$ and so R is recursive in H_a. Hence since M is a-r-maximal, $\overline{R} \cap \overline{M}$ is finite. Thus $R \cup M$ is cofinite. Hence $R \in U(M)$. So we have proved

$$R \in U(M) \equiv \overline{R} \notin U(M) \tag{6}$$

It follows ffom (3)-(6) that $U(M)$ is an ultrafilter in $B(a^*)$. To prove that $U(M)$ is non-principal we observe that cofinite sets are recursive and so are in $B(a^*)$ and thus $U(M)$ contains all cofinite sets.

(B) Suppose $U(M)$ is a non-principal ultrafilter in $B(a^*)$.

First $\overline{M}$ is infinite. For suppose to the contrary that $\overline{M}$ is finite. Then $\phi \cup M$ is cofinite. But $\emptyset$ is recursive and hence $\emptyset \in B(a^*)$. Hence $\emptyset \in U(M)$. This contradicts the supposition that $U(M)$ is an ultrafilter.

Next let $R \in B(a^*)$. Then R is recursive in H_a.

Hence

$\overline{R \cup M}$ infinite $\equiv R \notin U(M) \equiv \overline{R} \in U(M) \equiv \overline{R} \cup M$ cofinite

Thus

$\overline{R} \cap \overline{M}$ infinite $\equiv R \cap \overline{M}$ finite.

Hence either $\overline{R} \cap \overline{M}$ is finite, or $R \cap \overline{M}'$ is finite.

Thus M is a-r-maximal. qed.

Ultrafilters derived in the above manner from r-maximal sets are quite complicated. One can easily prove that if M is a-r-maximal then $\{w;\ w \in \mathrm{Ind}(a)\ \&\ M(w) \in U(M)\}$ is recursive in H_{a+4_0}. Since $U(M)$ is an ultrafilter in $B(a^*)$, if $b \leq_0 a$

$B(b^*) \cap U(M)$ is an ultrafilter in $B(b^*)$. However, we cannot get a better estimate of the complexity of $\{w;\ w \in \mathrm{Ind}(b)\ \&\ M(w) \in B(b^*) \cap U(M)\}$ than $a+4_0$. Our aim in the rest of this section is to construct ultrafilters U in $B(a^*)$ such that the ultrafilters $B(b^*) \cap U$ in $B(b^*)$ for $b \leq_0 a$ are recursive in H_c for $|c|$ less than the complexity of U. For this purpose we introduce

§ 4. Graded Ultrafilters

Definition 5.

Let Z be a path in <u>O</u>.

(i) A subset V of Ind(Z) is Z-graded if there exists a partial recursive function g - the <u>grading function</u> - defined in Z such that for all $b \in Z$, $g(b)$ is a recursive index of $\mathrm{Ind}(b) \cap V$ and $|\underline{1}.g(b)| < |b| + \omega$ if $|b|$ is a successor ordinal, and $\leq |b|$, if $|b|$ is a limit ordinal.

(ii) A set U of sets is Z-graded if every member of U has an index in Ind(Z) and $\{w; M(w) \in U\}$ is Z-graded.

Thus if U is a Z-graded family of sets and $b \in Z$ then the problem of deciding whether $M(\langle e, b\rangle) \in U$ is only a finite number of recursive jumps away from H_b

$$M(w) \in U \equiv \underline{1}.w \in Z\ \&\ w \in M(g(d)),$$

where d is any member of Z such that $d \geq_0 \underline{1}.w$.

Our construction of graded ultrafilters is based on the relation between ultrafilters and families of sets which are maximal with respect to the finite intersection property.

Definition 6.

(i) A family L of sets has the finite intersection property (FIP) if every finite collection of sets in L has non-zero intersection

(ii) A set L of indices has FIP if $\{M(w);\ w \in L\}$ has FIP.

Note (i) If $L = (m_0, m_1, \ldots)$ is a collection of sets then L has FIP if and only if for all n, $\bigcap_{i \leq n} m_i \neq \emptyset$.

(ii) Let $b \in \underline{O}$ and $D \subseteq \mathrm{Ind}(b)$. Then if D is maximal in Ind(b) with respect to FIP then $\{M(w):\ w \in D\}$ is maximal in $B^{\dagger}(b)$ with respect to FIP; but $\{M(w):\ w \in D\}$ may be maximal in $B^{\dagger}(b)$ with respect to FIP without D being maximal in Ind(b) with respect to FIP.

Lemma 4.

Let L be a family of sets including a field C of sets, and let K be a subclass of L which is maximal in L with respect to FIP. Then $K \cap G$ is an ultrafilter in C.

This lemma will be used in the following way. A uniform operation o will be defined below with the following property. Let Ω be a collection of sets in $B^{\dagger}(a)$ for any $a \in \underline{O}$, with FIP. Then $\Omega^{o} \supseteq \Omega$ and is maximal in $B^{\dagger}(a)$ with respect to FIP. Moreover the passage from $\{w; w \in \mathrm{Ind}(a^*)\ \&\ M(w) \in \Omega\}$ to $\{w; w \in \mathrm{Ind}(a^*)\ \&\ M(w) \in \Omega^{o}\}$ incurs only a finite number of recursive jumps. The operation can now be repeated. For Ω^{o} is a collection of sets in $B^{\dagger}(a^*)$ with FIP. Hence $\Omega^{oo} \supseteq \Omega^{o}$ and is maximal in $B^{\dagger}(a^*)$ with respect to FIP and again a finite number of recursive jumps are incurred in passing to $\{w;\ w \in \mathrm{Ind}(a^{**})\ \&\ M(w) \in \Omega^{oo}\}$. Hence, starting with a set of recursive sets with FIP and a recursive set of indices of Ω we can construct $U = \Omega \cup \Omega^{o} \cup \Omega^{oo} \cup \ldots$ which will be an ultrafilter in $B(\omega)$. And by the uniformity of o, U will be graded.

The operation o is obtained from a process for constructing paths in trees using "trial and error" predicates (Putnam, (14). By relativising Putnam's definition we have

Definition 7.

Let $a \in \underline{O}$. P is an a-trial and error predicate if there exists a function f recursive in H_a such that

$$P(\underline{x}) \equiv \lim_{y \to \infty} f(x,y) = 1$$

$$\neg P(\underline{x}) \equiv \lim_{y \to \infty} f(x,y) = 0$$

Note If $a, b \in \underline{O}$ and $|a| = |b|$ then P is an a-trial and error predicate if and only if it is a b-trial and error predicate.

An estimate of the complexity of an a-trial and error predicate is given by theorem 1, (13).

Theorem 5.

P is an a-trial and error predicate if and only if it is expressible in both the forms $(y)(Ez)R(\underline{x},y,z)$, $(Ey)(z)S(\underline{x},y,z)$ where R,S are a-recursive.

There easily follows

Corollary 6.

P is an a-trial and error predicate if and only if P is a*-recursive.

Corollary 7.

Let f be a function recursive in H_a such that $\lim_{y\to\infty} f(\underline{x}\ y)$ exists for all $\underline{x}$. Then the function F defined by $F(\underline{x}) = \lim_{y\to\infty} F(\underline{x}\ y)$ is recursive in H_{a^*}.

Proof Define

$$g(\underline{x},n,y) = \begin{cases} 1 & \text{if } f(\underline{x}\ y) = n \\ 0 & \text{if } f(\underline{x}\ y) \neq n \end{cases}$$

g is a*-recursive. Then $\lim g(\underline{x},n,y) = 0$ or 1 for all $\underline{x}$,n and so a trial and error predicate P can be defined by

$$P(\underline{x},n) \equiv \lim_{y\to\infty} g(\underline{x},n,y) = 1$$

$$\neg P(\underline{x}\ n) \equiv \lim_{y\to\infty} g(\underline{x},n,y) = 0$$

By corollary 12, P is a* recursive. It is easily seen that

$$F(x) \simeq \min n\ P(\underline{x},n)$$

Hence F is a*-recursive.

Note Actually, a stronger result can be read from Putnam's proof, namely there exists a recursive function h such that for all e and all $a \in \underline{O}$ if $\langle e,a\rangle$ is a recursive ind and if $\lim_{y\to\infty} \langle e,a\rangle : (x,y) = 0$ or 1 for all x, then $\langle h(e),a^*\rangle$ is a recursive index and $\langle h(e),a^*\rangle : (x) = \lim_{y\to\infty} \langle e,a\rangle : (x,y)$ for all x. Thus we can pass effectively from a recursive index of a function representing a trial and error predicat to a recursive index of the predicate itself.

Definition 8.

Let φ be a non decreasing function on the natural numbers such that $(n)(\varphi(n) > 0)$.

(i) A tree with modulus φ is a set Tr of sequences of natural numbers such that if $(x_1, \ldots, x_n) \in \mathrm{Tr}$ then $(r)(1 \leq r \leq n \rightarrow \varphi(r) > x_r \ \& \ (x_1, \ldots, x_r) \in \mathrm{Tr})$

(ii) A tree Tr is said to be a-recursive if φ and the set of Gödel numbers of members of Tr is a-recursive.

(iii) If $\varphi(n) \equiv 2$, Tr is a binary tree.

The following functions will be used in discussing finite or infinite sequences

(i) $$l(w) = \begin{cases} \text{length of } w \text{ if } w \text{ is finite,} \\ \infty \text{ if } w \text{ is not finite.} \end{cases}$$

(ii) $$[w]_n = \begin{cases} \text{the initial segment of } w \text{ of length } n \text{ if } n \leq l(w) \\ w \text{ if } n > l(w) \end{cases}$$

(iii) $$w(n) = \begin{cases} n\text{-th digit of } w \text{ if } n \leq l(w) \\ \varphi(n) \text{ if } n > l(w) \end{cases}$$

(iv) $$|w| = \sum_{n=1}^{l(w)} w(n) \cdot \varphi(n)^{-n}$$

These functions, restricted to finite sequences are recursive.

Our principal theorem concerning trees is reached by two lemmas.

Lemma 8.

Let Tr be an infinite tree with modulus φ and let, for $m > 0$, Tr(m) be the set of sequences of length m in Tr. For $m > 0$ define.

$$\beta^m = (\iota w \in \mathrm{Tr}(m))(|w| = \max\{|p|;\ p \in \mathrm{Tr}(m)\}).$$

Then

(i) For all $x > 0$, $\lim_{m\to\infty} \beta^m(x)$ exists and $< \varphi(x)$

(ii) if β is the infinite sequence such that for all x, $\beta(x) = \lim_{m\to\infty} \beta^m(x)$ (i.e. $\beta = \lim_{m\to\infty} \beta^m$), then $[\beta]_n \in \mathrm{Tr}$ for all n.

Proof

(i) β^m is well-defined. For Tr (m) is always finite, and as Tr is infinite Tr(m) is always non-empty for $m > 0$. Hence $\max\{|p|;\ p \in \mathrm{Tr}(m)\}$ is well-defined. So therefore, is β^m.

It is easy to establish the relations for sequences :

$$w \in Tr\ (m)\ \&\ n < m \rightarrow [w]_n \in Tr\ (n) \tag{7}$$

$$l(w_1) = l(w_2)\ \&\ |w_1| \leqslant |w_2| \rightarrow (m)_{\leqslant l(w_1)} (\ |[w_1]_m| \leqslant |[w_2]_m|) \tag{8}$$

Now by (7), if $r \geqslant s$, $[\beta^r]_s \in Tr\ (s)$. Hence $|[\beta^r]_s| \leqslant |\beta^s|$.

Then by (8) if $r \geqslant s \geqslant m$, $|[[\beta^r]_s]_m| \leqslant |[\beta^s]_m|$.

Thus

$$r \geqslant s \geqslant m \rightarrow |[\beta^r]_m| \leqslant |[\beta^s]_m| \tag{9}$$

By (9) for x fixed, $|[\beta^r]_x|$ is a decreasing function of r. But $[\beta^r]_x \in Tr(x)$ if $r \geqslant x$; and Tr (x) is finite. Hence the sequence $(\beta^x(x), \beta^{x+1}(x), \ldots, \beta^{x+n}(x) \ldots)$ is ultimately constant ($< \varphi(x)$). This proves (i).

(ii) In the sense of (i) the sequence $(\beta^1, \beta^2, \ldots\ldots)$ converges to an infinite sequence β such that $\beta(m) < \varphi(m)$ for $m > 0$. We show that for all n, $[\beta]_n \in Tr$.

Now $[\beta]_n = (\beta(1), \ldots\ \beta(n))$. Since $\lim_{n \to \infty} \beta^n = \beta$ for each n such that $1 \leqslant m \leqslant n$ there exists a number k (m) such that for all $n \geqslant k(m)$, $\beta^n(m) = \beta(m)$. Let $k = \max\{k(m) ;\ 1 \leqslant m \leqslant n\}$.

Then $\beta^k(m) = \beta(m)$ for all m such that $1 \leqslant m \leqslant n$.

Thus

$$[\beta]_n = (\beta^k(1), \ldots\ldots\ldots \beta^k(n)) = [\beta^k]_n.$$

But $\beta^k \in Tr$. Hence $[\beta^k]_n \in Tr$, i.e. $[\beta]_n \in Tr$.

q.e.d.

The infinite sequence β can be conceived as a path in the tree Tr;

<u>Definition 9.</u>

Let Tr be a tree:

(i) A <u>path</u> in Tr is an infinite sequence β such that $(n)_{>0}\ ([\beta]_n \in Tr)$.

(ii) If Tr is infinite, the <u>principal path</u> of Tr is the sequence $\beta = \lim_{m \to \infty} \beta^m$ of 8 (ii)

(iii) P (Tr) denotes the class of all paths of Tr.

(iv) Let β, γ be infinite sequences.

$$\beta \geqq \gamma \equiv (n)_{>0}\ (\beta(n) \geqslant \gamma(n)).$$

<u>Note</u> There is a natural one-one correspondence between sets X of natural numbers and infinite binary sequences :

$$x \in X \equiv \beta(x+1) = 1$$
$$x \notin X \equiv \beta(x+1) = 0.$$

Let X,Y be sets of natural numbers and β, γ the infinite binary sequences corresponding to them.

Then $X \geq Y \equiv \beta \geq \gamma$.

Lemma 9.

Let Tr be an infinite tree. The principal path in Tr is maximal in P(Tr) with respect to the ordering $\geq$.

Proof Let β be the principal path of Tr. Then in the notation of lemma 8, $\beta = \lim_{m\to\infty} \beta^m$. Suppose β is not maximal; we derive a contradiction.

There exists a $\gamma \in P(Tr)$ such that $\gamma > \beta$. Let $n = \min t(\gamma(t) > \beta(t))$.

Then

$$\gamma(n) > \beta(n) \quad \& \quad (m)(0 < m < n \to \gamma(m) = \beta(m)).$$

Hence

$$|[\gamma]_n| > |[\beta]_n| \tag{10}$$

Since $\beta = \lim_{m\to\infty} \beta^m$, $[\beta^k]_n = [\beta]_n$ for some $k \geq n$. So by (8), (10)

$$|[\gamma]_k| > |\beta^k| \tag{11}$$

But since $\gamma \in P(Tr)$, $[\gamma]_k \in$ Tr. This, together with (11) contradicts the definition of β^k.

By combining Lemmas 7,8 and 9, we have

Theorem 10.

Let Tr be an a-recursive infinite tree some $a \in \underline{O}$. Then P(Tr) has a maximal element a*-recursive.

Proof Let $a \in \underline{O}$. Suppose Tr is an a-recursive infinite tree. Then the a-recursive function f defined by

$$f(x,m) = \beta^m(x+1)$$

satisfies the conditions of corollary 7. Hence $\beta = \lim_{m\to\infty} \beta^m$ is a*-recursive. By Lemma 9, β is maximal in P(Tr).

q.e.d.

Note (i) It follows from 6 (note) that a recursive index of the principal path of an infinite binary tree Tr can be obtained effectively from a recursive index of Tr, that is there exists a recursive function g such that if $\langle e, a\rangle$ is a recursive index of Tr, then $\langle g(e), a^*\rangle$ is a recursive index of β i.e.

$$\beta = (\langle g(e), a^*\rangle : (0), \langle g(e), a^*\rangle : (1), \ldots\ldots)$$

Before proceeding to the construction of graded ultrafilters we mention an application of 9.

Consider the sets in $\bigcup_n B(n_0)$. It is well-known that these sets are arithmetical, that is definable by a formula with one free using constant and variable natural numbers, the functions + and $\cdot$, equality =, the operations $\rightarrow$, &, $\vee$, $\neg$ of the propositional calculus, and the quantifiers (x), (Ex) combined according to the usual syntactical rules. Let $\varphi_1(v_0)$, $\varphi_2(v_0)$.... be a recursive enumeration of the set Ar of all such formulae with one free variable v_0. We consider some provable versions of the finite intersection property. The sign " $\vdash$ " will be used to denote derivation from the usual first-order Peano axioms.

Definition 10.

Let $B \subseteq Ar$ and let $\bar{n}$ be the numeral denoting the number n

(i) B has PFIP(1) if for every finite sequence

$(\varphi_{n_1}(v_0), \ldots\ldots \varphi_{nt}(v_0))$ of members of B there exists an n such that

$\vdash \varphi_{n_1}(\bar{n}) \;\&\ldots\&\; \varphi_{nt}(\bar{n})$.

(ii) B has PFIP (2) if for every finite sequence

$(\varphi_{n_1}(v_0), \ldots\ldots\ldots \varphi_{nt}(v_0))$ of members of B

$\vdash (E\, v_0)(\varphi_{n_1}(v_0) \;\&\ldots\&\; \varphi_{nt}(v_0))$

(iii) B has PFIP(3) if for every finite sequence

$(\varphi_{n_1}(v_0), \ldots\ldots \varphi_{nt}(v_0))$ of members of B

$\nvdash \neg(E v_0)(\varphi_{n_1}(v_0) \;\&\cdots\&\; \varphi_{nt}(v_0))$

Theorem 11.

For i = 1,2,3, the statement s_i holds.

s_i: If B_i is a non-empty subset of Ar, has PFIP(i) and is 1_0-recursive, then there exists a set D_i such that $B_i \subseteq D_i \subseteq Ar$. D_i is maximal in Ar with respect of PFIP (i) and D_i is 2_0-recursive.

Proof For i = 1,2,3, let B_i be a non-empty subset of Ar with PFIP(i) and B_i recursive in H_{1_0}.

Let Tr be the set of all binary sequences $(\delta 1, \ldots\ldots \delta k)$ such that

a) $1 \leqslant r \leqslant k \;\&\; \varphi_r(v_0) \in B_1 \rightarrow \delta r = 1$, and

b) if 1,m..... are those numbers k such that $\delta 1 = \delta m = \ldots = 1$ then $(En)(\vdash \varphi_1(\bar{n}) \;\&\; \varphi_m(\bar{n}) \;\&\ldots\ldots)$

Let Tr^2 be the set of all binary sequences $(\delta 1, \ldots\ldots \delta k)$ such that

(c) $1 \leqslant r \leqslant k \;\&\; \varphi_r(v_0) \in B_2 \rightarrow r = 1$, and

(d) if 2,m,..... are those numbers $\leqslant k$ such that $\delta 1 = \delta m = \ldots\ldots\ldots = 1$ then $\vdash (E\, v_0)(\varphi_1(v_0) \;\&\varphi_m(v_0)\;\&$

Let Tr^3 be the set of all binary sequences $(\delta 1, \ldots\ldots \delta k)$ such that

(e) $1 \leq r \leq k$ & $\varphi_r(V_0) \in B_3 \rightarrow \delta r = 1$, and

(f) if 1,m,..... are those numbers $\leq k$ such that $\delta 1 = \delta m = ... = 1$

then $\nvdash \neg (E V_0)(\varphi_1(V_0)$ & $\varphi_m(V_0)$ &.....). Since B_i has PFIP(i), Tr^i is a binary tree for i = 1,2,3. Further Tr^1 is infinite - consider the infinite sequence β such that

$$\beta^{(i)}(r) = \begin{cases} 1 & \text{if } \varphi_r(V_0) \in B_i \\ 0 & \text{if } \varphi_r(V_0) \notin B_i. \end{cases}$$

Then $[\beta^{(i)}]_n \in Tr^i$ for all $n > 0$.

The theorem is established by applying theorem 10 to each of Tr^1, Tr^2, Tr^3. For this application we establish their complexity by using the fact that the set of theorems of first-order Peano arithmetic is recursively enumerable.

Since B_1, B_2, B_3 are recursive in H_{1_0} so are conditions (a),(b),(c). (b), (d) and the negation of (f) are partial recursive. Hence (b), (d) (f) are recursive in H_{1_0}. Therefore so are (a) & (b), (c) & (d), (e) & (f) and hence Tr^1, Tr^2, Tr^3 are recursive in H_{1_0}. Then by theorem 10, there are maximal paths B_1, B_2, B_3 in Tr^1, Tr^2, Tr^3 respectively, recursive in H_{2_0}. Then for i=1,2,3 define

$D_i = \{\varphi_r(V_0): \beta_i(r) = 1\}$. Conditions (a), (c),(e) ensure that

$D_i \supseteq B_i$ and (b), (d),(f) ensure that D_i has PFIP (i).

q.e.d.

We return to the construction of graded ultrafilters. The main step in this construction is analogous to theorem 11.

<u>Theorem 12.</u>

Let $b \in \underline{O}$, $|b| > 0$ and suppose B is b-recursive, $\phi \subset B \subseteq Ind(b)$. Then there exists a b*-recursive set D, $B \subseteq D \subseteq Ind(b)$, such that if B has FIP then D is maximal in Ind (b) with respect to FIP.

<u>Proof</u> Let Tr be the set of all binary sequences $(\delta 1, ..., \delta k)$ such that

(a) $1 \leq r \leq k$ & $r \in B \rightarrow \delta r = 1$, and

(b) either $\delta 1 = ... = \delta k = 0$, or if 1,m .... are those numbers k such that $\delta 1 = \delta m = .. = 1$

then

(b1) 1, m,.... $\in$ Ind (b), and

(b2) $M(1) \cap M(m)$ ..... $\neq \phi$

Tr is clearly a binary tree: it is infinite because B together with the infinite set of indices in Ind(b) of the set of all natural numbers satisfies (a) & (b).

Since B is b-recursive, so is condition (a).
Consider condition (b). We use the fact that there exists numbers $\mathbf{1}, \mathbf{1}'$ such that $\{\mathbf{1}\}(x) \equiv 1$ and $\{\mathbf{1}'\}(x) \equiv 0$ so that $M(\langle \mathbf{1},1\rangle) = \phi$ and $M(\langle \mathbf{1}',1\rangle)$ = all natural numbers. Since $C(b)$ is recursively enumerable, and $|b| > 0$, $C(b)$ is recursive in H_b. Hence there exists a b-recursive function t such that

$$t(w) = \begin{cases} \langle \mathbf{1}, m, \dots \rangle & \text{if } w \text{ is the Gödel-number of a binary sequence and } 0 < \mathbf{1} < m \dots \text{ where } \mathbf{1}, m, \dots \in \mathrm{Ind}(b) \\ & \text{and } \mathbf{1}, m, \dots \text{ are those numbers } r \text{ such that } 1 \leqslant r \leqslant \mathrm{lh} \ \& \ \underline{r-1}.w = 1. \\ \langle\langle \mathbf{1}',1\rangle\rangle & \text{if } w \text{ is the Gödel-number of a sequence of zeros.} \\ \langle\langle \mathbf{1},1\rangle\rangle & \text{otherwise.} \end{cases}$$

Next, there existsa partial recursive function v such that if w is the Gödel-number of a sequence of members of Ind(b), then

$$\left.\begin{aligned} M(v(w)) &= \bigcap_{i < \mathrm{lh}(w)} M(\underline{i}.w) \\ \underline{1}.v(w) &= \max\{\underline{1}.i;\ i < \mathrm{lh}(w)\} \leqslant_O b \end{aligned}\right\} \qquad (12)$$

Thus if w is the Gödel-number of a sequence of members of Ind(b), $(\underline{1}.v(w))^* \leqslant_O b$.

We shall now prove that

$$w \text{ satisfies (b)} \equiv (Ex)\ (vt(w):(x) = 0) \qquad (13)$$

Suppose w satisfies (b). Then either
(i) w is the Gödel-number of a sequence of zeros. Then $t(w) = \langle\langle \mathbf{1}',1\rangle\rangle$. Hence $M(vt(w)) = M(\langle \mathbf{1}',1\rangle)$ = all natural numbers. Thus $M(vt(w)) \neq \phi$ i.e. $(Ex)(vt(w):(x) = 0)$, or
(ii) w is the Gödel-number of a sequence satisfying (b1) & (b2). Then $t(w) = \langle \mathbf{1}, m \dots \rangle$ and so $M(vt(w)) = M(\mathbf{1}) \cap M(m) \cap \dots \neq \phi$.

Suppose w does not satisfy (b). Then neither (i) not (ii) above hold, Hence $t(w) = \langle\langle \mathbf{1}, 1\rangle\rangle$.
Thus $M(vt(w)) = M(\langle \mathbf{1},1\rangle) = \phi$ so that $\neg(Ex)(vt(w):(x) = 0)$.

Tr is b-recursive by (13) and lemma 1. By Theorem 12 there exists a set $D \subseteq \mathrm{Ind}(b)$, extending B and satisfying (b) i.e. having FIP (see def.6(Note)) q.e.d.

Note

The constructions involved in theorem 12 are effective in the sense that they are accomplished by recursive functions on indices. Thus there exists a recursive function g such that if u is a recursive index of a set B such that $\phi \subset B \subseteq \mathrm{Ind}(b)$ and

$\{M(w); w \in B\}$ has FIP then $g(u)$ is an index of the function vt(w) in (13). Hence, there exists a recursive function g' such that if u is an index of a set B satisfying the conditions of theorem 12 then $g'(u)$ is an index of Tr. Finally, from 7 (Note) the passage from $g'(u)$ to an index of D is effective. Hence there exists a recursive function k such that if $\langle j,b\rangle$ is a recursive index of the B of theorem 12 then $\langle k(j,b), b^*\rangle$ is a recursive index of D.

The passage from B to D in 12 is the operation 'o' p.22. This operation iterated transfinitely gives graded ultrafilters.

<u>Theorem 13.</u>

There exists a partial recursive function t such that if $b \in \underline{O}$, $|b| > 0$ and $\langle j,b\rangle$ is a recursive index of a non-empty set $B \subseteq \mathrm{Ind}(b)$ having FIP, then for any $c \in \underline{O}$ such that $c \geqslant_{o} b$, $\langle t(j,b,c), c\rangle$ is a recursive index of a set D such that $B \subsetneq D \subsetneq \mathrm{Ind}(\dot{c})$ and D is maximal in $\mathrm{Ind}(\dot{c})$ with respect to FIP.

<u>Proof</u> The folliwing functions will be used.

(i) k (as in previous note)

(ii) The recursive function f such that if $b \in \underline{O}$,

$$x \leqslant_{o} b \equiv \{f(b)\}(x) = 0$$

([10])

(iii) The function θ such that

$$\theta(x,b,e) = \{e\}(\min n (\{f(\{e\}^{\bullet}(n))\}(\underline{1}.x) = 0 \;\&\; \{f(\{e\}^{\bullet}(n))\}(b) = 0))$$

By (ii) θ is partial recursive. If $3.5^e \in \underline{O}$, $3.5^e \geqslant_{o} b$, $\langle z,d\rangle \in \mathrm{Ind}(3.5^e)$ then $\theta(\langle z,d\rangle, b, e)$ is the first notation $\{e\}(m)$ in the sequence $\{e\}(0) \leqslant_{o} \{e\}(1) \leqslant_{o} \cdots$ such that $\max(d,b) \leqslant_{o} \{e\}^{\bullet}(m)$

iv) The function $\mathfrak{l}$ such that if 3.5^e, $b \in \underline{O}$ and $3.5^e \geqslant_{o} b$ then $\langle \mathfrak{l}(i,e,b), 3.5^e\rangle$ is a recursive index of $\{x; x \in M(\langle \{i\}(j,b,\theta(x,b,e)), \theta(x,b,e)\rangle) \;\&\; \underline{1}.x <_{o} 3.5^e$

(v) primitive recursive functions pr, $\mathfrak{lm}$ such that

$$pr(x) = \begin{cases} \min a(a < x \;\&\; 2^a = x) & \text{if } (Ea)_{<x}(x = 2^a) \\ 0 & \text{otherwise} \end{cases}$$

$$\mathfrak{lm}(x) = \begin{cases} \min e(e < x \;\&\; 3.5^e = x) & \text{if } (Ee)_{<x}(x = 3.5^e) \\ 0 & \text{otherwise} \end{cases}$$

it follows from p.340 [14] and lemma 3[8] that $\mathfrak{l}$ is partial recursive.

Now we put $t(j,b,c) \simeq \{m^*\}(j,b,c)$ where m* is the solution form of the equation

$$\{m\}(j,b,c) = \begin{cases} k(j,b) & \text{if } c = 2^b \\ k(\{m\}(j,b,pr(c)),c) & \text{if } \begin{cases} pr(c) \neq b \;\& \\ pr(c) \neq 0 \;\&(b=c \vee \{f(c)\}(b)=0) \end{cases} \\ \mathfrak{l}(m, \mathfrak{lm}(c), b) & \text{if } \mathfrak{lm}(c) \neq 0 \;\&\; (b=c \vee \{f(c)\}(b) = 0) \end{cases}$$

Let $b \in \underline{O}$ and $\langle j,b\rangle$ a recursive index of a set $B \subseteq \mathrm{Ind}(b)$ having FIP. By induction on c we shall prove that for all $c \in \underline{O}$ such that $c \geq_{o} b$,

(a) $\langle t(j,b,c),c\rangle$ is a recursive index and $B \subseteq D(c) = M(\langle t(j,b,c),c\rangle) \subseteq \mathrm{Ind}(\mathring{c})$ and $D(c)$ is maximal in $\mathrm{Ind}(\mathring{c})$ with respect to FIP,

(b) if $b \leq_{o} d \leq_{o} c$ then $D(d) \subseteq D(c)$,

(c) if $b \leq_{o} c$ and $|c|$ is a limit ordinal, then $D(c) = \bigcup_{b \leq d < c} D(d)$ (14)

<u>Basis</u>

Suppose $c = 2^{b}$. Then $b^{*} = c$, $b = \mathring{c}$ and so $t(j,b,c) = k(j,b)$. By theorem 12 (note) $\langle k(j,b),b^{*}\rangle$ is a recursive index of a set $D \supseteq B$, maximal in Ind (b) with respect to FIP. But $\langle k(j,b),b^{*}\rangle = \langle t(j,b,c),c\rangle$ and Ind (b) = Ind $(\mathring{c})$. Hence 14 holds for $c = 2^{b}$.

<u>Inductive step</u>

As inductive hypothesis assume that (14) holds for all c such that $b \leq_{o} c <_{o} a$ where $a \in \underline{O}$. It will be shown that 2.12 holds also for $c = a$.

There are two cases to consider - $|a|$ a successor ordinal and $|a|$ a limit ordinal.

<u>Case 1.</u>

$|a|$ a successor ordinal. Then $a = 2^{c}$, say where $c <_{o} a$. So by the inductive hypothesis, (14) holds for c.

We also have $\mathring{a} = c$, $c^{*} = a$.

Now $t(j,b,a) = k(t(j,b,c)c)$. Then by the inductive hypothesis $\langle t(j,b,c)c\rangle$ satisfies 14(a). Then $B \subseteq D \subseteq \mathrm{Ind}(\mathring{c}) \subseteq \mathrm{Ind}(c)$ and $D(c)$ has FIP. Then by theorem 12 (note) $k(t(j,b,c),c^{*})$ is a recursive index of a set D such that $D(c) \subseteq D \subseteq \mathrm{Ind}(c)$ and D is maximal in Ind(c) with respect to FIP. But since $pr(a) = c$, $t(j,b,a) = k(t(j,b,c)c^{*})$. Hence $D = D(a)$ and 14(a) holds when $c = a$.

Since $D(c) \subseteq D = D(a)$ 14(b) holds when $c = a$.

14(c) holds vacuously for $c = a$.

<u>Case 2.</u>

$|a|$ is a limit ordinal. Then $a = 3.5^{e}$, say, and $(n)\ (\{e\}(n) <_{o} \{e\}(n+1) <_{o} 3.5^{e}$

So by the inductive hypothesis, (14) holds for $\{e\}(n)$ for all n. We also have $\mathring{a} = a$.

Now by 14 (b), (a), $D(c) = \bigcup\{D(\{e\}(n)) : \{e\}(n) \geq_{o} b\}$ is maximal in $\bigcup_{n \geq 0} \mathrm{Ind}(\{e\}(n))$. But since $\{e\}(n) <_{o} \{e\}(n+1)$, we have $\{e\}(n) \leq_{o} \{e\}(n+1)$. Hence $\bigcup_{n \geq 0} \mathrm{Ind}(\{e\}(n))$ $= \bigcup_{n \geq 0} \mathrm{Ind}(\{e\}(n)) = \mathrm{Ind}(\mathring{a})$. It easily follows that $D(c)$ satisfies 14(b),(c).

It remains to be proved that $\langle t(j,b,a),a\rangle$ is a recursive index of D(a).

We first observe that by the maximal property of the sets $D(\{e\}(n))$, $\langle z,d\rangle \in D($ if and only if $d <_{o} a$ and $\langle z,d\rangle \in D(\{e\}(m))$ where m is any number such that

$\{e\}(m) \geq_o d,b$. It is clear that we can put $\{e\}(m) = \theta(\langle z,d\rangle,b,e)$ here. Thus

$$x \in D(a) \equiv x \in D(\theta(x,b,e)) \ \& \ \underline{1}\cdot x \leq_o 3.5^e \tag{15}$$

But by inductive hypothesis 14a, if $\underline{1}\cdot x \leq_o 3.5^e$

$$D(\theta(x,b,e)) = M(\langle t(j,b,\theta(x,b,e)), \theta(x,b,e)\rangle)$$

$$= M(\langle\{m^*\}(j,b,\theta(x,b,e)), \theta(x,b,e)\rangle).$$

Further, $\langle\mathfrak{l}(m^*,e,b), 3.5^e\rangle$ is a recursive index of

$$\{x; x \in M(\langle\{m^*\}(j,b,\theta(x,b,e)), \theta(x,b,e)\rangle) \ \& \ \underline{1}\cdot x \leq_o 3.5^e\}$$

Hence by (15), $\langle\mathfrak{l}(m^*,e,b), 3.5^e\rangle$ is a recursive index of $D(a)$.
But $t(j,b,3.5^e) = \mathfrak{l}(m^*,e,b)$. Hence $\langle t(j,b,3.5^e),3.5^e\rangle$ is a recursive index of $D(a)$.
This completes the inductive step.

(14) holds for all $c \geq_o b$, $c \in \underline{O}$ by induction.

qed.

Then using the function $\ddot{}$ we have

<u>Corollary 14</u>

Let $\langle j, 1_o\rangle$ be a recursive index of a non-empty set with FIP. Then for all $b \in \underline{O}$ such that $b \geq_o 1_o$, $D(b) = M(\langle t(j,1_o,b),b\rangle)$ is $C(b)$-graded with grading function $g(c) = \langle t(j,1_o,\ddot{c}),\ddot{c}\rangle$ and

(i) if $|b|$ is a successor ordinal, $\{M(w); w \in D(b)\}$ is maximal in $B^+(b)$ with respect to FIP,

(ii) if $|b|$ is a limit ordinal, $\{M(w); w \in D(b)\}$ is maximal in $B(b)$ with respect to FIP.

<u>Proof</u> Let $1_o \leq_o c \leq_o b$. Then by the maximal property of $D(b)$, $D(c)$ we have

$$D(c) = \begin{cases} Ind(\dot{c}) \cap D(b) & \text{if } |c| \text{ is a successor} \\ Ind(c) \cap D(b) & \text{if } |c| \text{ is a limit.} \end{cases}$$

Hence

$$D(\ddot{c}) = Ind(c) \cap D(b).$$

By theorem 13, $g(c)$ is a recursive index of $D(c)$. g is partial recursive and further, $\underline{1}\cdot g(c) = \ddot{c} \leq_o c$. Hence by definition 5 $g(c)$ is a grading function.

If $|b|$ is a successor ordinal

$$\{M(w); w \in Ind(\dot{b})\} = B^+(\dot{b}) \tag{16}$$

It follows that if $|b|$ is a limit ordinal then $B(b) = \bigcup_{a \leq_o b} B^+(a)$. Hence

$$\{M(w); w \in Ind(\dot{b})\} = B(b). \tag{17}$$

(i) now follows from 14(a) and (ii) from 14(b).

qed

<u>Note</u>. It easily follows that if Z is a Π^1_1 path through $\underline{O}$ then $D(Z) = \bigcup\{D(c); c \in Z\}$ is Π^1_1 and $\{M(w); w \in D(Z)\}$ is a Z-graded ultrafilter in H.

References

1. S.Feferman "Systems of Predicative Analysis" JSL 29 (1964) 1-30

2. S.Feferman , C.Spector "Incompleteness along paths in progressions of Theories". JSL 27 (1962) 383-390

3. R.M.Friedberg "Three theorems on recursive Enumeration" JSL 23 (1958) 309-316

4. R.O.Gandy "Proof of Mostowski's Conjecture" Bull de L'Acad.Polonaise des Sciences 8 (1960) 571-575

5. A.Grzegorczyk "Elementarily Definable Analysis" Fund.Math. 41 (1955) 311-338

6. S.C.Kleene "On the Forms of Predicates in the theory of Constructive Ordinals" 2nd.paper - Amer J.Math 77 (1955) 405-428

7. S.C.Kleene "Arithmetical predicates and Quantifiers" Trans.A.M.S. 79 (1955) 312-

8. S.C.Kleene "Introduction to Metamathematics". North Holland Pub.Cp.Amsterdam.1962

9. G.Kreisel "La predicativité" Bull de la Societe. Math de France 88 (1960) 371-39

10. G.Kreisel G.Sacks "Metarecursive Sets". JSL 30 (1965) 318-338

11. A.Lachlan "On the Lattice of r.e.Sets" Trans. AMS (to appear)

12. A.Mostowski "A classification of Logical Systems". Studia Philosphica 4 (1951)231-274

13. A.Mostowski "On Various Degrees of Constructivity" pp.178-196 of "Constructivity in Mathematics" (A.Heyting) North Holland Pub.Co.Amsterdam.

14. H.Putnam "Trial and Error predicates and the Solution to a problem of Mostowski" JSL 30 (1965) 49-57.

15. R.W.Robinson "Simplicity of r.e. Sets" JSL 32 (1967) 162-172

16. D.Scott "On Constructing Models for Arithmetic" pp 235-255 of "Infinitistic Methods" Pergamon, 1961.

17. C.Spector "Recursive well-orderings" JSL 20 (1955) 151-163.

18. H.Weyl "Das Kontinuum. Kritische Untersuchung über die Grundlugen der Analysis" Leipzig 1918.

RECURSIVE EQUIVALENCE: A SURVEY

by

John N. Crossley, All Souls College, Oxford.

Recursion theory is "the theory of those properties of sets of natural numbers which are preserved under recursive permutations, where a recursive permutation is a one-one recursive function mapping the set of all natural numbers onto itself." Myhill [21] .

Given this premiss the theory of recursive equivalence types, which is one constructive analogue of cardinal number theory, can be developed. But if we seek an analogue of Myhill's condition for ordered sets it is not clear what the interpretation of "recursive permutation" should be; in particular what we mean by the action of such a function outside the actual ordered set considered. (Of course, if the ordering is an ordering of the set of all natural numbers this does not matter.) Further, unlike the case of sets, it is not at all clear why the points outside a set (which may have additional structure) should have any relevance. Later we shall see that we considered ordered sets as sets of rational numbers (suitably coded in the natural numbers) with the ordering induced by the ordering of the rationals. In this setting we might have asked about equivalence classes under recursive order automorphisms of the rationals. As a matter of fact we did not do this; and one reason for this was that it was not until the summer of 1965 that we even contemplated using the rationals as our basic setting. The hypothesis that we are dealing with recursive order automorphisms of the rationals is stronger than the one we use but we do not expect, though we have not verified this in detail, that the theory of constructive order types would be significantly affected by using the stronger hypothesis. It does turn out that if we consider one-one partial recursive onto maps p: $\alpha \to \beta$ where α, β are sets of natural numbers and we allow the domain of $\underline{p}$ possibly to include α <u>properly,</u> then a viable theory is possible in both cardinal and ordinal cases ($\underline{p}$ order preserving for the latter). Moreover, the following theorem shows that in the cardinal case Myhill's dictum is respected.

<u>Theorem</u> (Karp, Myhill [12]). There is a recursive permutation mapping α one-one onto β iff there are one-one, onto partial recursive functions p: $\alpha \to \beta$ and q: $\bar{\alpha} \to \bar{\beta}$ (where

$\bar{\alpha}$ denotes the complement of α in the set of natural numbers).

Recently (Nov.1967) R.I. Soare has shown that the order analogue of this theorem, (i.e. when p, q and the recursive permutation are all required to be order preserving) is false in the case of Dedekind sections.

2. Recursive Equivalence Types. Following Dekker [10] we now define $A = RET(\alpha) = \{\beta$: there exists a one-one partial recursive function p whose domain includes α and which maps α onto $\beta\}$, call A a recursive equivalence type (R.E.T.) and say α is recursively equivalent to β if $\alpha \in RET(\beta)$. It is easy to see that recursive equivalence is an equivalence relation, that finite cardinals coincide with R.E.T.s of finite sets (if we ignore all but sets of natural numbers) and that there are $2^{\aleph_0}$ R.E.T.s. Just as in the theory of cardinals the natural operations of addition and multiplication (+ and .) are introduced. We define

$$A + B = RET(\{2x : x \in \alpha\} \cup \{2x + 1 : x \in \beta\}),$$

$$A.B = RET j(\alpha, \beta)$$

where $A = RET(\alpha)$., $B = RET(\beta)$ and j is a recursive pairing function $j: N \times N \to N$ and N is the set of all natural numbers. We remark that in order to guarantee that addition is well-defined one has to have separability, i.e. the representatives of A and B must be contained in disjoint r.e. sets; we chose the even and odd numbers. Addition is commutative and associative, so is multiplication and the distributive laws hold. We can also define

$$A \leq B \leftrightarrow (\exists C)(A + C = B).$$

The proof that $\leq$ is a partial ordering uses the analogue of the Cantor-Bernstein theorem for whose proof we require the

Lemma (Myhill [12]). $B + A = A \leftrightarrow B.R \leq A$, where $R = RET(N)$.

The proof of this lemma is an interesting application of the well-known proof that every infinite recursively enumerable (r.e.) set contains an infinite recursive subset.

It also turns out that the following conditions are equivalent:

(1)

$A \neq A + 1$,

$A + B = A + C \quad B = C$,

$A \ngeq R = RET(N)$,

A is not recursively Dedekind infinite.

Dedekind defined "infinite set" by saying that such a set could be mapped one-one onto a proper subset of itself; "recursively Dedekind infinite set" is defined in the same way but also insisting that the mapping be partial recursive (and one-one on its domain).

We call R. E. T. s satisfying (1) isols and write Λ for the collection of all isols. There are $2^{\aleph_0}$ infinite isols which are mutally incomparable under $\leq$. All R. E. T. s of finite sets are isols too. Isols are closed under + and . and because of the commutativity, etc. noted above, the isols can be extended to a ring Λ^* containing the ring of integers just as the natural numbers are extended to the ring of integers.

Theorem (Myhill [23]). The first order theory of Λ^* with + as the only function letter is the theory of a free abelian group on an infinite set of generators and is complete (but Λ^* is not a free abelian group).

Exponentiation (corresponding to weak cardinal exponentiation) can be defined in a natural way and we can show that cancellation laws (involving only +, ., exponentiation and equality) hold for isols just as they do for natural numbers, e. g. $A.B = A.C \ \& \ A \neq O \rightarrow B = C$, but inequalities often fail. In general cancellation laws do not hold for R. E. T. s which are not isols, but Nerode [25] and Friedberg [15] proved cancellation laws involving addition. Indeed Nerode [25] has shown that the theory of universal Horn sentences involving only addition and equality for R. E. T. s is decidable. The basic example of such cancellation laws is Friedberg's

Theorem [15]. For arbitrary R. E. T. s A, B and natural number $n > 0$,

$$A.n = B.n \rightarrow A = B.$$

Myhill [20, 21] and Nerode [25] have extended these laws for isols to a large class of functions, containing +, . and exponentiation (to a base > 1), called recursive combinatorial functions. Every one-place number-theoretic function can be expressed as $f(n) = \sum_{i \leq n} {}_nC_i . r_i$ where the r_i are the Stirling coefficients and ${}_nC_i = n!/(n-i)!\,i!$. If all the $r_i \geq 0$, then f is combinatorial; similarly when f has more than one argument.

There is no hope of extending cancellation laws beyond Firedberg's result above since we have

Theorem (Ellentuck, Myhill [22, 23]). There exists an R. E. T. A such that

$f(A) = f(A^2)$ for every (extension of any) non-linear recursive combinatorial function f.

In order to study R. E. T. s in general new techniques are required and a start has been made by Nerode. However, further analysis suggests that Nerode's combinatorial series [26] are merely one special case of the generalization we mention in section 5.

There is one part of the theory of R. E. T. s we have completely neglected, that is the theory of regressive isols and infinite sums and products. For this we refer the reader to Dekker's survey [11] where an extensive bibliography of works on isols can also be found.

3. Constructive Order Types. We first introduced constructive order types in 1962. There were also considered independently by A. Manaster (unpublished) at about the same time. In direct analogy to R. E. T. s we define $A = COT(\alpha) = \{\beta :$ there exists a one-one partial recursive order preserving function p whose domain includes α and maps α onto $\beta\}$ where α, β are ordered sets ; and call A a constructive order type (C. O. T.). Originally we did not assume there was any significant connection between the set α and its ordering [4, 5, 3] and some anomalies arose in the theory. This point is also (forcibly) made by Kreisel in his review [18] of [4] (see the beginning and remark at the end of [18]). For example there were constructive order types whose (classical) ordinal was $\omega . 2$ but which had no predecessor of type $\omega + n$ ($n \geq 0$). Consideration of losols and Morley's lemma (see below) convinced us that we should restrict ourselves to sub-orderings of r. e. orderings. In fact as far as constructive order types are concerned this is just the same (by Cantor's proof that any linearly ordered countable set is order embeddable in the rationals) as looking at C. O. T. s of sets of rationals with the induced ordering (and some standard coding of the rationals in the natural numbers). It is in this setting that the monograph [6] is presented.

In commencing the study of C. O. T. s our first objective was to obtain analogues of Dekker and Myhill's work (especially [12]) and we used Tarski's Ordinal Algebras [29] as a guidebook. But there are other reasons for studying C. O. T. s. It was felt that the theory might throw light on the question of what is a "natural" well-ordering. We have given an answer to this for ordinals less that ε_ω, the ω-th solution of $\omega^\alpha = \alpha$ (namely, any well-ordering whose C. O. T. is $\leq E(n)$ for some n, see below). But although it is possible to extend the methods used to larger ordinals (cf. [27, 8, 1, 28, 14]) nevertheless we have no general method or answer for arbitrarily large recursive ordinals and this despite the fact that the first significant attack on the problem was (essentially) made by Veblen in 1908 [30] . (We note that it is certainly not true that the well-orderings given by Kleene's O [17] are "natural" for any ordinal $\geq \omega^2$ under any reasonable criteria see, e.g. [2, 9].) Moreover, we have felt for some time that constructing actual well-orderings is an ultimately fruitless way of achieving a definition of "natural" or of showing that there is a bound strictly less that the first non-recursive ordinal to all ordinals which have natural well-orderings. The reason for this view may be expressed roughly as follows: if one can construct well-orderings for all ordinals $< \alpha$ in a uniform way then one can construct a well-ordering of type α which is still "natural" (because of the uniformity). Hence one can progress further. Moreover if the well-ordering of type α has been constructed by an explicit recursive process then $\alpha + \omega_1 = \omega_1$ (the first non-recursive ordinal) so α is "very small" compared with ω_1.

Because of this we feel that there is very little point in trying to construct more and more functions, though this can be continued along the lines sketched in [8] . The most up-to-date treatment of the problem of natural well-orderings may be found in [14] where certain principles, generalizing those of [8] are given. Kreisel in an early version of [18] claimed that these are "far reaching extensions" but exactly the same objections as above can be raised. If one has specific projects, e.g. consistency proofs, for which one requires natural well-orderings of a particular ordinal then there is no problem. But equally, given explicit descriptions like those in [14] one can perform exactly the same sort of process again and continue the Sisyphian task. In this survey we restrict ourselves to the theory of C.O.T's as a part of recursion theory (which is how we originally envisaged it, though not how Kreisel did or does, see [18]).

For C.O.T.'s we can again imitate classical constructions and define +, . and exponentiation (to a base > 1): separability has to be borne in mind just as for addition of R.E.T.s. We also write $A \subsetneq B$ if there exist $\alpha \in A$, $\beta \in B$ with $\alpha \subsetneq \beta$ and $A \leq B$ ($A \leq {}^*B$) if there exist $\alpha \in A$ and $\beta \in B$ with α an initial (final) segment of β, $A < B$ if $A \leq B$ and $A \neq B$. Recently the following analogue of the Lindenbaum-Tarski theorem has been obtained.

Lemma (Morley [6]). If $A \leq B$ and $B \leq {}^*A$ then $A = B$.

Corollary. $A.n = B.n$ & $n > 0$ $A = B$.

We also have

Theorem (Aczel & Crossley [6]). $n.A = n.B$ & $n > 0 \rightarrow A = B$.

Definition. A co-ordinal is the constructive order type of a well-ordering. It is easy to see that there are $2^{\aleph_0}$ co-ordinals containing only well-orderings of a given countable ordinal.

Theorem. $\leq$ is a partial well-ordering of the co-ordinals.

That $\leq$ is very definitely not a linear ordering is clear from the following

Theorem (Crossley [4]). A non-empty collection of co-ordinals (of power $\leq 2^{\aleph_0}$) has a least upper bound iff it has a maximum; otherwise there are $2^{\aleph_0}$ mutually incomparable (under $\leq$) upper bounds which are minimal.

Co-ordinals, like ordinals, have one-sided cancellation properties,

e.g. $A + B = A + C \rightarrow B = C$

but on the other hand we have $B \leq C \not\rightarrow B + A \leq C + A$

and similar results hold for multiplication and exponentiation (with trivial side conditions). In studying analogues of classical laws the most interesting are those concerning principal (or main) numbers. An ordinal α is said to be a principal number for a function f if it is infinite and

$$n < \beta < \alpha \rightarrow f(\beta, \alpha) = \alpha.$$

(The "n" is included to avoid anomalies like $0^{\alpha} = 0$.) We extend the definition to co-ordinals in the obvious way. We also write $|A|$ for the ordinal of (any) well-ordering in A.

Theorem (Crossley [4]). (i) If $|A| < \omega^{\omega}$ then A is a principal number for addition iff $A = W^n$ for some finite co-ordinal n,

(ii) If $|A| < \omega^{\omega^{\omega}}$ then A is a principal number for multiplication iff $A = W^{W^n}$ for some n; where W is the C. O. T. of N with the standard ordering.

Every co-ordinal of the form W^A is a principal number for addition but the converse is false (Aczel [6]) and likewise every co-ordinal of the form W^{W^A} is a principal number for multiplication but there are some principal numbers for multiplication which are not even of the form W^A let alone W^{W^A} (Hamilton [6]) When we turn to exponentiation the situation is very close to the classical one.

Theorem (Aczel & Crossley [3]) The following are equivalent:

A is a principal number for exponentiation,

$2^A = A$,

$A = W$ or $W^A = A$,

$A = W$ or $A = E(B)$ for some co-ordinal B, where E is a specific constructive function on co-ordinals (defined in [5]).

We can also prove a version of the Cantor Normal Form Theorem.

Theorem (Crossley [5]). If $A \leq W^B$ then

$$A = W^{A_1}.n_1 + \ldots + W^{A_r}.n_r$$

where $B \geq A_1 > \ldots > A_r$ and the n_i are finite; similarly with W replaced throughout by any other infinite co-ordinal, C, say.

The above theorems rely heavily on the lemma

Lemma (Crossley [4]). $B + A = A \longleftrightarrow B.W \leq A$ (when A is a co-ordinal)

We do not know whether this result is true for arbitrary C. O. T. s. An even simpler (to express), but related open question is

(2) does $2 + A = A \rightarrow 1 + A = A$ for A an arbitrary C. O. T. ?

Recently Hamilton [16] has shown that the answer is "no" with the old definition of C. O. T. (see the beginning of this section) but the question is wide open with the new definition.

Now we turn to losols. A losol is the C. O. T. of a set (of rationals) whose R. E. T. is an isol. (The name "losol" is the joint responsibility of the author and Joseph Rosenstein: it is hoped it will suggest "l(inear) o(rdering i)sol".) At first the author erroneously thought that losols were C. O. T. s of linear orderings with no recursive ascending or descending chains but this latter class properly contains the losols (Tennenbaum; a proof based on Jockusch's ideas is in [6]). The theory of losols is quite close to that of isols but life is made a little easier by the fact that there is a unique order isomorphism between finite linearly ordered sets of the same cardinality.

Theorem (Crossley [6]). $\subseteq$ is a partial order on losols (and so are $\leq$, $\leq^*$).

Theorem (Crossley [6]). For losols A, B, C,

$A + B = A + C \rightarrow B = C$,

$A.B = A.C \ \& \ A \neq 0 \rightarrow B = C$ (similarly for $B.A = C.A$),

$A^B = A^C \ \& \ A > 1 \rightarrow B = C$ (and similarly for $B^A = C^A$ if $A \neq 0$),

similarly with "$\leq$" or "$\leq^*$" replacing "=" throughout.

In fact one can develop a theory of combinatorial functions on C. O. T. s [7].

4. Connections between R. E. T. s and C. O. T. s. Because classically cardinals and ordinals have clear connections (apart from the G. C. H. !) it is natural to ask what are the relations between R. E. T. s and C. O. T. s. The answer is very simple: Every infinite R. E. T. contains sets α such that α is the underlying set of a well-ordering of type β for any denumerably infinite ordinal β, and hence we can obtain a co-ordinal A with $|A| = \beta$ from α. Now let us suppose we revert to the old definition of C. O. T. ([4, 5, 3]) and restrict ourselves to co-ordinals A which are full in the sense that if $\beta < |A|$ then there exist co-ordinals B, C with $|B| = \beta$ and $B + C = A$. (Not all co-ordinals are full: see the first paragraph of section 3 and recall that we have to have separability). Then we can define $\rho(A)$, for A an R. E. T., to be the set of all ordinals of full co-ordinals whose underlying set is an element of A.

Theorem (Manaster [19]). If A is infinite then either $\rho(A)$ is the set of all denumerably infinite ordinals or, for some ordinal α,

(3) $$\rho(A) = \{\beta : \omega \leq \beta < \omega^{\alpha}.n\} \quad \text{where } n > 1.$$

Conversely, given a denumerable ordinal α and a finite $n > 1$ there exists A with $\rho(A)$ as in (3); $\rho(R)$ is the set of all denumerable ordinals.

5. Further extensions. All the theory of R. E. T. s we have discussed (and of regressive isols too) can be extended to an arbitrary recursive set of generators; for R. E. T. s the set is $\{1\}$ and we identify sets with their characteristic functions. For extensions to C. O. T. s and beyond we have to consider characteristic functions whose domains are ordered sets (e.g. the rationals) but this causes only rather more complication and a lack of nice

abelian properties (recall that even for order types $\alpha + \beta \neq \beta + \alpha$ in general). This work is in progress and will appear in [7].

Dekker has commenced a study of R. E. T. s of certain vector spaces and Hassett has recently announced results for R. E. T. s of groups in a rather different setting from ours. It would appear that Hassett's work is more closely related to work problems than to constructive versions of purely algebraic results, though his abstract (Journal of Symbolic Logic 1967) includes specializations to groups which constitute particular cases of the structures considered in [7].

Finally, suppose we wish to leave recursion theory and work in set theory. Corresponding to the theory of isols Ellentuck [13] has developed the theory of Dedekind finite cardinals using the axiom of choice for finite sets but not the full axiom of choice. We conjecture that the theory of Dedekind finite order types will not require even this axiom but will correspond directly to the theory of losols. Here again work is in progress.

References.

[1] P.H.G. Aczel, D.Phil. Thesis, Oxford (1966).

[2] ———— Paths in Kleene's <u>0</u>, Archiv für math. Logik und Grundlagenforschung, <u>10</u> (1967), 8 - 12.

[3] ———— & J.N. Crossley, Constructive Order Types, III, Archiv für math. Logik und Grundlagenforschung, <u>9</u> (1967), 112 - 116.

[4] J.N. Crossley, Constructive Order Types, I, in Formal Systems and Recursive Functions, Amsterdam (1965), 189 - 264.

[5] ———— Constructive Order Types, II, Journal of Symbolic Logic, <u>31</u> (1966), 525 - 538.

[6] ———— Constructive Order Types, monograph to be published by North Holland Pub. Co., Amsterdam.

[7] ———— & A. Nerode, Effective Dedekind Types, monograph in preparation.

[8] ———— & R.J. Parikh On Isomorphisms of Recursive Well-orderings (abstract), Journal of Symbolic Logic, <u>28</u> (1963), 308. (See correction

in [5]).

[9] ——— & K. Schütte Non-uniqueness at ω^2 in Kleene's $\underline{\mathcal{O}}$, Archiv für math. Logik und Grundlagenforschung, 9 (1967), 95-101.

[10] J.C.E. Dekker A non-constructive extension of the number system (abstracts), Journal of Symbolic Logic, 20 (1955) 204-205.

[11] ——— Regressive Isols, in Sets, Models and Recursion Theory, Amsterdam (1967), 272-296.

[12] ——— & J. Myhill Recursive Equivalence Types, University of California Publications in Mathematics (new series) 3 (1960), 67-214.

[13] E. Ellentuck The Universal Properties of Dedekind Finite Cardinals, Annals of Mathematics, 82 (1965), 225-248.

[14] S. Feferman Systems of Predicative Analysis, II: Representations of Ordinals, to appear in Journal of Symbolic Logic.

[15] R.M. Friedberg The Uniqueness of finite division for Recursive Equivalence Types, Math. Zeitschrift, 75 (1961) 3-7.

[16] A.G. Hamilton An Unsolved Problem in the theory of Constructive Order Types, to appear in Journal of Symbolic Logic.

[17] S.C. Kleene On the forms of Predicates in the theory of Constructive Ordinals (second paper), American Journal of Mathematics, 77 (1955), 405-428.

[18] G. Kreisel review of [4] to appear in Zentralblatt für Mathematik.

[19] A. Manaster Full Co-ordinals of RETs, submitted to Pacific Journal of Mathematics.

[20] J. Myhill Recursive Equivalence Types and Combinatorial Functions, Bulletin of the American Mathematical Society, 64 (1958), 373-376.

[21] ——— Recursive Equivalence Types and Combinatorial Functions, Proceedings of the 1960 International Congress in Logic, Methodology and Philosophy of Science, Stanford (1962),

46-55.

[22] ——— $\Omega - \Lambda$ in Recursive Function Theory, American Mathematical Society Proceedings of Symposia in Pure Mathematics 5 (1962), 97-104.

[23] ——— Elementary Properties of the Group of Isolic Integers, Math. Zeitschrift, 78 (1962), 126-130.

[24] A. Nerode Extensions to Isols, Annals of Mathematics, 73 (1961), 362-403.

[25] ——— Additive relations among Recursive Equivalence Types, Mathematische Annalen, 159, 329-343 (1965).

[26] ——— Combinatorial Series and Recursive Equivalence Types, Fundamenta Mathematicae, 58 (1966), 113-141.

[27] A. J. Page Ph.D. Thesis, Cambridge (1966)

[28] K. Schütte Predicative Well-orderings, in Formal Systems and Recursive Functions, Amsterdam (1965), 280-303.

[29] A. Tarski Ordinal Algebras, Amsterdam (1956)

[30] O. Veblen Continuous Increasing Functions of Finite and Transfinite Ordinals, Transactions of the American Mathematical Society, 9 (1908), 280-292.

Half-Ring Morphologies

A. S. Davis

Every language consists of 1) a class of things called "expressions" which combine with one another in various ways to form more complex utterances from simpler ones, 2) rules by which certain categories and relations among expressions are defined, such as the category of sentences or the noun-phrase to sentence relation, and 3) rules by which certain expressions are in some sense said to follow from or be consequences of others. In short, every language has a morphology, a grammar, and a logical syntax. The morphology tells us what manner of stuff the language is made of and how to build with it; the grammar tells us what a sentence is and how to parse a given sentence into its constituent phrases; and the syntax tells us how several sentences may be strung together in a coherent way to form a proof in mathematics, a poem or story in literature, a conversation between people, and other such linguistic structures that go beyond the boundaries of a single sentence. (Hence "logical" rather than "grammatical" syntax is intended here.)

The grammar and syntax of a language are based upon its morphology. Thus any mathematical theory of language must presuppose or provide some sort of underlying calculus of expressions. The grammars currently studied in mathematical linguistics generally assume no more than a semigroup for their underlying morphology: only a single binary operation, called "concatenation", is used and expressions are combined by simply writing (uttering, thinking) one right after the other, an expression being just a string of lexical items. [Cf. e.g. 5]

Now it appears that even the best systems of grammatical rules must be quite complicated in order to provide good approximations to the actual grammar of a natural language. [1, 4] Perhaps a richer morphology would enable one to get by with much simpler rules.

One way of enriching the morphology on which grammatical rules are to be based is to allow as morphemes so-called "discontinuous strings" or expressions with "blank places" in them and to provide for substituting phrases for the blanks in such expressions. This seems also a reasonable thing to do from considering what should be counted as the morphemes in a language. It seems more natural, for instance, to list the expression

$$\text{if} \_\_\__1 \text{ then} \_\_\__2$$

as a single lexical item than to count only 'if' and 'then' as formatives. One should think that the sentence

if it rains then it pours

has but five unanalyzable parts rather than six. But then a different kind of operation than concatenation must be used to compose a sentence from such formatives.

Motivated by these considerations, we shall sketch here an algebraic theory of morphologies with two binary operations, one for substitution, the other for concatenation. The algebraic systems appropriate for our theory turn out to be essentially the "transformation algebras" introduced for other purposes in [3]. Our point of view here also seems to be closely related to that of Curry in [2].

The author gratefully acknowledges his debt to Mrs. Lelia Chance for corrections to the statement of some of the theorems in this paper.

1. The abstract algebraic approach to the morphology of a language. By a "language" we shall mean anything that has essentially the same structure as what is ordinarily referred to as "language". That is to say, the theory of languages we have in mind is not only intended to be quite general, it is abstract: to the category of systems to be called "languages" belongs any system which is isomorphic to a member of that category. This means that a lot of things not ordinarily thought of as languages will be included. For instance, to the extent that a given language (in the usual sense of the word) mirrors the system of thought it is supposed to express, the conceptual system itself will be a language (in our abstract sense of the word). A by-product of this point of view is that dual or triple terminologies, usually encountered in semantics, can be slighted. (For example, a sentence is said to express a proposition and a proposition conceptualizes a possible state of affairs. In our abstract approach, sentence, proposition, and state of affairs may all be called "sentences", each in a different language, for we may speak not only of a written or spoken language but of "a language of ideas" or of "the language of experience" as well. Of course there is nothing to

hinder us from using the other terminologies whenever we wish to make the distinctions intended by them.)

The elements of which a language consists will be called "expressions". An expression may be made of almost anything, depending on the language considered and the medium or channel by which it is conveyed. Modulated sound waves, magnetized spots on steel tape, brush strokes on rice paper, neural impulses in a cerebral cortex, body motions on a dance floor - any of these may form the substance of expressions in some language or other.

An expression may be long or short, simple or complex. A single noun and an entire Shakespearian play both count as expressions in the English language.

An expression may be "discontinuous" in that it may have one or more places" or "blanks" which can be filled by other expressions.

What things are to be included or excluded as expressions depends on the level of analysis at which one is working. In studying French grammar words and phrases of the vocabularly of French are expressions, whereas the letters of the French alphabet are not. On the other hand, if we are concerned with spelling, then letters as wëll as words are expressions, whereas the individual strokes which make up a letter are not.

Expressions possess three notable properties: 1) They are replicable; any expression can be duplicated any number of times so that an unlimited store of copies is available for constructing other expressions. 2) They may be concatenated, one after another in time, space, or some other dimension, to form longer expressions. 3) They consist of interchangeable parts; replicas of one expression can be substituted for the occurences of another in a larger expression, the result being again an expression.

These properties suggest an algebraic approach to the morphology of a language. Corresponding to the last two, we shall consider two operations in an algebra of expressions. One of them - concatenation - is easy to characterize. (The system of expressions under concatenation alone is just a free semigroup with identity, at least for the finitary languages with which we are primarily concerned.) The other one - substitution - is more involved. We shall treat it as a binary operation, which we prefer to call "composition", and seek a characterization of it by means of a suitable set of postulates.

2. Concatenation and composition. The constituent parts of an expression will be called "phrases". Every expression is either a single phrase or a (possibly empty) sequence of phrases. (A phrase by itself is counted as a sequence of length one.)

A phrase can be quite complex or "deep" and be composed of other phrases, not in a sequence, but nested within each other in various ways. The nesting is accomplished by substituting phrases for blanks. For example the sentence

necessity is the mother of invention

is a phrase of English composed from the more basic phrases

$\_\_\__1$ is $\_\_\__2$,

necessity ,

the mother of $\_\_\__1$,

and

invention .

(Let it be emphasized that the above sentence is not to be thought of as a concatenation of shorter expressions, at least not at the level of analysis we have in mind, as it would be according to the more usual concatenative morphology.)

Blanks will be treated on a par with ordinary expressions. They are phrases. We assume a denumerable infinity of them, distinguished by numbering, and each capable of being replicated like any other expression. Blanks will be denoted by numerals in parentheses (rather than by dashes). Thus

(1), (2), (3), etc.,

are the first blank, the second blank, and so on. But this numbering is not to have any necessary connexion with the spatial or temporal position of blanks as they occur in a larger expression. For instance, in the phrase

(2) loves (1) and (1) is hateful
but (2) doesn't care even if (3) is
troubled ,

the numbering is used only to distinguish one blank from another and to indicate that there happens to be more than one occurrence of the same blank in the expression.

Now we are ready to describe the operations of concatenation and composition. These will be denoted by an asterisk and by juxtaposition, respectively. The concatenation $x*y$ of two expressions x and y is the sequence consisting of the phrases of x (as a sequence of phrases) immediately followed by those of y, all in the given order. The composition xy of x with y is the expression obtained by substituting the first phrase of y (as a sequence) for the first blank in x, the second phrase of y for the second blank in x, and so on until all the blanks in x have been filled with phrases of y, or until the last phrase of y has been used, in which case the substitution begins again with the first phrase of y for the next blank in x. (If y is the empty string $\emptyset$, then put $xy = \emptyset$.) In performing either of these operations, any punctuation (marks, pauses, etc.) which might be used to prevent ambiguity in the resulting expression is to be considered as part of the glue which holds an expression together and not as consisting of expressions in their own right.

Here is indicated a concatenation of an expression of length two with an expression of length one:

(Jack, Jill) $*$ (her best friend) =
(Jack, Jill, her best friend).

The result is an expression of length three. As an example of composition, let y be the string

Jack, Jill, her best friend

and let x be

(2) loves (1) and (1) is hateful
but (2) doesn't care even if (3) is
troubled.

Then xy is

Jill loves Jack and Jack is hateful
but Jill doesn't care even if her best friend is
troubled.

3. <u>Half-ring morphologies</u>. By a half-ring we mean an algebraic system $(E, \emptyset, *, \cdot)$ with a distinguished element $\emptyset$ and binary operations $*$ and $\cdot$ such that

i) $x(yz) = (xy)z$,
ii) $x*(y*z) = (x*y)*z$,
iii) $x*y = x*z$ implies $y = z$,
iv) $(x*y)z = xz * yz$,
v) $x*\emptyset = x$ and $x\emptyset = \emptyset$,

for all x, y, z in E. (Notationally, $*$ takes precedence over $\cdot$, so that $xy * z$ is $(xy)*z$, not $x(y*z)$.) The operations $*$ and $\cdot$ will be called *concatenation* and *composition*, respectively, and $\emptyset$ the *null* element.

One is easily convinced that the concatenation and composition of expressions described in the previous section satisfy the postulates for a half-ring. There are many other kinds of half-rings, however. The arithmetic of ordinal numbers (up to an appropriate limit-ordinal) or the system of real functions under point-wise addition and ordinary composition provide examples. Of course every ring is a half-ring.

Consider the half-ring generated by a denumerable sequence of elements (1), (2), etc., subject to just these defining-relations:

$$(1)((m) * x) = (m)$$

and

$$(n+1)((m) * x) = (n)(x * (m)) ,$$

for all $m, n = 1, 2$, etc., and all x. It is easily seen that such a half-ring does exist, is unique up to isomorphism, admits only the trivial automorphism, and is essentially the system of all finite sequences of natural numbers under ordinary concatenation and a composition defined by

$$(a_1, \ldots , a_n)(b_1, \ldots , b_m) = (b_{\bar{a}_1}, \ldots , b_{\bar{a}_n}) ,$$

where $\bar{a}_i \equiv a_i \pmod{m}$, for each $i = 1, \ldots , n$. Call this, or any system isomorphic to it, "the blank-morphology", for it summarizes the algebra of blanks implicit in the previous section.

Now morphologies in general are half-rings in which the blank-morphology is embedded in a suitable manner. This could easily be made precise, but the following definition will do just as well, is more elegant, and more easily used with universal algebraic concepts since it avoids mentioning generators or assuming an infinity of distinguished elements.

A *morphology* is a system $(E, \emptyset, \pi, {}', *, \cdot)$ consisting of a half-ring $(E, \emptyset, *, \cdot)$ whose elements are called *expressions* (or *strings*), among which π is distinguished as a *first blank*, and a unary *shift* operation ' such that

vi) $(xy)' = xy'$,

vii) $\pi\pi = \pi$,

viii) if $x \neq \emptyset$, then $\pi(x * y) = \pi x$,

ix) if $y \neq \emptyset$, then $x'(\pi y * z) = x(z * \pi y)$,

for all x, y, z in E, and

x) $\pi \neq \emptyset$.

The elements π, π', π'', etc., called <u>blanks</u>, will be denoted by

(1), (2), (3), etc.,

respectively. Further notation: $(j_1 \ldots j_n) = (j_1) * \ldots * (j_n)$, and $(\ldots 0) = \emptyset$, $(\ldots 1) = (1)$, $(\ldots 2) = (12)$, $(\ldots 3) = (123)$, etc. An expression x is <u>closed</u> if $xy = x$ for all $y \neq \emptyset$. The <u>degree</u> of a closed expression is zero; otherwise the degree of x is either infinite or is the least n such that $x(\ldots n) = x$. The <u>length</u> of x, if not infinite, is the least $m \geq 0$ such that $(\ldots m)x = x$. Expressions of length one are <u>phrases</u>. Closed phrases are <u>formulas</u>. If all expressions are of finite length and degree, that is, if

xi) for each x there are numbers m and n such that $(\ldots m)x = x = x(\ldots n)$,

then the morphology is <u>locally finite</u>. If in addition it is finitely generated, it is called <u>finitary</u>. A minimal set of phrases which generates the morphology is a <u>vocabulary</u>, whose members are called <u>morphemes</u> (or <u>primitives</u>, or <u>formatives</u>).

Although languages in which infinite strings are permitted as expressions have been used in connexion with foundational studies in mathematics, natural languages are always finitary. In this paper, "morphology" will mean a locally finite morphology. Also, to rule out one trivial case for which some of the theorems below fail, we shall assume

xii) $\pi' \neq \pi$

in the following.

These definitions and terminology were chosen to suggest the intended interpretation: the morphology of some language in the sense of Section 1. The extent to which the above postulates characterize this notion remains to be considered. Apparently Postulates i through xii are not too much, since the calculus of expressions sketched in Section 2 satisfies them. Later we shall show that every system which satisfies these postulates is a homomorphic image of the morphology of a certain canonical kind of language and adduce reasons why this justifies our feeling that we have postulated just enough.

<u>4</u>. <u>Bracketing</u>. When one presents a proof in mathematics one customarily marks off the boundaries of the sequence of expressions constituting the proof by writing "Proof:" at the beginning and "QED" or "This completes the proof." or "∎" at the end. At the very least

the proof is enclosed in a paragraph by indenting. One way or another the sequence is bracketed so that it will be seen as a single unit - which is to say, an expression of length one, hence a phrase, rather than a string of phrases.

A piece of literature is usually not so much a sequence of sentences as it is a string of larger linguistic units which are themselves strings of sentences or other phrases. A play is a sequence of acts, which in turn are sequences of scenes, which in turn are sequences of speeches, which in turn are sequences of sentences (and other utterances or gestures) by individual characters in the play. Various devices are used to denote or display these units in the script and in the performance of a play.

The act of marking the boundaries of an expression in order to form a single phrase let us call a "bracketing operation". Another instance of bracketing is the use of quotation marks to transform an expression into a noun of the language, thereby allowing the language the ability to make statements about itself.

What seem to be the crucial properties of a bracketing operation are summarized by the following definition:

A morphology with bracketing is a system $(E, \emptyset, \pi, ', [\ ], *, \cdot)$ such that $(E, \emptyset, \pi, ', *, \cdot)$ is a morphology and $[\ ]$ is a unary operation over E, called <u>bracketing</u>, such that

xiii) if $[x] = [y]$, then $x = y$,

xiv) $\pi[x] = [x]$,

xv) $[x]y = [xy]$,

for all x, y in E.

Note that the function of bracketing cannot be served by a special expression b, such that if $[\ ]$ is defined by

$$[x] = bx ,$$

then $[\ ]$ satisfies the above three properties, at least not in a locally finite morphology, since such an expression b would have to be of infinite degree. Note also that only one bracketing operation is ever needed, since other "labeled bracketing" operations can be defined in terms of it: If $[\ ]_c$ is defined by

$$[x]_c = [c * x], \text{ if } x \neq \emptyset, \text{ and } [\emptyset]_c = \emptyset ,$$

where c is any closed expression, then $[\ ]_c$ satisfies the bracketing properties, and if $d \neq c$ then $[x]_d \neq [x]_c$ for each $x \neq \emptyset$.

It is straightforward but rather tedious to see that every morphology can be enlarged to a morphology with bracketing.

Since bracketing operations in a language properly belong to the morphology of the language, this brief discussion of them is included to show how they may be easily incorporated into the present algebraic approach. No further mention of bracketing will be made in this paper.

5. Linear morphologies. Morphemes, hence expressions in general, can be two- or three-dimensional objects with blanks positioned in them like atoms in an organic molecule. The radicals and characters of Chinese calligraphy are good examples of fundamentally two-dimensional, non-linear phrases. On the other hand the morphemes of a language (like English) can be essentially one-dimensional or linear in that each consists of alphabetical symbols (or phonemes) and blanks written (or spoken) in sequence.

Let S be an arbitrary set of things called an alphabet of symbols. Let $N = \{\underline{1}, \underline{2}, \ldots\}$ be a denumerable set of numerals, disjoint from S. Let W be the set of all non-null finite sequences in $S \cup N$, and let E be the set of all finite sequences in W. Define, for x and y in E, the sequences x', $x * y$, and xy in E as follows:

x' = the result of replacing each numeral $\underline{n}$ in x by $\underline{n+1}$,

$$(x * y)_i = \begin{cases} x_i, & \text{if } 0 < i \leq m, \\ y_{i-m}, & \text{if } m < i \leq m+n, \end{cases}$$

for $i = 1, \ldots, m+n$, where m and n are the lengths of x and y, respectively, and

$$(xy)_i = \begin{cases} \emptyset, & \text{if } y = \emptyset, \\ \text{the result of substituting } y_k \text{ for } \underline{k} \text{ in } x_i \text{ modulo the length of } y, \text{ for each } \underline{k}, & \text{if } y \neq \emptyset, \end{cases}$$

for each i. Then $(E, \emptyset, \pi, {}', *, \cdot)$, where $\emptyset$ is the null sequence and π is the sequence whose only term is the sequence whose only term is $\underline{1}$, is called the total linear morphology over S. Any submorphology of this is a linear morphology over S.

An important class of linear morphologies is the so-called "Polish notation". Let us define a Łukasiewicz morphology to be a linear morphology over a set S, generated by a vocabulary each of whose members is of the form

(s)

or of the form

$(s\underline{1}\ldots\underline{n})$

with s in S. (We are denoting a sequence in $S \cup N$ by a string of juxtaposed symbols and a sequence of sequences by means of parentheses and commas.) For example, the morphollogy underlying the propositional calculus expressed in Polish notation is a Łukasiewicz morphology over {N, K, A, C, p, q, ...} whose vocabulary is $\{(N\underline{1}), (K\underline{1}\underline{2}), (A\underline{1}\underline{2}), (C\underline{1}\underline{2}), (p), (q), \ldots\}$. In this morphology the phrase

(CNKpqANpNq)

expresses one of DeMorgan's Laws and factors into morphemes like this:

$(C\underline{1}\underline{2})((N\underline{1})(K\underline{1}\underline{2})(p*q)*(A\underline{1}\underline{2})((N\underline{1})(p)*(N\underline{1})(q)))$.

Lukasiewicz morphologies are "monotectonic" in that the factorization of an expression into morphemes is essentially unique. Presently we shall see that, up to isomorphism, these are the only morphologies which are monotectonic. First let us provide a precise definition of this notion and of factorization.

6. Factorization of phrases. Given a set V of phrases in a morphology, we define the set of V-factorizations recursively by

1) if x is a closed member of V or is a blank, then the one-tuple (x) is a V-factorization;

2) if x in V is of positive degree and $F_1, \ldots, F_n$ are V-factorizations, then the tuple $(x, F_1, \ldots, F_n)$ is a V-factorization.

The product $\overline{F}$ of a V-factorization F is defined recursively by

3) if x is a closed member of V or is a blank, then $\overline{(x)} = x$;

4) if $(x, F_1, \ldots, F_n)$ is a V-factorization, then

$$\overline{(x, F_1, \ldots, F_n)} = x(\overline{F_1}*\ldots*\overline{F_n}).$$

If $\overline{F} = x$, then F is said to be a V-factorization of x. It is easy to see that if V is a vocabulary for a morphology, then every phrase has at least one V-factorization. If each phrase has just one V-factorization, call V and the morphology monotectonic. Otherwise V is polytectonic, as is a morphology all of whose vocabularies are polytectonic.

The fact that a morphology is polytectonic may be associated with "structural ambiguity" in the language based on that morphology, in the sense that one cannot tell exactly which lexical items were used in the construction of an expression and how they were used. For example,

the union of the union of A and B

in the mathematical English of set theory could refer to $(\bigcup A)\cup B$ or to $\bigcup(A\cup B)$, depending on how the phrase is parsed. The formatives involved are

x = the union of (1) and (2),
y = the union of (1),
a = A, and b = B,

all distinct. The morphology is a linear morphology. The ambiguity is that $x(ya*b) = yx(a*b)$. The vocabulary is polytectonic since this phrase has two different factorizations: (x, ((y, (a)), (b))) and (y, (x, ((a), (b)))).

But the morphology of a language may be polytectonic without there being structural ambiguity in any real sense. For instance the morphology might have a vocabulary such that any permutation of the component factorizations in any factorization yields a factorization with the same product. This would be the case if the blanks bore no particular relationship, spatial or otherwise, to one another in a morpheme. (Think of a morpheme of degree n as a recepticle or box to contain an unordered set of no more than n phrases.) If this be the case, one would not want to say that there is structural ambiguity on account of the permutability of components.

In order to state some facts about monotectonic vocabularies, we need a few more definitions.

If an expression x is such that, for some n sufficiently large, $x((\ldots i-1)*y*(i+1\ldots n)) = x$ for every phrase y, then x will be said to be <u>free of the i-th blank</u>. The <u>number of blanks</u> in an expression is $n-k$, where n is the degree and k is the number of blanks, among the first n, of which the expression is free. (So the degree of an expression need not be the same as the number of blanks in it.) An expression is <u>reduced</u> if the number of blanks in it is the same as its degree.

<u>Theorem</u>: For every expression x there exist elements y and z of the blank-morphology such that xy is reduced and $(xy)z = x$. Hence each vocabulary for a morphology may be replaced in a one-one fashion by

a vocabulary whose members are reduced.

Theorem: Every member of a monotectonic vocabulary is already reduced.

Since there is no real loss of generality in assuming that morphemes are always reduced (by the first theorem) let us mean henceforth by "vocabulary" a vocabulary all of whose members are reduced.

Call a member $(j_1 \ldots j_n)$ of the blank-morphology a permutation if the string $(j_1 \ldots j_n)$ is obtained from $(\ldots n)$ by a permutation of this string of blanks.

Theorem: Given two (reduced) vocabularies for a monotectonic morphology, corresponding to each morpheme v in one of them there is a unique morpheme w in the other and a permutation p such that $v = wp$. Thus a monotectonic morphology has essentially just one vocabulary (modulo permutations on the right).

The main result of this section is the fact that Łukasiewicz morphologies are, up to isomorphism, the only ones that are monotectonic:

Theorem: If a morphology with vocabulary V is monotectonic, then it is isomorphic to a Łukasiewicz morphology over V as a set of symbols. Conversely, every morphology isomorphic to a Łukasiewicz morphology is monotectonic.

The proofs of these theorems are not immediate but require nothing devious.

7. First-degree morphologies.

As remarked in the introduction, most grammars studied in mathematical linguistics have been based on a semigroup of expressions rather than on a half-ring or other kind of system. We wish now to include these semigroup-morphologies as morphologies in our present sense of the word. Since we are thinking of sentences - the main concern of grammar - as being built primarily by composition rather than by concatenation, we shall want these semigroups to be composition-semigroups in the appropriate half-rings. That this is always possible is the point of the next theorem.

By a first-degree morphology we mean a morphology whose reduced phrases are all of degree no greater than one.

Theorem: Every semigroup with unity is the semigroup of reduced phrases in some first-degree morphology. (π is the unity element.)

In particular, the concatenative semigroup underlying traditional grammars can be enlarged to a half-ring which is a monotectonic morphology. But what was called "concatenation" becomes composition, what was the "null string" becomes the first blank, and each morpheme is to be thought of as having a single blank where the "next" morpheme goes as bigger phrases are constructed. And one may wish to add a formative of degree zero, called a "period" or "stop" or "boundary marker", with which to close a phrase that is to be a sentence. This too can be included in the enlargement.

8. Interpretations. Let us denote morphologies by merely naming their sets of expressions, always using the same notation to denote the algebraic structure (π, ', $*$, $\cdot$).

By an interpretation of a morphology A in a morphology B we shall mean nothing more nor less than a homomorphism of A into B, i.e., a mapping $\Theta: A \longrightarrow B: x \longrightarrow x^{\Theta}$ which preserves operations:

$$\pi^{\Theta} = \pi, \quad x'^{\Theta} = x^{\Theta}{}', \quad (x * y)^{\Theta} = x^{\Theta} * y^{\Theta}, \text{ and } (xy)^{\Theta} = x^{\Theta}y^{\Theta},$$

for all x, y in A. We shall refer to the image of A in B also as "the interpretation of A"(under Θ) and will call A a formulation of of its image. Thus to say that one morphology can be formulated in another is to say that there is an interpretation mapping the latter onto the former.

It is also suggestive to say that a morphology A is "expressed in" or is "translated into" a morphology B when we are speaking of an interpretation of A in B.

Here are some facts about interpretations:

Theorem: Under any interpretation of one morphology in another, the blank-morphology of the first maps isomorphically onto that of the other.

Theorem: The length of an expression is always preserved under an interpretation, and the degree is never increased. (It may decrease.)

Call a mapping of a subset of one morphology into another conservative if it preserves length and does not increase degree. A morphology is freely generated by a vocabulary if every conservative mapping of that vocabulary can be extended to an interpretation of the whole morphology. Call a morphology <u>free</u> if it possesses a vocabulary by which it is freely generated.

<u>Theorem</u>: The free morphologies are precisely those which are isomorphic to Łukasiewicz morphologies. Hence a morphology is free if and only if it is monotectonic.

<u>Corollary</u>: Every morphology is the interpretation of some Łukasiewicz morphology. Thus every morphology has a monotectonic formulation.

The proofs are straightforward but a bit tedious.

Suppose now that A is the morphology of some "language of ideas". That is, suppose the concatenating and composing of concepts in some system of thought takes place in A. Suppose B is the morphology of a natural language which expresses those ideas, i.e., that there is a homomorphism of the conceptual system into the language so that B is an interpretation of A. If the homomorphism is not an isomorphism, then detail is lost, structure becomes simplified, and ambiguity of some kind may be introduced. In particular, to the extent that the language harbours structural ambiguities of a morphological nature, they may be viewed as the result of a blurring of distinctions made in A when this system is translated into B. (B is closer to what Chomsky and others have called the "surface structure of a language" and A is closer to the "deep structure".) It may be that structural ambiguity can always be accounted for in terms of an interpretation which maps the more monotectonic morphology of concepts into a polytectonic morphology of utterances or writing. That is, it may be reasonable to locate the source of an ambiguity always in the morphology of a language rather than in its grammar.

Proceeding now in the other direction, structural ambiguity inherent in a system of thought can be resolved by formulating the system in, say, a language based on the first-order predicate calculus in Polish notation. If we assume that every language expresses some system of thought and that every conceptual system can be formulated in a language based on a monotectonic morphology, and if we assume that the translations involved are indeed homomorphisms of the underlying algebras of expressions, then we may conclude that every language is based on a morphology in the sense of our definition. On the other hand, if what we mean by "language" is such that every homomorphic image of a first-order predicate calculus in Polish notation is a language, then we may conclude that every system satisfying the postulates of our definition is at least a part of the morphology of some language. Thus Postulates i through xii appear to be just enough to characterize the intuitive (but abstract) notion of the morphology of a language with which we began.

9. Grammars based on half-ring morphologies. The analogue of a context-free grammar (in the sense of Chomsky *et al*) for half-ring morphologies is a system consisting of a number of variables α, β, ..., σ, ..., to stand for various grammatical categories (with σ to stand for the category of sentences), and a list of rules of the form

$$\xi \longrightarrow P\,,$$

where ξ is a single variable and P is a polynomial made up of variables and vocabulary symbols combined by indicated compositions and concatenations (to be performed in the given morphology). And just as every "first degree" context-free grammar is equivalent to one whose rules are all of the form

$$\alpha \longrightarrow \beta\gamma,$$

α, β, and γ being single variables, or of the form

$$\alpha \longrightarrow v\,,$$

where v is a morpheme, every "context-free grammar" based on an arbitrary half-ring morphology is equivalent to one with rules of the above form plus rules of the form

$$\alpha \longrightarrow \beta \cdot \gamma.$$

Sentences are the formulas generated by the rules, starting with σ. If the morphology is a first-degree linear morphology, then the sets generated by such grammars are just sets of strings of symbols from the alphabet, and the additional power of a grammar due to rules involving both composition and concatenation can be compared with that of a grammar with rules involving composition (= "concatenation") alone.

The power of context-free rules based on a half-ring morphology is yet to be investigated. It is known that many more sets can be generated by context-free rules involving both operations than by rules based only on composition. (For instance it is easy to write a list of rules which will generate the non-"context-free" set $\{a^n b^n a^n : n = 1, 2, \text{etc.}\}$ in the linear morphology with vocabulary consisting of (a1) and (b1).)

Context-free grammars based on the simple "concatenative" morphologies used in the past are inadequate to handle the grammar of natural languages. [6] The extra power of "context-free" rules which are based on the kind of morphology introduced in this paper supports the hope that these rules may succeed in many places where the earlier ones have failed.

REFERENCES

1. Chomsky, N.: Aspects of the theory of syntax. Cambridge: M.I.T. Press, 1965.
2. Curry, H.B.: Some logical aspects of grammatical structure. In R. Jakobson (ed.), Structure of language and its mathematical aspects, Proceedings of the Twelfth Symposium in Applied Mathematics, pp. 56-68. Providence: Amer. Math. Soc., 1961.
3. Davis, A.S.: An axiomatization of the algebra of transformations over a set. Math. Annalen 164, 372-377 (1966).
4. Fodor, J.A.; Katz, J.J. (eds.): The structure of language: readings in the philosophy of language. Englewood Cliffs, N.J.: Prentice-Hall, 1964.
5. Ginsburg, S.: The mathematical theory of context-free languages. New York: McGraw-Hill, 1966.
6. Postal, P.M.: Limitations of phrase structure grammars. In Fodor and Katz (eds.), The structure of language. 1964.

Department of Mathematics
The University of Oklahoma
Norman, Oklahoma, U. S. A.

FORMALISATIONS OF SOME $\aleph_0$-VALUED ŁUKASIEWICZ PROPOSITIONAL CALCULI

ALAN ROSE

It has been shown[1] that the $\aleph_0$-valued Łukasiewicz propositional calculus[2] with one designated truth-value may be formalised completely by means of the axiom schemes

Ax 1 CPCQP,

Ax 2 CCPQCCQRCPR,

Ax 3 CCCPQQCCQPP,

Ax 4 CCNPNQCQP

and the rule of modus ponens. The object of the present paper is to formalise, for all rational numbers r such that $0<r<1$, the corresponding propositional calculus whose designated truth-values are the rational numbers x such that $r \leqslant x \leqslant 1$.

We define the functors K, L, B as in the previous paper[1] and then make the definitions

$M_{\alpha-1,\alpha}P =_{df} KB^{\alpha-2}P^{\alpha-1}NP \quad (\alpha=2, 3, \dots),$

$M_{\alpha-\beta,\alpha}P =_{df} L^{\beta-1}(M_{\alpha-1,\alpha}P)^{\beta} \quad (2 \leqslant \beta \leqslant \alpha-1, (\alpha, \beta)=1; \alpha=2, 3, \dots).$

Thus, if P is capable of taking all truth-values, the highest truth-value taken by $M_{\beta\alpha}P$ is β/α $(1 \leqslant \beta \leqslant \alpha-1, (\alpha, \beta)=1; \alpha=2, 3, \dots)$.

Next we define functors $G_{\beta\alpha}$ such that if P, $G_{\beta\alpha}P$ take the truth-values x, $g_{\beta\alpha}(x)$ respectively then $g_{\beta\alpha}(x)=1$ if and only if $x \geqslant \beta/\alpha$ $(1 \leqslant \beta \leqslant \alpha-1, (\alpha, \beta)=1; \alpha=2, 3, \dots)$. We note that there is an effective method[1] of constructing a formula $H_{\beta\alpha}P$ such that if P, $H_{\beta\alpha}P$ take the truth-values x, $h_{\beta\alpha}(x)$ respectively then

$h_{\beta\alpha}(x)=\max(0, \min(1, \alpha x-\beta+1)) \quad (1 \leqslant \beta \leqslant \alpha-1, (\alpha, \beta)=1; \alpha=2, 3, \dots).$

Since $\alpha x-\beta+1 \geqslant 1$ if and only if $x \geqslant \beta/\alpha$, it follows at once that we may make the definitions

$G_{\beta\alpha}(P) =_{df} H_{\beta\alpha}(P) \quad (1 \leqslant \beta \leqslant \alpha-1, (\alpha, \beta)=1; \alpha=2, 3, \dots).$

---

[1] A. Rose and J. B. Rosser, Fragments of many-valued statement calculi, Trans. Amer. Math. Soc., 87 (1958), 1-53; C. A. Meredith, The dependence of an axiom of Łukasiewicz, Ibid, 54; C. C. Chang, Proof of an axiom of Łukasiewicz, Ibid., 55-56.

[2] J. Łukasiewicz and A. Tarski, Untersuchungen über den Aussagenkalkül, Comptes rendus (Warsaw) Classe III, 23 (1930), 30-50.

Let us now suppose that $r=\beta/\alpha$ $((\alpha, \beta)=1)$ and consider the formalisation which uses the axiom schemes A1-4 and primitive rules of procedure R1, R2 given below, where P, Q, R are formulae, S is a propositional variable, for A1, A3, A4 and R1 S does not occur in P, Q and for A2 S does not occur in P, Q, R and for R2 S does not occur in P.

A1 $NCCPCQPM_{\alpha-\beta,\alpha}S$

A2 $NCCCPQCCQRCPRM_{\alpha-\beta,\alpha}S$

A3 $NCCCCPQQCCQPPM_{\alpha-\beta,\alpha}S$

A4 $NCCCNPNQCQPM_{\alpha-\beta,\alpha}S$

R1 If $NCPM_{\alpha-\beta,\alpha}S$ and $NCCPQM_{\alpha-\beta,\alpha}S$ then $NCQM_{\alpha-\beta,\alpha}S$.

R2 If $NCG_{\beta\alpha}PM_{\alpha-\beta,\alpha}S$ then P.

In order to establish the plausibility of our formalisations we first establish that, if P is a formula and S is a propositional variable not occurring in P then P takes the truth-value 1 always if and only if the truth-value of $NCPM_{\alpha-\beta,\alpha}S$ is always designated. If P always takes the truth-value 1 then, if $M_{\alpha-\beta,\alpha}S$ takes the truth-value z, $NCPM_{\alpha-\beta,\alpha}S$ takes the truth-value 1-z. Since

$$z\leq(\alpha-\beta)/\alpha,$$

$$1-z\geq\beta/\alpha$$

and the truth-value of $NCPM_{\alpha-\beta,\alpha}S$ is always designated. If, conversely, the truth-value of $NCPM_{\alpha-\beta,\alpha}S$ is always designated and P takes the truth-value x then, with the above notation,

$$1-\min(1,\ 1-x+z)\geq\beta/\alpha.$$

Hence

$$\max(0,\ x-z)\geq\beta/\alpha$$

and it follows at once that

$$x-z\geq\beta/\alpha.$$

Thus, if the truth-value of S is such that $z=(\alpha-\beta)/\alpha$, then, whatever the truth-values of the propositional variables occurring in P (since S does not occur in P),

$$x-(\alpha-\beta)/\alpha\geq\beta/\alpha.$$

Hence $x\geq1$ and it follows at once that P always takes the truth-value 1.

It follows at once from the above result that all instances of A1-4 take designated truth-values exclusively. It is therefore now sufficient for us to establish, for both primitive rules of procedure, that if the premiss(es) take(s) designated truth-values exclusively, so does the conclusion.

For R1 we note that, if $NCPM_{\alpha-\beta,\alpha}S$ and $NCCPQM_{\alpha-\beta,\alpha}S$ take designated truth-values exclusively then P, CPQ always take the truth-value 1. Hence Q always takes the truth-value 1 and the truth-value of $NCQM_{\alpha-\beta,\alpha}S$ is always designated.

For R2 we note that, if the truth-value of $NCG_{\beta\alpha}PM_{\alpha-\beta,\alpha}S$ is always designated then $G_{\beta\alpha}P$ always takes the truth-value 1. Hence, if P takes the truth-value x,

$$x \geqslant \beta/\alpha = r.$$

Thus the truth-value of P is always designated.

In order to establish completeness we first prove a lemma.

Lemma. *If P always takes the truth-value 1 then, for any propositional variable S which does not occur in the proof of P, the formula $NCPM_{\alpha-\beta,\alpha}S$ is provable in the formalisation.*

Since P always takes the truth-value 1 it is provable by means of a finite number of instances of Ax 1-4 and the rule of *modus ponens*. Let us suppose that, in this proof, the rule of *modus ponens* is used exactly l times. We shall prove the lemma by strong induction on l.

If $l = 0$ then P is an instance of one of Ax1-4 and S does not occur in P. Hence $NCPM_{\alpha-\beta,\alpha}S$ is an instance of one of A1-4. We now assume the lemma for all non-negative integers less than l and deduce it for l.

The last step in the proof of P is, for some formula Q, an application of the rule of *modus ponens* to the formulae Q, CQP. Since S does not occur in either of the proofs of Q, CQP it follows from our induction hypothesis that the formulae $NCQM_{\alpha-\beta,\alpha}S$, $NCCQPM_{\alpha-\beta,\alpha}S$ are provable in the present formalisation. Applying R1 to these formulae we derive the formula $NCPM_{\alpha-\beta,\alpha}S$. Thus the lemma is proved.

Proof of the Main Theorem. Let R be a formula whose truth-value is always designated. Thus the formula $G_{\beta\alpha}R$ always takes the truth-value 1 and, by the Lemma, the formula $NCG_{\beta\alpha}RM_{\alpha-\beta,\alpha}S$ is, for some propositional variable S not occurring in R, provable in the formalisation. Applying R2 we infer the formula R. Thus the formalisation is complete.

The University,
Nottingham.

THEORIES WHICH ARE NOT $\aleph_0$-CATEGORICAL

by Joseph G. Rosenstein

A well-known theorem (Engeler, Ryll-Nardzewski, Svenonius) characterizes the $\aleph_0$-categorical theories T as those for which $B_n(T)$ is finite for each n (where $B_n(T)$ is the Boolean algebra of equivalence classes of well-formed formulas in the variables $x_1,\ldots,x_n$ with respect to the equivalence relation

$$T \vdash (x_1)\ldots(x_n)(\varphi(x_1,\ldots,x_n) \longleftrightarrow \psi(x_1,\ldots,x_n)) .$$

One immediately asks whether a better result is possible--namely, whether there is an m such that T is $\aleph_0$-categorical whenever $B_m(T)$ is finite. To deny this, it is sufficient to exhibit for each n a system M_n for which $B_n(T(M_n))$ is finite but $B_{n+1}(T(M_n))$ is infinite. That is precisely the purpose of this paper.

The system M_n is a partial ordering in which transitivity is vacuously satisfied--i.e., no element satisfies both $(Ey)(x<y)$ and $(Ey)(y<x)$, so that M_n consists of a top set and a bottom set with each element on the bottom being less than certain elements on the top. Thus the universe of M_n can be written as the disjoint union $\{a_i\}_{i\in N} \cup \{b_j\}_{j\in N}$, and if $d<c$ then d is b_j for some (unique) j and c is a_i for some (unique) i. We shall denote $\{b_j \mid b_j < a_i\}$ by $\underline{a}_i$.

To guarantee that $B_{n+1}(T(M_n))$ is infinite, we construct M_n so that for each k the well-formed formula $(E_k z)(z<x_1 \wedge \ldots \wedge z<x_{n+1})$ is satisfied in M_n. ($(E_k z)(\ldots)$ is of course an abbreviation for a well-formed formula which would normally be interpreted "there are exactly k distinct z's such that...")

The same shouldn't happen to $B_n(T(M_n))$ so we will be particularly careful and have $\underline{a}_{i_1} \cap \underline{a}_{i_2} \cap \ldots \cap \underline{a}_{i_n}$ infinite for any choice of $i_1, i_2, \ldots, i_n$. Of course more than that will be necessary in order to make sure that $B_n(T(M_n))$ is finite; we shall in fact arrange matters so that whenever two n-tuples $\langle r_1,r_2,\ldots,r_n\rangle$ and $\langle s_1,s_2,\ldots,s_n\rangle$ have the same "configuration" (i.e., for each i and j, $r_i = r_j$ if and only if $s_i = s_j$, $r_i < r_j$ if and only if $s_i < s_j$, and r_i is a top (bottom) element if and only if s_i is a top (bottom) element, then there is an automorphism π of M_n such that $\pi(r_i) = s_i$ for $1 \leq i \leq n$. This symmetry guarantees that the well-formed formula which describes the "configuration" of the n-tuple is an atom of $B_n(T(M_n))$, and, since there are but a finite number of "configurations" of n elements, the number of such atoms is finite; but since any n-tuple has a configuration, it satisfies some atom, so we may conclude that $B_n(T(M_n))$ is generated by these atoms, hence is finite. This symmetry will be realized by guaranteeing that for any set of top elements and any coherent way of choosing for each n-element subset $a_{i_1},\ldots,a_{i_n}$ of these a finite subset B of $\underline{a}_{i_1} \cap \ldots \cap \underline{a}_{i_n}$ there are top elements a which simultaneously realize all of these intersections in the sense that

$$\underline{a} \cap (\underline{a}_{i_1} \cap \ldots \cap \underline{a}_{i_n}) = B$$

for each $i_1,\ldots,i_n$ and corresponding B.

Thus we shall define a relation $<$ on $\{a_i\}_{i\in N} \cup \{b_j\}_{j\in N}$ so that

(i) if $d < c$ then d is b_j for a unique j and c is a_i for a unique i,

(ii) $\underline{a}_{i_1} \cap \underline{a}_{i_2} \cap \ldots \cap \underline{a}_{i_n}$ is infinite for each $i_1,i_2,\ldots,i_n$,

(iii) $a_{i_1} \cap \underline{a}_{i_2} \cap \ldots \cap \underline{a}_{i_n} \cap \underline{a}_{i_{n+1}}$ is finite if $i_1,i_2,\ldots,i_{n+1}$ are all different,

(iv) if B is a finite subset of $\{b_j\}_{j\in N}$ then there are infinitely many i's for which $B \subseteq \underline{a}_i$,

(v) if K is finite and for each n-element subset L of K we are given a finite set $B_L \subseteq (\bigcap_{i\in L} \underline{a}_i)$ such that

$$(\bigcup_{L\in\mathcal{L}} B_L) \cap (\bigcap_{i\in\mathcal{L}} \underline{a}_i) = B_L$$

for each $L \in \mathcal{L}$ (where $\mathcal{L}$ is the set of all n-element subsets of K), then there are infinitely many distinct i's for which $\underline{a}_i \cap (\bigcap_{i\in L} \underline{a}_i) = B_L$ for each $L \in \mathcal{L}$.

Before defining the relation $<$ we shall see why the above properties do guarantee that M_n serves our purposes.

<u>Theorem</u>: Let $\langle r_1,r_2,\ldots,r_n\rangle$ and $\langle s_1,s_2,\ldots,s_n\rangle$ be two n-tuples of elements of M_n such that for each i and j, $r_i = r_j$ if and only if $s_i = s_j$, $r_i < r_j$ if and only if $s_i < s_j$ and r_i is a top (bottom) element of M_n if and only if s_i is a top (bottom) element of M_n. Then there is an automorphism π of M_n such that $\pi(r_i) = s_i$ for $1 \leq i \leq n$.

__Proof:__ Let $\langle a_{i_1},\dots,a_{i_t}\rangle$ be an arrangement of the distinct top elements among $\{r_1,\dots,r_n\}$ and let $\langle a_{j_1},\dots,a_{j_t}\rangle$ be the corresponding arrangement of the distinct top elements among $\{s_1,\dots,s_n\}$. Let $\langle b_{i_1},\dots,b_{i_c}\rangle$ be an arrangement of the distinct bottom elements among $\{r_1,\dots,r_n\}$ and let $\langle b_{j_1},\dots,b_{j_c}\rangle$ be the corresponding arrangement of the distinct bottom elements among $\{s_1,\dots,s_n\}$.

So that we can apply (v), we select $a_{i_{t+1}},\dots,a_{i_n}$ and $a_{j_{t+1}},\dots,a_{j_n}$. Using (iv) let $a_{i_{t+1}},\dots,a_{i_n}$ be $n-t$ distinct a's for which $\{b_{i_1},\dots,b_{i_c}\}\subseteq \underline{a}$ and let $a_{j_{t+1}},\dots,a_{j_n}$ be $n-t$ distinct a's for which $\{b_{j_1},\dots,b_{j_c}\}\subseteq \underline{a}$. We shall construct an automorphism π of M_n such that $\pi(a_{i_x})=a_{j_x}$ for $1\le x\le n$ and $\pi(b_{i_x})=b_{j_x}$ for $1\le x\le c$.

The construction of π will be done in stages. We assume that at the end of stage k we have defined two sequences $\langle a_{i_1},a_{i_2},\dots,a_{i_{n+k}}\rangle$ and $\langle a_{j_1},a_{j_2},\dots,a_{j_{n+k}}\rangle$. Let $R_k=\bigcup_{K\in\mathcal{K}}(\bigcap_{x\in K}\underline{a}_{i_x})\cup\{b_{i_1},\dots,b_{i_c}\}$ and $S_k=\bigcup_{K\in\mathcal{K}}(\bigcap_{x\in K}\underline{a}_{j_x})\cup\{b_{j_1},\dots,b_{j_c}\}$ where $\mathcal{K}$ is the set of all $n+1$-element subsets of $\{1,\dots,n+k\}$. (Note that R_k and S_k are finite.) We assume also that we have defined $\pi_k\colon R_k\cup\{a_{i_1},\dots,a_{i_{n+k}}\}\to S_k\cup\{a_{j_1},\dots,a_{j_{n+k}}\}$ so that π_k is 1-1 onto, $\pi_k(a_{i_x})=a_{j_x}$ for $1\le x\le n+k$, $\pi_k(b_{i_x})=b_{j_x}$ for $1\le x\le c$, and that if $b\in R_k$ then $b<a_{i_x}$ if and only if $\pi_k(b)<a_{j_x}$.

At stage $k+1$ we will define $a_{i_{n+k+1}}$, $a_{j_{n+k+1}}$, R_{k+1}, S_{k+1} and we will extend π_k to $\pi_{k+1}\colon R_{k+1}\cup\{a_{i_1},\dots,a_{i_{n+k+1}}\}\to S_{k+1}\cup\{a_{j_1},\dots,a_{j_{n+k+1}}\}$. It will be clear from our definitions that $\bigcup_k R_k=\bigcup_k S_k=\{b_j\}_{j\in N}$, that every a_i is some a_{i_x} and is some a_{j_y}, and that $\bigcup_k \pi_k$ is indeed an automorphism of M_n with the required properties.

__Stage $k+1$:__ Assume that k is even and let $a_{i_{n+k+1}}$ be the first a_i not occurring among $a_{i_1},\dots,a_{i_{n+k}}$. (If k is odd, we let $a_{j_{n+k+1}}$ be the first a_i not occurring among $a_{j_1},\dots,a_{j_{n+k}}$ and proceed analogously to find $a_{i_{n+k+1}}$.)

Let $A_L=\underline{a}_{i_{n+k+1}}\cap(\bigcap_{x\in L}\underline{a}_{i_x})$ for each $L\in\mathcal{L}$ (where $\mathcal{L}$ is the set of all n-element subsets of $\{1,\dots,n+k\}$) and let $A_L'=A_L-R_k$. Note that the A_L' are pairwise disjoint and that the A_L are finite.

Let B_L' be a subset of $(\bigcap_{x\in L}\underline{a}_{j_x}-S_k)$ with exactly $|A_L'|$ elements (this is possible because of (ii)) and let $B_L=B_L'\cup\pi_k(A_L\cap R_k)$. Then $B_L\subseteq\bigcap_{x\in L}\underline{a}_{j_x}$ and

$$(\bigcup_{L\in\mathcal{L}} B_L) \cap \bigcap_{x\in L_o} \underline{a}_{j_x} = \bigcup_{L\in\mathcal{L}} (B_L \cap \bigcap_{x\in L_o} \underline{a}_{j_x}) = \bigcup_{L\in\mathcal{L}} ([B'_L \cup \pi_k(A_L \cap R_k)] \cap \bigcap_{x\in L_o} \underline{a}_{j_x})$$

$$= \bigcup_{L\in\mathcal{L}} (B'_L \cap \bigcap_{x\in L_o} \underline{a}_{j_x}) \cup \bigcup_{L\in\mathcal{L}} (\pi_k(A_L \cap R_k) \cap \bigcap_{x\in L_o} \underline{a}_{j_x}) = B'_{L_o} \cup \pi_k(A_{L_o} \cap R_k) = B_{L_o}$$

for each $L_o \in \mathcal{L}$. Hence by (v) we can find an $a_{j_{n+k+1}}$ distinct from $a_{j_1},\ldots,a_{j_{n+k}}$ so that

$$\underline{a}_{j_{n+k+1}} \cap (\bigcap_{c\in L} \underline{a}_{j_c}) = B_L$$

for each $L \in \mathcal{L}$. Let

$$R_{k+1} = \bigcup_{K\in\mathcal{K}} (\bigcap_{c\in K} \underline{a}_{i_c}) \cup \{r_{i_1},\ldots,r_{i_t}\}$$

$$S_{k+1} = \bigcup_{K\in\mathcal{K}} (\bigcap_{c\in K} \underline{a}_{j_c}) \cup \{s_{i_1},\ldots,s_{i_t}\}$$

where $\mathcal{K}$ is now the set of all n+1-element subsets of $\{1,\ldots,n+k+1\}$ and define

$$\pi_{k+1}: R_{k+1} \cup \{a_{i_1},\ldots,a_{i_{n+k+1}}\} \to S_{k+1} \cup \{a_{j_1},\ldots,a_{j_{n+k+1}}\}$$

by $\pi_{k+1} | D\pi_k = \pi_k$, $\pi_{k+1}(a_{i_{n+k+1}}) = a_{j_{n+k+1}}$, and $\pi_{k+1}: A'_L \to B'_L$ any old 1-1 way (for each $L \in \mathcal{L}$.)

We leave it to the reader to verify that the induction hypotheses are preserved and that the claims concerning the result of the construction are true.

We claim that it follows that $B_n(T(M_n))$ is finite and that $B_{n+1}(T(M_n))$ is infinite. The latter claim is easily substantiated; indeed, by (ii) and (v) the formulas

$$(E_k z)(z < x_1 \wedge \ldots \wedge z < x_n)$$

are satisfied by distinct n-tuples of elements of M_n.

To see that $B_n(T(M_n))$ is finite, we note that any n-tuple has a unique configuration and hence satisfies a unique "configuration formula" φ. Now if ψ is any element of $B_n(T(M_n))$ then because of the theorem above, either

$$T(M_n) \vdash (x_1)\ldots(x_n)(\varphi \to \psi) \quad \text{or}$$

$$T(M_n) \vdash (x_1)\ldots(x_n)(\varphi \to \sim \psi) ;$$

hence ψ is equivalent (in $T(M_n)$) to the disjunction of some subset of the set of configuration formulas. Hence $B_n(T(M_n))$ is finite.

It remains only to define $<$ so that (i)-(v) hold. This will be done by stages--at stage k we will define $\underline{a}_k$. We assume before we start stage $k+1$ that (ii) and (iii) hold when $i_1,\ldots,i_{n+1}$ come from $1,\ldots,k$ and also that infinitely many b's have been placed in none of $\underline{a}_1,\ldots,\underline{a}_k$. At the end of stage $k+1$ (ii) and (iii) will continue to hold and in addition we will have met one of the requirements in (iv) or (v). In other words, we assume that we have listed all the requirements of (iv) and (v) so that each appears infinitely many times. Thus at stage $k+1$, we will try to do one of two things.

The first possibility is that we are presented with a finite subset B of $\{b_j\}_{j\in N}$ and we want $B \subseteq \underline{a}_{k+1}$; in this case, noting that by the induction hypothesis if $N \subseteq \{1,2,\ldots,k\}$ with $\leq n$ elements then $(\bigcap_{x\in N} \underline{a}_x) \cap (\bigcap_{\substack{x\notin N \\ x\leq k}} \overline{\underline{a}_x})$ has infinitely many elements, for each $N \subseteq \{1,2,\ldots,k\}$ with $\leq n$ elements we divide $(\bigcap_{x\in N} \underline{a}_x) \cap (\bigcap_{\substack{x\notin N \\ x\leq k}} \overline{\underline{a}_x})$ into two infinite pieces N_1 and N_2, and we set $\underline{a}_{k+1} = (\bigcup_{N\in\mathfrak{n}} N_1) \cup B$ (where $\mathfrak{n}$ is the set of all subsets of $\{1,2,\ldots,k\}$ with fewer than n elements.) Noting that for distinct N's in $\mathfrak{n}$ the sets above are disjoint, we can conclude that (ii) and (iii) continue to hold. Of course, we have succeeded in getting $B \subseteq \underline{a}_{k+1}$.

The second possibility is that we are presented with a case (v) requirement. (Note that if some of the a_i mentioned in the case (v) requirement have not yet been defined, we should just attack the next requirement.) We proceed as above to divide $(\bigcap_{x\in N} \underline{a}_x) \cap (\bigcap_{\substack{x\notin N \\ x\leq k}} \overline{\underline{a}_x})$ into two infinite pieces N_1 and N_2 for each $N \in \mathfrak{n}$ ($\mathfrak{n}$ as above) and we set $\underline{a}_{k+1} = (\bigcup_{N\in\mathfrak{n}} N_1) \cup (\bigcup_{L\in\mathcal{L}} B_L)$. Again (ii) and (iii) continue to hold, and again we have succeeded in meeting the requirement that presented itself at this stage.

This completes the construction of the systems M_n. I was unable to find in nature a system M_n with these properties for any $n \geq 2$.

After my talk at the Leeds Summer School, J. V. Howard pointed out examples for $n=2$ and $n=3$ which use infinitely many relations, but, except for that, do occur in nature. (There are other examples which involve an infinite number of relations.) For $n=2$, we define relations R_{ab} on the (rational) plane (a,b natural numbers) by $R_{ab}(x,y,z)$ if and only if $ax+by=(a+b)z$; it is then clear that $B_3(T(M))$ is infinite and that $B_2(T(M))$ is finite. For $n=3$, we define relations R_z on the (rational) complex plane (z (rational) complex) by $R_z(z_1,z_2,z_3,z_4)$ if and only if the cross ratio (z_1,z_2,z_3,z_4) is z ; since the cross ratio is invariant under linear transformations, and since any three distinct points can be carried to any other three distinct points by a linear transformation, it follows that $B_3(T(M))$ is finite and that $B_4(T(M))$ is infinite.

BIBLIOGRAPHY

[1] E. Engeler, "A characterization of theories with isomorphic denumerable models," Amer. Math. Soc. Notices, vol. 6 (1959), p. 161.

[2] C. Ryll-Nardzewski, "On the categoricity in power $\leq \aleph_0$," Bull. Acad. Polon. Sci. Sér. Sci. Math. Astro. Phys., vol. 7 (1959), pp. 545-548.

[3] J. G. Rosenstein, "Theories which are not $\aleph_0$-categorical," Abstract to appear in Journal of Symbolic Logic.

[4] L. Svenonius, "$\aleph_0$-categoricity in first-order predicate calculus," Theoria (Lund), vol. 25 (1959), pp. 82-94.

The Monadic Fragment of Predicate Calculus with the Chang Quantifier and Equality[1]

A. Slomson

Introduction

Let L be the language of predicate calculus with equality, which we suppose is formulated without constant or function symbols. L_Q is the language obtained from L by adding the Chang or Equicardinal quantifier Q. Q is defined by the satisfaction clause:

$$\underline{\underline{A}} \models_x Qv_n\phi \quad \text{iff} \quad \text{card}(\{a\in A : \underline{\underline{A}} \models_{x(n/a)} \phi\}) = \text{card}(A),$$

where we have used the usual notation of model theory.

Thus, intuitively, $Qx\phi(x)$ is true, iff the set of x's satisfying ϕ has the same cardinal as the whole domain.

In a finite model Q behaves like the universal quantifier, so to avoid trivial complications, in what follows by "model" we shall mean "infinite model". L_{Q_α} is the language obtained from L by adding the quantifier Q_α which is interpreted as saying "there are at least $\aleph_\alpha$". In a model of cardinal $\aleph_\alpha$, Q_α behaves like Q and so many of the results about Q and Q_α are inter-translatable. In particular it follows from Theorem 10 of Mostowski [1957] that the monadic fragment of L_Q without equality is decidable. In §1 we extend this result to the monadic fragment of L_Q with equality.

Yasuhara [1966] has provided an explicit axiomatization for the fragment of L_Q without equality. In §2 we provide an axiomatization for the monadic fragment of L_Q with equality. In effect our axiomatization is obtained by adding a single axiom for equality to Yasuhara's system. In §3 we extend the results

---

1. The results described here are contained in the author's D.Phil. Thesis, Oxford, September 1967, which was prepared while the author held an S.R.C. research studentship.

of §2 to arbitrary sets of sentences of the monadic fragment of L_Q.

§4 is devoted to pointing out that recent results of Fuhrken [1964] and Keisler [1967] imply that, assuming a certain hypothesis about limit cardinals, the Compactness Theorem holds for countable sets of sentences of L_Q. Assuming the G.C.H. we use the ultraproduct construction to extend this result to arbitrary sets of sentences of L_Q. Finally we remark that the results of Vaught [1964] and Keisler [1967] imply, given the same assumption about limit cardinals, that L_Q is recursively axiomatizable.

§1. The Decision Procedure

Let L_Q^1 be the language obtained from L_Q by dropping all predicate letters other than monadic predicate letters and equality. For the time being we restrict our attention to $L_Q^{1,k}$, the fragment of L_Q^1 whose only monadic predicate letters are $P_0,\ldots,P_{k-1}$.

A <u>type</u> is an element of 2^k. If $\alpha \in 2^k$ is a type, then $\Phi_\alpha(x)$ is the formula

$$R_0(x) \; \& \; \ldots \; \& \; R_{k-1}(x)$$

where, for $0 \leqslant i < k$, R_i is P_i if $\alpha(i) = 0$ and is $\neg P_i$ if $\alpha(i) = 1$.
If $\underline{\underline{A}}$ is a realization of $L_Q^{1,k}$ with domain A, an element $a \in A$ is said to be <u>of type α</u> iff $\underline{\underline{A}} \models \Phi_\alpha[a]$. Thus a is of type α iff

$$\text{for} \quad 0 \leqslant i < k, \quad \underline{\underline{A}} \models P_i[a] \quad \text{iff } \alpha(i) = 0.$$

$\alpha(A)$ is the set of all elements of A of type α.

Let $l = 2^k$. A <u>genus</u> is an element of ω^l which takes the value 0 for some $\alpha \in 2^k$. A realization $\underline{\underline{A}}$ of $L_Q^{1,k}$ is said to be <u>of genus β</u> if for each $\alpha \in 2^k$,

$$\text{if} \quad \text{card}\,(\alpha(A)) = \text{card}\,(A), \quad \text{then} \quad \beta(\alpha) = 0,$$

$$\text{if} \quad \aleph_0 \leqslant \text{card}\,(\alpha(A)) < \text{card}\,(A), \text{ then } \quad \beta(\alpha) = 1,$$

$$\text{and if} \quad \text{card}\,(\alpha(A)) = p < \aleph_0 \quad \text{then} \quad \beta(\alpha) = p + 2.$$

Thus β is the function which tells us "how many" elements there are of each type.

Let $\underline{\underline{A}}$, $\underline{\underline{B}}$ be realizations of $L_Q^{1,k}$ and let $a_0,\ldots,a_n$ be elements of the domain A of $\underline{\underline{A}}$, and $b_0,\ldots,b_n$ elements of the domain B of $\underline{\underline{B}}$. We write $(a_0,\ldots,a_n) \simeq (b_0,\ldots,b_n)$ iff

(i) for $0 \leqslant i \leqslant n$, a_i is of the same type as b_i

and (ii) for $0 \leqslant i < j \leqslant n$, $a_i = a_j$ iff $b_i = b_j$.

<u>Lemma 1·1</u>: Let $\underline{\underline{A}}$, $\underline{\underline{B}}$ be realizations of $L_Q^{1,k}$ of the same genus. If $a_0,\ldots,a_n \in A$, $b_0,\ldots,b_n \in B$ and $(a_0,\ldots,a_n) \simeq (b_0,\ldots,b_n)$ then for any formula $\phi(v_0,\ldots,v_n)$ of $L_Q^{1,k}$

$$\underline{\underline{A}} \models \phi[a_0,\dots,a_n] \quad \text{iff} \quad \underline{\underline{B}} \models \phi[b_0,\dots,b_n]. \tag{1}$$

Proof. The proof is by induction on the number of logical symbols in ϕ. The proof of (1) for atomic formulas and of the induction step for the logical connectives is obvious. We prove the induction step for the quantifier Q; that for $\exists$ is similar.

So suppose (1) holds for $\psi(v_0,\dots,v_n)$, $\phi = Qv_0\psi$, $\underline{\underline{A}} \models \phi[a_1,\dots,a_n]$ and $(a_1,\dots,a_n) \simeq (b_1,\dots,b_n)$. Let $X = \{a \in A : \underline{\underline{A}} \models \psi[a, a_1,\dots,a_n]\}$. By the definition of Q, card (X) = card (A) which is infinite. There are only finitely many types, hence there is some type $\alpha \in 2^k$ such that card $(\alpha(A))$ = card (A) and there is an element, a, in $X - \{a_1,\dots,a_n\}$ of type α. Since $\underline{\underline{A}}$ is of the same genus as $\underline{\underline{B}}$, card $(\alpha(B))$ = card (B). Suppose $b \in B' = B - \{b_1,\dots,b_n\}$. Then $(a, a_1,\dots,a_n) \simeq (b, b_1,\dots,b_n)$ and so, by the induction hypothesis $\underline{\underline{B}} \models \psi[b, b_1,\dots,b_n]$. But card (B') = card (B) and therefore $\underline{\underline{B}} \models \phi[b_1,\dots,b_n]$. The proof of the converse implication is similar and hence the proof of the induction step for the quantifier Q is completed.

If $\beta \in \omega^1$ is a genus and p is a positive integer β/p is the genus defined thus:

$$\text{for } \alpha \in 2^k, \text{ if } \beta(\alpha) = 0 \text{ or } 2 \leqslant \beta(\alpha) \leqslant p+2, \text{ then } \beta/p(\alpha) = \beta(\alpha)$$
$$\text{and if } \beta(\alpha) = 1 \text{ or } p+2 \leqslant \beta(\alpha), \text{ then } \beta/p(\alpha) = p+2.$$

A p-genus is a genus β such that for all $\alpha \in 2^k$, $\beta(\alpha) = 0$ or $2 \leqslant \beta(\alpha) \leqslant p+2$. Thus a p-genus is a genus β such that in any realization $\underline{\underline{A}}$ of genus β for each type $\alpha \in 2^k$, either there are at most p elements in $\underline{\underline{A}}$ of type α or there are card (A) elements in $\underline{\underline{A}}$ of type α. Clearly β/p is always a p-genus.

Lemma 1.2: If $\underline{\underline{A}}$ is a realization of $L_Q^{1,k}$ of genus β and $\underline{\underline{B}}$ is a subrealization of $\underline{\underline{A}}$ of genus β/p then for any formula $\phi(v_0,\dots,v_n)$ of $L_Q^{1,k}$ containing less than p distinct variables and any $b_0,\dots,b_n \in B$

$$\underline{\underline{B}} \models \phi[b_0,\dots,b_n] \quad \text{iff} \quad \underline{\underline{A}} \models \phi[b_0,\dots,b_n]. \tag{2}$$

Proof. The proof is by induction on the number of logical symbols in ϕ. Again the only non-trivial part is the proof of the induction step for the quantifiers.

Suppose (2) holds for $\psi(v_o,\ldots,v_n)$ and let $\phi = \exists v_o\psi$. If $\underline{\underline{B}} \models \phi[b_1,\ldots,b_n]$ than clearly $\underline{\underline{A}} \models \phi[b_1,\ldots,b_n]$. Suppose $\underline{\underline{A}} \models \phi[b_1,\ldots,b_n]$. Then for some $a \in A$, $\underline{\underline{A}} \models \psi[a,b_1,\ldots,b_n]$. If $a \in B$ then $\underline{\underline{B}} \models \psi[a,b_1,\ldots,b_n]$, by the induction hypothesis, and so $\underline{\underline{B}} \models \phi[b_1,\ldots,b_n]$. If $a \notin B$, then a must be of some type α such that card $(\alpha(A)) > p$ and hence such that card $(\alpha(B)) \geqslant p > n$. Hence there is some $b \in B-\{b_1,\ldots,b_n\}$ of the same type as a. $(a,b_1,\ldots,b_n) \simeq (b,b_1,\ldots,b_n)$ and so by Lemma 1·1, $\underline{\underline{A}} \models \psi[b,b_1,\ldots,b_n]$. Therefore, by the induction hypothesis, $\underline{\underline{B}} \models \psi[b,b_1,\ldots,b_n]$ whence $\underline{\underline{B}} \models \phi[b_1,\ldots,b_n]$.

This completes the proof of the induction step for $\exists$, that for Q is similar.

We use $\exists^{>r} x\, \phi(x)$ as an abbreviation for

$$\exists x_o \ldots \exists x_r[x_o \neq x_1 \,\&\, x_o \neq x_2 \,\&\, \ldots . \,\&\, x_{r-1} \neq x_r \,\&\, \phi(x_o) \,\&\, \ldots . \,\&\, \phi(x_r)].$$

If ϕ is a formula of L_Q^1, $\phi^{>r}$ is the formula obtained from ϕ by replacing each occurrence of Qx by $\exists^{>r}x$.

<u>Lemma 1·3</u>: Let p,q be positive integers and let $r = p \cdot 2^k + q$. Let $\underline{\underline{A}}$ be a realization of $L_Q^{1,k}$ of some p-genus. For any formula $\phi(v_o,\ldots,v_n)$ of $L_Q^{1,k}$ containing $\leqslant$ q distinct variables and any $a_o,\ldots,a_n \in A$

$$\underline{\underline{A}} \models \phi[a_o,\ldots,a_n] \text{ iff } \underline{\underline{A}} \models \phi^{>r}[a_o,\ldots,a_n]. \tag{3}$$

<u>Proof</u>: Again the proof is by induction on the number of logical symbols in ϕ. This time the only non-trivial part is the induction step for Q.

Suppose then that (3) holds for $\psi(v_o,\ldots,v_n)$. We show it holds also for $\phi = Qv_o\psi$.

Note first that $\phi^{>r} = \exists^{>r} x\, \psi^{>r}$. Suppose $\underline{\underline{A}} \models \phi[a_1,\ldots,a_n]$. Then if $X = \{a \in A: \underline{\underline{A}} \models \psi[a,a_1,\ldots,a_n]\}$, by the definition of Q, card $(X) =$ card $(A) \geqslant \aleph_o > r$. By the induction hypothesis

$X = \{a \in A: \underline{\underline{A}} \models \psi^{>r}[a,a_1,\ldots,a_n]\}$. So $\underline{\underline{A}} \models \exists^{>r} x\, \psi^{>r}[a_1,\ldots,a_n]$, that is, $\underline{\underline{A}} \models \phi^{>r}[a_1,\ldots,a_n]$.

Conversely, suppose $\underline{\underline{A}} \models \phi^{>r}[a_1,\ldots,a_n]$. Let $Y = \{a \in A: \underline{\underline{A}} \models \psi^{\;r}[a,a_1,\ldots a_n]\}$.

Then card (Y) $>$ r and, by the induction hypothesis,

$$Y = \{a \in A: \underline{\underline{A}} \models \psi[a,a_1,\ldots,a_n]\}.$$

Let $Y' = Y - \{a_1,\ldots,a_n\}$. $q \geqslant n$ and so card $(Y') > p\cdot 2^k$. Hence there is some $a_o \in Y'$ of a type α such that card $(\alpha(A)) > p$. Since $\underline{\underline{A}}$ is of some p-genus it follows that card $(\alpha(A))$ = card (A). Since $\underline{\underline{A}} \models \psi[a_o,a_1,\ldots,a_n]$, it follows from Lemma 1·1 that for any $a \in \alpha(A) - \{a_1,\ldots,a_n\}$, $\underline{\underline{A}} \models \psi[a,a_1,\ldots,a_n]$. Therefore $\underline{\underline{A}} \models \phi[a_1,\ldots,a_n]$.

This completes the proof.

<u>Theorem 1·4</u>: A sentence σ of $L_Q^{1,k}$ containing less than p variables has a model iff $\sigma^{>r}$ has a model of some p-genus, where $r = p(2^k + 1)$.

<u>Proof</u>: This is an immediate consequence of Lemmas 1·2 and 1·3.

$\sigma^{>r}$ is a sentence of $L_Q^{1,k}$ not containing the quantifier Q. The proof that for each p there is a decision procedure for deciding whether a sentence of $L_Q^{1,k}$, not containing Q, has a model of some p-genus is essentially the same as the proof that there is a decision procedure for monadic predicate calculus with equality and so is omitted here. Since any sentence of L_Q^1 can be regarded as a sentence of $L_Q^{1,k}$ for some calculable k, and given any sentence σ of $L_Q^{1,k}$ we can effectively find the sentence $\sigma^{>r}$ of Theorem 1·4, we can conclude that

<u>Theorem 1·5</u>: There is a decision procedure for determining whether or not a sentence of L_Q^1 has a model.

Let L^{MF} be the language obtained from L^1 by dropping equality but allowing monadic function letters and let L_Q^{MF} be the result of adding Q to L^{MF}. It has been shown by Eichholz [1957] that there is a decision procedure for L^{MF}. Another proof of this result is due to Löb [1967] where it is shown that the decision problem for L^{MF} can be reduced to that for L^1, the monadic fragment of predicate calculus with equality. It is easy to see that Löb's reduction procedure can be applied also to reduce the decision problem for L_Q^{MF} to that for L_Q^1. Hence we have:

<u>Theorem 1·6</u>: There is a decision procedure for determining whether or not a sentence of L_Q^{MF} has a model.

§2. An Axiom System for L_Q^1

Let V_Q^1 be the set of universally valid sentences of L_Q^1. It follows from Theorem 1·5 that V_Q^1 is recursive. A fortiori there is a recursive axiomatization of V_Q^1. We present such an axiomatization in this section.

Let P be some usual axiomatization for first order predicate calculus with equality. Let P_Q^1 be the system obtained from P by adding the following axiom schemas:

Q1 $\quad Qx\phi(x) \rightarrow \exists x\phi(x)$

Q2 $\quad \forall x\phi(x) \rightarrow Qx\phi(x)$

Q3 $\quad Qx(\phi(x) \vee \psi(x)) \rightarrow Qx\phi(x) \vee Qx\psi(x)$

Q4 $\quad [Qx\phi(x) \,\&\, \forall x(\phi(x) \rightarrow \psi(x))] \rightarrow Qy\psi(y)$

Q5 $\quad \forall y \neg Qx(x = y)$

where ϕ, ψ are any formulas and x,y are any variables (distinct in the case of Q5). The change of variable in Q4 is to enable the equivalence $Qx\phi(x) \leftrightarrow Qy\phi(y)$ to be provable. It should be noted that Q3 is only valid in infinite models. Our original axiomatization had in place of Q5 the axiom schema which is the case n = 1 of Lemma 2·1(ii). I am grateful to Professor Löb for pointing out that this schema can be derived from the single axiom Q5 in P_Q^1.

The system P_Q^- obtained from P_Q^1 by dropping Q5 is equivalent to the system shown by Yasuhara [1967] to be sufficient for proving all the universally valid sentences of L_Q not containing equality.

We use '$\vdash$' for 'provable in P_Q^1' in the usual way. The following Lemma is easily proved by induction on n.

Lemma 2·1: For any formula ϕ and all positive integers n

(i) $\vdash Qx\phi(x) \rightarrow \exists^{>n} x\phi(x)$

and (ii) $\vdash Qx\phi(x) \rightarrow \forall y_1 \ldots \forall y_n\, Qx[\phi(x) \,\&\, x \neq y_1 \,\&\, \ldots \,\&\, x \neq y_n]$.

For the time being we restrict our attention to $L_Q^{1,k}$ the fragment of L_Q^1 whose only monadic predicate letters are $P_0, \ldots, P_{k-1}$.

$\bigwedge_{1 \leq i \leq n} \phi_i$ is an abbreviation for $\phi_1 \,\&\, \ldots \,\&\, \phi_n$. We often extend this notation in the obvious way, e.g. immediately below.

We let $\mathrm{Iso}(x_0,\ldots,x_n;\ y_0,\ldots,y_n)$ be an abbreviation for the formula

$$\bigwedge_{0\leqslant i<j\leqslant n} [(x_i = x_j) \leftrightarrow (y_i = y_j)] \quad \& \bigwedge_{0\leqslant i\leqslant n} \left[\bigwedge_{0\leqslant j<k} [P_j(x_i) \leftrightarrow P_j(y_i)]\right]$$

<u>Lemma 2·2</u>: For any formula $\phi(v_0,\ldots,v_n)$ with free variables among $v_0,\ldots,v_n$

$$\vdash \forall x_0 \ldots \forall x_n\ \forall y_0 \ldots \forall y_n \Big(\mathrm{Iso}(x_0,\ldots,x_n;\ y_0,\ldots,y_n) \rightarrow$$
$$[\phi(x_0,\ldots,x_n) \leftrightarrow \phi(y_0,\ldots,y_n)]\Big) .$$

<u>Proof</u>: The proof is by induction on the number of logical symbols in ϕ. The proof for atomic formulas is obvious as is also that of the induction step for the logical connectives. The proof of the induction step for $\exists$ is a straightforward exercise in predicate calculus and is omitted here. We complete the proof by showing that if the Lemma holds for ϕ then it holds also for $Qv_1\phi$. For simplicity we assume that ϕ contains only the variables v_0 and v_1 free.

Clearly $\mathrm{Iso}(x_0;\ y_0),\ x_0 \neq v_1,\ y_0 \neq v_1 \vdash \mathrm{Iso}(x_0,v_1;\ y_0,\ v_1)$

and hence by the induction hypothesis

$$\mathrm{Iso}(x_0;\ y_0),\ x_0 \neq v_1,\ y_0 \neq v_1,\ \phi(x_0,v_1) \vdash \phi(y_0,v_1)$$

whence $$\mathrm{Iso}(x_0;\ y_0) \vdash \forall v_1\Big([x_0 \neq v_1\ \&\ y_0 \neq v_1\ \&\ \phi(x_0,v_1)] \rightarrow \phi(y_0,v_1)\Big). \quad (1)$$

By Lemma 2·1 (ii)

$$Qv_1\phi(x_0,v_1) \vdash Qv_1[x_0 \neq v_1\ \&\ y_0 \neq v_1\ \&\ \phi(x_0,v_1)] \quad (2)$$

and so from (1), (2) and Q4

$$\mathrm{Iso}(x_0;\ y_0),\ Qv_1\phi(x_0,v_1) \vdash Qv_1\phi(y_0,v_1). \quad (3)$$

Clearly (3) also holds with $\phi(x_0,v_1)$ and $\phi(y_0,v_1)$ interchanged and so

$$\vdash \forall x_0\ \forall y_0\Big(\mathrm{Iso}(x_0;\ y_0) \rightarrow [Qv_1\ \phi(x_0,v_1) \leftrightarrow Qv_1\ \phi(y_0,v_1)]\Big).$$

This completes the induction step.

We associate with each genus $\beta \in \omega^l$, where $l = 2^k$, the set Δ_β of sentences $L_Q^{1,k}$, defined as follows:

For $\alpha \in 2^k$, if $\beta(\alpha) = 0$ then we put $\Delta_\beta^\alpha = \{Qx\Phi_\alpha(x)\}$,

if $\beta(\alpha) = 1$ then we put $\Delta_\beta^\alpha = \{\neg Qx\Phi_\alpha(x)\} \cup \{\exists^{>n} x\Phi_\alpha(x) : n < \omega\}$,

and if $\beta(\alpha) = p+2$ with $p \geqslant 0$, we put $\Delta_\beta^\alpha = \{\exists!^p x\Phi_\alpha(x)\}$

where $\exists^{>n} x$ says "there are more than n" and $\exists!^p x$ "there are exactly p". Both these quantifiers can be defined in terms of $\exists$.

Finally we let $\Delta_\beta = \bigcup_{\alpha\in 2^k} \Delta_\beta^\alpha$. Obviously a realization $\underline{A}$ of $L_Q^{1,k}$ is of genus β iff it is a model of Δ_β. Hence each Δ_β has a model and is consistent. Lemma 1.1 implies that each Δ_β is semantically complete. The main step in our proof is to show that each Δ_β is also syntactically complete.

Let $C = \{c_{n\alpha} : \alpha\in 2^k, n<\omega\}$ be a set of constant symbols and let L' be the language obtained from $L_Q^{1,k}$ by adding all the constants in C. We associate with each genus $\beta\in 2^l$ a set Γ_β of sentences of L' in the following manner.

If $\alpha\in 2^k$ and $\beta(\alpha) \leqslant 1$, we let $\Lambda_\beta^\alpha = \{\Phi_\alpha(c_{n\alpha}) : n<\omega\}$ while if $\beta(\alpha) = p+2$ with $p \geqslant 0$ we let $\Lambda_\beta^\alpha = \{\Phi_\alpha(c_{n\alpha}) : n< p\}$. We put $\Lambda_\beta = \bigcup_{\alpha\in 2^k} \Lambda_\beta^\alpha$ and let C_β be the set of all constants in Λ_β. L_β is the language obtained by adding the constants of C_β to $L_Q^{1,k}$. A constant in C_β is said to be of type α if it occurs in Λ_β^α.

Then $\Gamma_\beta = \Delta_\beta \cup \Lambda_\beta \cup \{c_{n\alpha} \neq c_{n'\alpha'} : c_{n\alpha}, c_{n'\alpha'} \in C_\beta, \langle n,\alpha\rangle \neq \langle n',\alpha'\rangle\}$.

Thus Γ_β is a set of sentences of L_β. It is clear that for any sentence σ of $L_Q^{1,k}$, if $\Gamma_\beta \vdash \sigma$ then $\Delta_\beta \vdash \sigma$. It follows at once that Γ_β is consistent. Our main result is

<u>Theorem</u> 2.3: For any sentence σ of L_β

(i) $\Gamma_\beta \cup \{\sigma\}$ is consistent iff $\Gamma_\beta \vdash \sigma$

(ii) $\Gamma_\beta \vdash \sigma$ or $\Gamma_\beta \vdash \neg\sigma$.

<u>Proof</u>. The proof is by induction on the number of logical symbols in σ. We note first that since Γ_β is consistent, $\Gamma_\beta \vdash \sigma$ implies $\Gamma_\beta \cup \{\sigma\}$ is consistent and (i) implies (ii) in each case.

The proof for atomic formulas is obvious. Suppose the Theorem holds whenever σ contains less than μ logical symbols and σ is a sentence which contains μ logical symbols. The proof in the case σ is of the form $\neg\phi$ or $\phi \& \psi$ is straightforward.

Suppose now σ is $\exists x\psi(x)$. Assume that the constants in $\psi(x)$ are $c_0,\dots,c_{k-1}$ and let C^{ψ} be the set of these constants.

Assume $$\text{not } \Gamma_\beta \vdash \exists x\psi(x). \tag{1}$$

Then for each $c \in C_\beta$, not $\Gamma_\beta \vdash \psi(c)$, whence by the induction hypothesis

for each $c \in C_\beta$, $$\Gamma_\beta \vdash \neg\psi(c). \tag{2}$$

We divide the types $\alpha \in 2^k$ into two classes. T is the set of all those types α such that $C_\beta - C^{\psi}$ contains a constant of type α. U is the set of all other types. For each $\alpha \in T$ we let c_α be a constant of type α in $C_\beta - C^{\psi}$.

By (2), for each $\alpha \in T$, $\Gamma_\beta \vdash \neg\psi(c_\alpha)$ (3)
and also for each $c \in C^{\psi}$, $\Gamma_\beta \vdash \neg\psi(c)$. (4)

Clearly for each $\alpha \in 2^k$,

$$\Gamma_\beta,\ \Phi_\alpha(x),\ \bigwedge_{i\in k} x \neq c_i \ \vdash \ \mathrm{Iso}(c_\alpha, c_0,\dots,c_{k-1};\ x, c_0,\dots,c_{k-1})$$

and hence, by Lemma 1·2 and (3),

for each $\alpha \in T$, $$\Gamma_\beta,\ \Phi_\alpha(x),\ \bigwedge_{i\in k} x \neq c_i \ \vdash \neg\psi(x). \tag{5}$$

Therefore, by (4) and (5)

$$\Gamma_\beta \vdash \forall x\Big[\Big(\bigvee_{\alpha\in T} \Phi_\alpha(x) \ \vee \ \bigvee_{i\in k} x=c_i\Big) \rightarrow \neg\psi(x)\Big]. \tag{6}$$

For each $\alpha \in U$, evidently

$$\Gamma_\beta \vdash \neg\exists x\Big(\Phi_\alpha(x) \ \& \ \bigwedge_{i\in k} x \neq c_i\Big). \tag{7}$$

Since $\vdash \forall x \bigvee_{\alpha\in 2^k} \Phi_\alpha(x)$ it follows from (7) that

$$\Gamma_\beta \vdash \forall x\Big[\bigvee_{\alpha\in T} \Phi_\alpha(x) \ \vee \ \bigvee_{i\in k} x=c_i\Big]$$

and hence we can conclude from (6) that

$$\Gamma_\beta \vdash \forall x \neg \psi(x).$$

This shows that $\Gamma_\beta \cup \{\exists x\psi(x)\}$ is not consistent and so concludes the proof of the induction step for $\exists$.

We suppose finally that σ is $Qx\psi(x)$ and again assume that C^ψ is the set of constants in ψ.

Suppose $$\text{not } \Gamma_\beta \vdash Qx\psi(x). \tag{8}$$

By Lemma 2·1,

$$Qx\psi(x) \vdash Qx\Big(\psi(x) \;\&\; \bigwedge_{i\epsilon k} x\neq c_i\Big), \quad \text{and hence}$$

$$Qx\psi(x) \vdash Qx \bigvee_{\alpha\epsilon 2^k} \Big(\psi(x) \;\&\; \bigwedge_{i\epsilon k} x\neq c_i \;\&\; \Phi_\alpha(x)\Big)$$

and so, by Q3,

$$Qx\psi(x) \vdash \bigvee_{\alpha\epsilon 2^k} \Big[Qx\Big(\psi(x) \;\&\; \bigwedge_{i\epsilon k} x\neq c_i \;\&\; \Phi_\alpha(x)\Big)\Big]. \tag{9}$$

Clearly,

$$\text{for each } \alpha\epsilon 2^k, \text{ either } \Gamma_\beta \vdash Qx\Phi_\alpha(x) \text{ or } \Gamma_\beta \vdash \neg Qx\Phi_\alpha(x). \tag{10}$$

Using Q4,

$$\text{if } \Gamma_\beta \vdash \neg Qx\Phi_\alpha(x) \text{ then } \Gamma_\beta \vdash \neg Qx\Big(\psi(x) \;\&\; \bigwedge_{i\epsilon k} x\neq c_i \;\&\; \Phi_\alpha(x)\Big). \tag{11}$$

Suppose $\Gamma_\beta \vdash Qx\Phi_\alpha(x)$. Then by Lemma 1·1, $\Gamma_\beta \vdash Qx\Big(\Phi_\alpha(x) \;\&\; \bigwedge_{i\epsilon k} x\neq c_i\Big)$ (12)
and therefore there is some $c_\alpha \epsilon C_\beta - C^\psi$ such that $\Gamma_\beta \vdash \Phi_\alpha(c_\alpha)$.
By the induction hypothesis

$$\text{either } \Gamma_\beta \vdash \psi(c_\alpha) \text{ or } \Gamma_\beta \vdash \neg\psi(c_\alpha). \tag{13}$$

If $\Gamma_\beta \vdash \psi(c_\alpha)$ then since $\Gamma_\beta, \Phi_\alpha(x), \bigwedge_{i\epsilon k} x\neq c_i \vdash \mathrm{Iso}(c_\alpha, c_0, \ldots, c_{k-1};\; x, c_0, \ldots, c_{k-1})$ it follows from Lemma 2·2 that

$$\Gamma_\beta \vdash \forall x\Big[\Big(\Phi_\alpha(x) \;\&\; \bigwedge_{i\epsilon k} x\neq c_i\Big) \rightarrow \psi(x)\Big]. \tag{14}$$

Applying Q4 to (13) and (14) we can now conclude that

$$\Gamma_\beta \vdash Qx\psi(x)$$

which contradicts our assumption (8). Hence we can conclude from (13) that

$\Gamma_\beta \vdash \neg\psi(c_\alpha)$ and hence, just as in our proof of (14) that

$$\Gamma_\beta \vdash \forall x\left[\left(\Phi_\alpha(x) \ \& \ \bigwedge_{i\in k} x \neq c_i\right) \rightarrow \neg\psi(x)\right] . \tag{15}$$

This finishes our proof that

$$\text{If} \quad \Gamma_\beta \vdash Qx\Phi_\alpha(x) \quad \text{then} \quad \Gamma_\beta \vdash \neg Qx\left(\Phi_\alpha(x) \ \& \ \bigwedge_{i\in k} x\neq c_i \ \& \ \psi(x)\right) . \tag{16}$$

From (9), (10), (11) and (16) it follows that $\Gamma_\beta \vdash \neg Qx\psi(x)$ which shows that $\Gamma_\beta \cup \{Qx\psi(x)\}$ is not consistent. This completes the induction step for Q and hence the proof of Theorem 2·3.

<u>Theorem 2·4:</u> <u>The Completeness Theorem for $L_Q^{1,k}$</u>

If Σ is a consistent set of sentences of $L_Q^{1,k}$ then Σ has a model. (Where by "consistent" we mean "consistent relative to the system P_Q^1").

<u>Proof</u>: Clearly if Σ is consistent there is some genus β such that $\Delta_\beta \cup \Sigma$ is consistent. Any realization of genus β is a model of Δ_β. By Theorem 3, for each $\sigma \in \Sigma$, $\Gamma_\beta \vdash \sigma$, and therefore, since σ contains no constants, $\Delta_\beta \vdash \sigma$. Thus any model of Δ_β is a model of Σ.

§3. The Completeness Theorem for L_Q^1

We now set about extending Theorem 2·4 to L_Q^1. We suppose that the monadic predicate letters of L_Q^1 can be well-ordered as a sequence $\{P_\xi : \xi < \underline{\underline{m}}\}$ where $\underline{\underline{m}}$ is a cardinal. We let $S(\underline{\underline{m}})$ be the set of all non-empty finite subsets of $\underline{\underline{m}}$. If $\chi \in S(\underline{\underline{m}})$, by a χ-*type* we mean an element of 2^χ. If α is a χ-type $\Phi_\alpha^\chi(x)$ is the formula $\bigwedge_{\xi\in\chi} R_\xi(x)$, where, for $\xi \in \chi$, R_ξ is P_ξ if $\alpha(\xi) = 0$ and is $\neg P_\xi$ if $\alpha(\xi) = 1$. If $\underline{\underline{A}}$ is a realization of L_Q^1 an element $a \in A$ is said to be of χ-type α iff $\underline{\underline{A}} \models \Phi_\alpha^\chi[a]$. A *full-type* is an element of $2^{\underline{\underline{m}}}$. If γ is a full-type an element $a \in A$ is said to be of *full-type* γ iff for each $\xi < \underline{\underline{m}}$, $\underline{\underline{A}} \models P_\xi[a]$ iff $\gamma(\xi) = 0$.

A χ-genus is an element of $\omega^{(2^\chi)}$ which takes the value 0 for some $\alpha \in 2^\chi$. If β is a χ-genus the set Δ_β^χ of sentences of L_Q^1 is defined in the same way as Δ_β was defined in §2, except that "Φ_α" is replaced throughout by "Φ_α^χ".

If α is a χ-type and γ is a full type we write $\alpha \prec \gamma$ iff $\alpha = \gamma \upharpoonright \chi$. If α' is a χ'-type we write $\alpha \subset \alpha'$ if $\chi \subseteq \chi'$ and $\alpha = \alpha' \upharpoonright \chi$.

To prove the Completeness Theorem for L_Q^1 we need only prove

Theorem 3·1: If for each $\chi \in S(\underline{\underline{m}})$, β_χ is a χ-genus and

$$\Delta = \bigcup_{\chi \in S(\underline{\underline{m}})} \Delta_{\beta_\chi}^\chi$$

is consistent, then Δ has a model.

Proof: For each $\chi \in S(\underline{\underline{m}})$ we define the function μ_χ on 2^χ as follows:

for $\alpha \in 2^\chi$, if $\beta_\chi(\alpha) = 0$ then $\mu_\chi(\alpha) = \underline{\underline{m}}^+$

if $\beta_\chi(\alpha) = 1$ then $\mu_\chi(\alpha) = \underline{\underline{m}}$,

and if $2 \leqslant \beta_\chi(\alpha) = p + 2$, then $\mu_\chi(\alpha) = p$.

Then we define λ on $2^{\underline{\underline{m}}}$ by,

$\lambda(\gamma) = \min\{\mu_\chi(\alpha) : \text{for some } \chi \in S(\underline{\underline{m}}),\ \alpha \text{ is a } \chi\text{-type and } \alpha \prec \gamma\}$.

Thus $\lambda(\gamma)$ equals $\underline{\underline{m}}^+$, or $\underline{\underline{m}}$ or is finite.

A χ-type α is said to be finite if $\mu_\chi(\alpha)$ is finite, and otherwise it is said to be infinite. Note that $\mu_\chi(\alpha) = p$ iff $\Delta \vdash \exists!^p x \Phi_\alpha^\chi(x)$. α is said to be fixed if for each $\xi < \underline{\underline{m}}$ either

$$\text{(i)} \quad \Delta \vdash \neg \exists x \left[\Phi_\alpha^\chi(x) \ \& \ P_\xi(x) \right]$$

$$\text{or (ii)} \quad \Delta \vdash \neg \exists x \left[\Phi_\alpha^\chi(x) \ \& \ \neg P_\xi(x) \right].$$

If $\mu_\chi(\alpha) \leqslant 1$, α is obviously fixed. With this basis it is easily proved by induction on $\mu_\chi(\alpha)$ that

I. If α is a finite χ-type, there is some $\chi' \epsilon S(\underline{\underline{m}})$ with $\chi \subseteq \chi'$, and a set $\alpha_1, \ldots, \alpha_k$ of fixed finite χ'-types such that, for $1 \leqslant i \leqslant k$, $\alpha \subset \alpha_i$ and

$$\mu_\chi(\alpha) = \mu_{\chi'}(\alpha_1) + \ldots . + \mu_{\chi'}(\alpha_k).$$

Next we prove

II. If α is a fixed finite χ-type or an infinite χ-type, there is a full type γ_α such that $\alpha \prec \gamma_\alpha$ and $\mu_\chi(\alpha) = \lambda(\gamma_\alpha)$.

<u>Proof</u>: Suppose first α is a fixed finite χ-type. γ_α is defined as follows. For each $\xi < \underline{\underline{m}}$ either (i) or (ii) of the definition of 'fixed' holds. If (i) holds we put $\gamma_\alpha(\xi) = 1$ and if (ii) holds we put $\gamma_\alpha(\xi) = 0$. Clearly $\alpha \prec \gamma_\alpha$ and $\mu_\chi(\alpha) = \lambda(\gamma_\alpha)$.

Next suppose α is an infinite χ-type. There are two cases. Assume first that $\mu_\chi(\alpha) = \underline{\underline{m}}^+$, hence $\Delta \vdash Qx \Phi_\alpha^\chi(x)$.

Let $X = \{P'_\xi : \xi < \underline{\underline{m}}\}$ be an enumeration of all the monadic predicate letters of L_Q^1 not in $\{P_\xi : \xi \epsilon \chi\}$. We will define a set $Y = \{R'_\xi : \xi < \underline{\underline{m}}\}$ such that, for $\xi < \underline{\underline{m}}$, R'_ξ is P'_ξ or $\neg P'_\xi$ and for each finite subset $\{\xi_1, \ldots, \xi_s\}$ of $\underline{\underline{m}}$

$$\Delta \vdash Qx \left[\Phi_\alpha^\chi(x) \ \& \ R'_1(x) \ \& \ \ldots \ \& \ R'_s(x) \right]. \tag{1}$$

Suppose $\zeta < \underline{\underline{m}}$ and we have defined R'_ξ for $\xi < \zeta$ so as to satisfy (1), but that (1) is not satisfied if we put $R'_\zeta = P'_\zeta$. We show that in this case it is satisfied if we put $R'_\zeta = \neg P'_\zeta$.

By hypothesis, there is some finite subset ζ_0 of ζ such that

$$\text{not} \quad \Delta \vdash Qx\Big[\Phi^{\chi}_{\alpha}(x) \ \& \ \bigwedge_{\xi\in\zeta_0} R'_{\xi}(x) \ \& \ P'_{\zeta}(x)\Big]. \tag{2}$$

Let η_0 be any finite subset of ζ. It is sufficient to show that

$$\Delta \vdash Qx\Big[\Phi^{\chi}_{\alpha}(x) \ \& \ \bigwedge_{\xi\in\eta_0} R'_{\xi}(x) \ \& \ \neg P'_{\zeta}(x)\Big]. \tag{3}$$

Now $\tau_0 = \zeta_0 \cup \eta_0$ is a finite subset of ζ and so by hypothesis

$$\Delta \vdash Qx\Big[\Phi^{\chi}_{\alpha}(x) \ \& \ \bigwedge_{\xi\in\tau_0} R'_{\xi}(x)\Big]$$

whence, using Q3, either

$$\Delta \vdash Qx\Big[\Phi^{\chi}_{\alpha}(x) \ \& \ \bigwedge_{\xi\in\tau_0} R'_{\xi}(x) \ \& \ P'_{\zeta}(x)\Big] \tag{4}$$

or

$$\Delta \vdash Qx\Big[\Phi^{\chi}_{\alpha}(x) \ \& \ \bigwedge_{\xi\in\tau_0} R'_{\xi}(x) \ \& \ \neg P'_{\zeta}(x)\Big]. \tag{5}$$

If (4) holds, then since $\zeta_0 \subseteq \tau_0$, by Q4 it follows that it holds also with 'τ_0' replaced by ζ_0, contradicting (2). Therefore (5) holds. But $\eta_0 \subseteq \tau_0$, so, using Q4 again, (3) holds.

Thus we can define Y so as to satisfy (1). The full type $\gamma_\alpha \in 2^{\underline{\underline{m}}}$ is defined as follows:

$$\text{if} \quad \xi \in \chi \quad \text{then} \quad \gamma_\alpha(\xi) = \alpha(\xi)$$

if $\xi \in \underline{\underline{m}} - \chi$, P'_{ζ} is P_{ξ} and R'_{ζ} is P'_{ζ} then $\gamma_\alpha(\xi) = 0$, while if R'_{ζ} is $\neg P'_{\zeta}$ then $\gamma_\alpha(\xi) = 1$. Then $\alpha \prec \gamma_\alpha$ and by our construction $\lambda(\gamma_\alpha) = \underline{\underline{m}}^+ = \mu_\chi(\alpha)$.
[The argument we have used to find Y is essentially the same as that which can be used to prove the Compactness Theorem for propositional calculus, see, e.g. Kreisel and Krivine [1967] p. 7.]

A similar proof works if $\mu_\chi(\alpha) = \underline{\underline{m}}$, so our proof of II is completed.

Let $T = \{\alpha :$ for some $\chi \in S(\underline{\underline{m}})$, α is a fixed finite χ-type or an infinite χ-type$\}$. Card (T) $\leqslant \underline{\underline{m}}$. For each $\alpha \in T$, let γ_α be a full-type such that $\alpha \prec \gamma_\alpha$ and $\mu_\chi(\alpha) = \lambda(\gamma_\alpha)$. The existence of such a γ_α is guaranteed by II. Let $\Gamma = \{\gamma_\alpha : \alpha \in T\}$.

There is a realization $\underline{\underline{A}}$ of L^1_Q which for each $\gamma \in \Gamma$ contains precisely $\lambda(\gamma)$

elements of full type γ and contains no elements of any other full type. We conclude our proof of Theorem 3·1 by showing that $\underline{\underline{A}}$ is a model of Δ.

If α is a fixed finite χ-type, say $\mu_\chi(\alpha) = p$, then there are precisely p elements in $\underline{\underline{A}}$ of full-type γ_α. Since $\alpha \prec \gamma_\alpha$ all these elements are of χ-type α. Because α is fixed if $\alpha \prec \gamma'$ and $\gamma' \neq \gamma_\alpha$, $\lambda(\gamma') = 0$ and hence $\underline{\underline{A}}$ contains precisely p elements of χ-type α. If α is finite but not fixed we can draw the same conclusion using I.

If α is an infinite χ-type and $\mu_\chi(\alpha) = \underline{\underline{m}}$ then for any full type γ such that $\alpha \prec \gamma$, $\lambda(\gamma) \leqslant \underline{\underline{m}}$. Hence $\underline{\underline{A}}$ contains at most $\underline{\underline{m}}$ elements of χ-type α. But by construction $\underline{\underline{A}}$ contains $\underline{\underline{m}}$ elements of full-type γ_α and $\alpha \prec \gamma_\alpha$. So $\underline{\underline{A}}$ contains precisely $\underline{\underline{m}}$ elements of χ-type α.

If $\mu_\chi(\alpha) = \underline{\underline{m}}^+$, $\underline{\underline{A}}$ contains $\underline{\underline{m}}^+$ elements of full-type γ_α with $\alpha \prec \gamma_\alpha$. Since for each χ, there must be some $\alpha \in 2^\chi$ such that $\mu_\chi(\alpha) = \underline{\underline{m}}^+$, card (A) $\geqslant \underline{\underline{m}}^+$ and by construction card (A) $\leqslant \underline{\underline{m}}.\underline{\underline{m}}^+ = \underline{\underline{m}}^+$. Thus card (A) $= \underline{\underline{m}}^+$.

These remarks show that $\underline{\underline{A}}$ is a model of Δ.

<u>Theorem 3·2</u>: <u>The Completeness Theorem for L_Q^1</u>

If Σ is a consistent set of sentences of L_Q^1 then Σ has a model.

<u>Proof</u>: If Σ is consistent we can obviously find, for each $\chi \in S(\underline{\underline{m}})$, a χ-genus β_χ such that $\Sigma \cup \Delta$ is consistent, where

$$\Delta = \bigcup_{\chi \in S(\underline{\underline{m}})} \Delta^\chi_{\beta_\chi}.$$

By Theorem 3·1, Δ has a model, and from Theorem 1·3 it follows that any model of Δ is a model of Σ.

<u>Theorem 3·3</u>: <u>The Lowenheim-Skolem Theorem for L_Q^1</u>

Let Σ be a set of sentences of L_Q^1 containing $\underline{\underline{m}}$ distinct monadic predicate letters. If Σ has a model then Σ has a model of any cardinal $\underline{\underline{n}} \geqslant \underline{\underline{m}}^+$.

<u>Proof</u>: If Σ has a model, Σ is consistent. It follows from the construction used in the proof of Theorem 3·1 that Σ has a model of cardinal $\underline{\underline{m}}$. By replacing "$\underline{\underline{m}}^+$" by "$\underline{\underline{n}}$" throughout this construction can be modified to show that

Σ has a model of any cardinal $\underline{\underline{n}} \geqslant \underline{\underline{m}}^+$.

<u>Theorem 3·4</u>: If Σ is a set of sentences of L_Q^1 containing only finitely many distinct monadic predicate letters and variables, then if Σ has a model it has a model of any infinite cardinal.

<u>Proof</u>: We can regard Σ as a set of sentences of $L_Q^{1,k}$ for some k. Suppose there are p+1 distinct variables in Σ, $p \geqslant 0$. By Lemma 1·2, if Σ has a model, Σ has a model of some p-genus, say β. Hence $\Sigma \cup \Delta_\beta$ is consistent. Clearly Δ_β has models of all infinite cardinals, and by Theorem 2·3, any model of Δ_β is a model of Σ. This completes the proof.

It is easily seen that $\underline{\underline{m}}^+$ cannot be replaced by any smaller cardinal in Theorem 3·3. Theorem 3·2 could be deduced more directly from Theorems 2·4, 3·4 and Corollary 4·7·1, but in this case we would not obtain the best possible lower bound $\underline{\underline{m}}^+$ in Theorem 3·3, and indeed we would have to replace $\underline{\underline{m}}^+$ by $(2^{\underline{\underline{m}}})^+$.

§4. Compactness and Completeness Results for L_Q

A cardinal $\underline{\underline{m}}$ is said to be a strong limit cardinal if whenever $\underline{\underline{n}} < \underline{\underline{m}}$, also $2^{\underline{\underline{n}}} < \underline{\underline{m}}$. By the Limit Cardinal Hypothesis (L.C.H.) we mean the hypothesis that each limit cardinal is a strong limit cardinal. The L.C.H. is clearly implied by the G.C.H. The results of Easton [1964] show that it is strictly weaker than the G.C.H. and is not provable in set theory even using the axiom of choice.

In this section we point out that, assuming the L.C.H., the Compactness Theorem for L_Q, with a countable set of predicate letters, can be deduced from some recent results of Fuhrken [1964] and Keisler [1967]. We also show, that assuming the G.C.H., this result can be extended to uncountable sets of sentences by a straightforward application of the ultraproduct construction. The results that we quote were originally stated in terms of the quantifiers Q_α, but by our introductory remarks they can be translated into Theorems about L_Q.

Theorem 4·1: (Keisler [1967] Corollary 3·5(v)). If Σ is a set of sentences of L_Q and each finite subset of Σ has a model of some singular strong limit cardinal, then Σ has a model.

Theorem 4·2: If Σ is a countable set of sentences of L_Q and each finite subset of Σ has a model of some regular cardinal, then Σ has a model.

Proof: By Theorem 3·1 of Fuhrken [1964] if each finite subset of Σ has a model of some regular cardinal, each finite subset of Σ has a model of cardinal $\aleph_1$. The result now follows from the Compactness Theorem for Countable sets of sentences of L_{Q_1} (Theorem 3·4 of Fuhrken [1964]).

By making use of various two cardinal results, Fuhrken obtains some transfer theorems for the languages L_{Q_α}. These results translate into the following Löwenheim-Skolem Theorems for L_Q.

Theorem 4·3: (Fuhrken, MacDowell & Specker). If Σ is a countable set of sentences of L_Q with a model of cardinal $\aleph_0$, then Σ has a model of each infinite cardinal.

Theorem 4·4: (Fuhrken). If Σ is a countable set of sentences of L_Q with a model of some regular cardinal, then Σ has a model of cardinal $\aleph_1$.

Theorem 4·5: (Fuhrken and Chang). Assuming the G.C.H., if Σ is a countable set of sentences of L_Q with a model of some non-limit cardinal, then Σ has a model of cardinal $\aleph_{\alpha+1}$ for each regular cardinal $\aleph_\alpha$.

Theorem 4·3, which depends on a result of MacDowell & Specker [1961] is just the translation of Theorem 3·2 of Fuhrken [1964], Theorem 4·4 is the translation of Theorem 3·1 in the same paper, and Theorem 4·5 which uses a result of Chang [1965] is the translation of (15) of Fuhrken [1965].

We now show how Theorem 4·2 can be extended to uncountable sets of sentences of L_Q.

If $\underline{\underline{m}}$, $\underline{\underline{n}}$ are two cardinals, we say that $\underline{\underline{m}}$ is $\underline{\underline{n}}$-normal if given any collection $\{\underline{\underline{m}}_\xi : \xi < \underline{\underline{n}}\}$ of cardinals each less than $\underline{\underline{m}}$,

$$\prod_{\xi<\underline{\underline{n}}} \underline{\underline{m}}_\xi < \underline{\underline{m}}^{\underline{\underline{n}}} .$$

If X is a set, $S_\omega(X)$ is the set of all finite subsets of X. An ultrafilter F on I is said to be regular if there is a one-one map f, from I onto $S_\omega(I)$ such that for all $j \in I$, $\{i \in I : j \in f(i)\} \in F$. This notion is due to Keisler [1964].

Theorem 4·6: Łoš's Theorem for L_Q.

Suppose $\underline{\underline{m}}$ is $\underline{\underline{n}}$-normal. Let card (I) = $\underline{\underline{n}}$ and for each $i \in I$, let $\underline{\underline{A}}_i$ be a realization of L_Q of cardinal $\underline{\underline{m}}$. If F is a regular ultrafilter on I, then for any formula ϕ of L_Q and any $x/F \in (\Pi A_i/F)^\omega$

$$\Pi\underline{\underline{A}}_i/F \models_x \phi \text{ iff } \{i \in I : \underline{\underline{A}}_i \models_{x_i} \phi\} \in F.$$

Proof: The proof is by induction on the number of logical symbols in ϕ. The proof for atomic formulas and that of the induction steps for $\neg$, & and $\exists$ is the same as usual (see e.g. Frayne, Morel & Scott [1962] p. 213).

To see that the induction step works for the quantifier Q it is sufficient to notice that because F is regular, card $(\Pi A_i/F) = \underline{\underline{m}}^{\underline{\underline{n}}}$ (see Chang [1967] p. 97) and therefore since $\underline{\underline{m}}$ is $\underline{\underline{n}}$-normal subsets of $\Pi A_i/F$ which are "small" in "almost all" factors are also small in $\Pi A_i/F$. The details are left to the reader.

Theorem 4·7: Let Σ be an infinite set of sentences of L_Q of cardinal $\underline{\underline{n}}$. If $\underline{\underline{m}}$ is $\underline{\underline{n}}$-normal and each finite subset of Σ has a model of cardinal $\underline{\underline{m}}$, then Σ has a model of cardinal $\underline{\underline{m}}^{\underline{\underline{n}}}$.

Proof: For $\Delta \in S_\omega(\Sigma)$, let $\underline{\underline{A}}_\Delta$ be a model of Δ of cardinal $\underline{\underline{m}}$. Let $I = S_\omega(\Sigma)$ and for $\Delta \in I$ let

$$\Delta^* = \{\Delta' \in I : \Delta \subseteq \Delta'\}.$$

It is easily seen that $\{\Delta^* : \Delta \in S_\omega(\Sigma)\}$ has the finite intersection property and so can be extended to an ultrafilter F on I.

Let g be any one-one map of Σ onto I. We define the map $f : I \to S_\omega(I)$ by $f(\Delta) = \{g(\sigma) : \sigma \in \Delta\}$ for each $\Delta \in I$. f is one-one onto. Suppose $\Delta \in I$, and let $\Delta^+ = \{\Delta' \in I : \Delta \in f(\Delta')\}$. Since f is onto for some $\Delta_0 \in I$, $\Delta \in f(\Delta_0)$. Clearly

$$\Delta_0{}^* = \{\Delta' \in I : \Delta_0 \subseteq \Delta'\} \subseteq \Delta^+.$$

By construction $\Delta_0{}^* \in F$. Therefore $\Delta^+ \in F$.

We have thus shown that F is regular. Hence Theorem 4·6 holds for the ultraproduct $\Pi \underline{\underline{A}}_\Delta/F$, and so, just as in the usual proof of the Compactness Theorem using ultraproducts (see e.g. Chang [1967] p. 93), it follows that $\Pi \underline{\underline{A}}_\Delta/F$ is a model of Σ. Clearly card $(\Pi A_\Delta/F) = \underline{\underline{m}}^{\underline{\underline{n}}}$.

Corollary 4·7·1: Let Σ be a set of sentences of L_Q of cardinal $\underline{\underline{n}}$. If each finite subset of Σ has a model of cardinal $(2^{\underline{\underline{n}}})^+$ then so too does Σ.

Proof: It is sufficient to notice that $(2^{\underline{\underline{n}}})^+$ is $\underline{\underline{n}}$-normal and that $((2^{\underline{\underline{n}}})^+)^{\underline{\underline{n}}} = (2^{\underline{\underline{n}}})^+$.

Corollary 4·7·2: Let Σ be a set of sentences of L_Q each finite subset of which has a countable model, then Σ has a model.

Proof: Immediate from Theorem 4·3 and the previous Corollary.

Theorem 4·8: Assuming the G.C.H., if Σ is a set of sentences of L_Q and each finite subset of Σ has a model of some regular cardinal, then Σ has a model.

Proof: We can assume Σ is infinite, say card $(\Sigma) = \aleph_\xi$. By Theorem 4·4 each finite subset of Σ has a model of cardinal $\aleph_1$, and hence, by Theorem 4·5 of cardinal $\aleph_{\xi+2}$. Therefore by Corollary 4·7·1, Σ has a model of cardinal $\aleph_{\xi+2}$.

Theorem 4·9: The Compactness Theorem for L_Q.

Let Σ be a set of sentences of L_Q every finite subset of which has a model. If Σ is countable, then assuming the L.C.H., Σ itself has a model, while if Σ is uncountable this conclusion follows from the G.C.H.

Proof: All singular cardinals are limit cardinals, hence the L.C.H. implies that the collections of singular and singular strong limit cardinals coincide. Notice now that if each finite subset of Σ has a model then either each finite subset of Σ has a model of some regular cardinal or each finite subset of Σ has a model of some singular cardinal. The Theorem now follows from Theorems 4·1, 4·2 and 4·8.

It would be interesting to know how essential the L.C.H. and the G.C.H. are to these results. In this connection it is interesting to note that assuming the L.C.H., the Completeness Theorem for L_Q can be deduced from some results of Vaught [1964] and Keisler [1967]. Let V_α, V_R, V_S, V be, respectively, the sets of sentences of L_Q valid in all models of cardinal $\aleph_\alpha$, in all models of some regular cardinal, in all models of some singular cardinal and in all models. By Theorem 4·4, $V_R = V_1$, and Vaught's result [1964] shows that V_R is recursively enumerable. Assuming the L.C.H., it follows from Corollary 3·7(i) of Keisler [1967], that $V_S = V_\omega$ and from Corollary 3·5(iv) of the same paper that V_ω is recursively enumerable. Thus assuming the L.C.H., $V = V_S \cap V_R$ is recursively enumerable.

Again it would be interesting to know if the use of the L.C.H. can be eliminated in this result. Fuhrken has provided an example of a sentence σ of L_Q which is only valid in models of some singular cardinal. We can take for σ the sentence

$$Qy\ \exists x\ P(x,y)\ \&\ \neg Qx\ \exists y\ P(x,y)\ \&\ \forall x\ \neg Qy\ P(x,y).$$

Conversely the sentence τ which says that "R is a total ordering of the domain and D is a dense subset of smaller cardinal", namely the sentence

$$\forall x \neg R(x,x)\ \&\ \forall x\, \forall y\, \forall z[R(x,y)\ \&\ R(y,z) \rightarrow R(x,z)]\ \&\ \forall x\, \forall y(R(x,y) \vee R(y,x) \vee y = x)$$
$$\&\ \forall x\, \forall y\{R(x,y) \rightarrow \exists z[D(z)\ \&\ R(x,z)\ \&\ R(x,y)]\}\ \&\ \neg Qx\ D(x)$$

does not have models of any strong limit cardinals. Thus assuming the L.C.H., $\sigma \rightarrow \neg\tau$ is a universally valid sentence. If however, say $2^{\aleph_n} \geqslant \aleph_\omega$ for some $n < \omega$, then $\sigma \& \tau$ has a model of cardinal $\aleph_\omega$ and hence $\sigma \rightarrow \neg\tau$ is not universally valid.

The remark above shows that even if it is possible to show that V is recursively enumerable without using the L.C.H. the problem of finding an explicit axiomatization for V cannot be solved without making some assumptions about cardinal exponential arithmetic which are independent of ZF set theory.

References

C. C. Chang

[1965] A Note on the Two Cardinal Problem, Proc. Amer. Math. Soc. 16, 1148-1155

[1967] Ultraproducts and Other Methods of Constructing Models, in Sets, Models and Recursion Theory, edited by J. N. Crossley, North-Holland, Amsterdam, pp. 85-121.

W. B. Easton

[1964] Powers of Regular Cardinals, Princeton University Dissertation.

Th. Eichholz

[1957] Semantische Untersuchungen zur Entscheidbarkeit im Prädikatenkalkül mit Funktionsvariablen, Archiv für Mathematische Logik und Grundlagenforschung, 3, 19-28.

T. Frayne, A. Morel & D. Scott

[1962] Reduced Direct Products, Fund. Math. 51, 195-228.

G. Fuhrken

[1964] Skolem-type Normal Forms for First Order Languages with a Generalized Quantifier. Fund. Math. 54, 291-302.

[1965] Languages with the Added Quantifier "There Exist at Least $\aleph_\alpha$" in The Theory of Models, edited by J. Addison, L. Henkin and A. Tarski, North-Holland, Amsterdam, 121-131.

H. J. Keisler

[1964] On Cardinalities of Ultraproducts, Bull. Amer. Math. Soc. 70, 644-647.

[1967] Models with Orderings, duplicated typescript (Abstract: Weakly Well-Ordered Models, Notices Amer. Math. Soc. 14, p. 414).

G. Kreisel & J. L. Krivine

[1967] Elements of Mathematical Logic, North-Holland, Amsterdam.

M. H. Löb

[1967] Decidability of the Monadic Predicate Calculus with Unary Function Symbols, J.S.L. 32, 563.

R. MacDowell & E. Specker

[1961] Modelle der Arithmetik, in Infinitistic Methods, Pergamon, Oxford. 257-263.

A. Mostowski

[1957] On a Generalization of Quantifiers, Fund. Math. 44, 12-36.

R. L. Vaught

[1964] The Completeness of Logic with the Added Quantifier, 'There are uncountably many', Fund. Math. 54, 303-4.

M. Yasuhara

[1966] An Axiomatic System for the First Order Language with an Equicardinality Quantifier, J.S.L., 31, 635-640.

The Π^1_1-comprehension schema and ω-rules

Gaisi Takeuti*

Many theorems in the first order proof-theory follow from Gentzen's Hauptsatz, i.e., the cut-elimination theorem. This is still true even for higher order proof-theory in which the cut-elimination theorem is proved constructively. (Cf. [2], [6], [7]). However if one wishes to consider an extension of arithmetic, it is impossible to eliminate all the cuts because proof-figures contain mathematical inductions. Schütte has introduced the ω-rule and eliminated all cuts and mathematical inductions in first order arithmetic (cf. [4]). This is an excellent idea and can be considered a nicer form of the cut-elimination when mathematical induction is involved. However, since our main object is a finite proof-figure but not an infinite proof-figure, it is better if we can restrict the ω-rule so that the infinite proof-figure considered is close to the finite proof-figure. In this sense we consider the following constructive ω-rule. We define simultaneously the constructive ω-rule and the Gödel number of an infinite proof-figure containing the constructive ω-rule. We assume a standard method of Gödel numbering for axioms and for finite inferences. Our ω-rule is expressed by the following.

$$\frac{\overset{P_0}{\Gamma \longrightarrow \Delta, A(0)} \;\cdots\cdots \qquad \cdots\; \overset{P_n}{\Gamma \longrightarrow \Delta, A(n)} \;\cdots\cdots}{\Gamma \longrightarrow \Delta, \forall x A(x)}.$$

Here P_n is defined for every natural number n and is a proof-figure ending with $\Gamma \longrightarrow \Delta, A(n)$. Let $\ulcorner P_n \urcorner$, a Gödel number of P_n, be already assigned to P_n. If there exists a recursive function such that $f(n) = \ulcorner P_n \urcorner$ for every n, then this ω-rule is said to be constructive and 3.5^e is assigned to the whole proof-figure, where e is a Gödel number of f, i.e. $\{e\}(n) = \ulcorner P_n \urcorner$ holds. Let T be any logical system. A proof-figure of the system obtained from T by adjoining the constructive ω-rule to it is called an ω-proof in T.

* Work partially supported by National Science Foundation grant GP-6132.

In [6], we proved the consistency of the Π^1_1-comprehension schema by using ordinal diagrams. The following remarks about [6] may orient the reader sufficiently to understand the present paper.

The class of all semi-isolated formulas and varieties is defined to be the least class K satisfying the following conditions:

(1) $\forall\psi A(\psi, a_1, \ldots, a_n)$ and $A(a_o, a_1, \ldots, a_n)$ are in K, if $A(\psi, a_1, \ldots, a_n)$ is arithmetical in ψ, $a_1, \ldots, a_n$.

(2) A variety $\{x_1, \ldots, x_n\}A(x_1, \ldots, x_n)$ is in K, if $A(a_1, \ldots, a_n)$ is in K.

(3) If $A(\alpha)$ is in K and V is a variety with the same number of argument-places as α and is in K, then $A(V)$ is in K.

The system SINN is obtained from G^1LC by modifying it as follows:

(1) Every beginning sequence is either of the form $D \longrightarrow D$ or $a = b$, $A(a) \longrightarrow A(b)$, or a mathematical beginning sequence.

(2) The inference 'induction' is added:

$$\frac{A(a),\ \Gamma \longrightarrow \Delta,\ \ A(a+1)}{A(0),\ \Gamma \longrightarrow \Delta,\ \ A(t)}\ ,$$

where a is not contained in any of $A(0)$, Γ, Δ and t is an arbitrary term. A(a) is called the <u>induction formula</u> and a is called the <u>eigen-variable</u>.

(3) The inference schema $\forall$ left on an f-variable of the form

$$\frac{F(V),\ \Gamma \longrightarrow \Delta}{\forall\psi\ F(\psi),\ \Gamma \longrightarrow \Delta}$$

is restricted by the condition that V be semi-isolated.

SINN is equivalent to the system obtained from second order pure logic by adjoining arithmetic with full induction and the Π^1_1-comprehension schema. In this paper we shall prove the following theorems (more precise statements will be given later). Let S be the system of ordinal diagrams used to prove the consistency of SINN and $\prec$ be the ordering of S.

THEOREM. There exists a $\prec$-recursive function f such that for every proof-figure P ending with a sequence $\Gamma \longrightarrow \Delta$ without any free t-variable, $f(\ulcorner P \urcorner)$ is an ω-proof ending with $\Gamma \longrightarrow \Delta$, which has no mathematical inductions or cuts.

THEOREM. If $\lessdot$ is a provable well-ordering in our system, then there exists a recursive function which is a $\lessdot - \prec$ order-preserving map from the integers into S.

Before proceeding let us recall the definition of $\lessdot$ -recursive.

DEFINITION. Let S(a) and $a \lessdot b$ be primitive recursive predicates such that $\lessdot$ is a well-ordering of $\{a: S(a)\}$, whose first element is 0. A number-theoretic function ψ is called $\lessdot$-recursive, if and only if one of the following holds.

(i) $\psi(a) = a+1$

(ii) $\psi(a_1, \ldots, a_n) = 0$

(iii) $\psi(a_1, \ldots, a_n) = a_i \ (1 \leq i \leq n)$

(iv) $\psi(a_1, \ldots, a_n) = \varphi(\chi_1(a_1, \ldots, a_n), \ldots, \chi_m(a_1, \ldots, a_n))$, where φ and χ_i $(1 \leq i \leq m)$ are $\lessdot$-recursive.

(v) $$\begin{cases} \psi(0, a_2, \ldots, a_n) = \varphi(a_2, \ldots, a_n) \\ \psi(a+1, a_2, \ldots, a_n) = \chi(a, \psi(a, a_2, \ldots, a_n), a_2, \ldots, a_n), \end{cases}$$
where φ and χ are $\lessdot$-recursive.

(vi) $$\begin{cases} \psi(0, a_2, \ldots, a_n) = \varphi(a_2, \ldots, a_n) \\ \psi(a+1, a_2, \ldots, a_n) = \chi(a, \psi(\tau^*(a, a_2, \ldots, a_n), \\ \qquad\qquad a_2, \ldots, a_n), a_2, \ldots, a_n), \end{cases}$$
where φ, χ and τ are $\lessdot$-recursive and
$$\tau^*(a, a_2, \ldots, a_n) = \begin{cases} \tau(a, a_2, \ldots, a_n) & \text{if } \tau(a, a_2, \ldots, a_n) \lessdot a+1, \\ 0 & \text{otherwise.} \end{cases}$$

Some acquaintance with [6] is assumed throughout this paper.

Chapter I. ω-proofs and cut-elimination

§1. We shall transform a proof-figure in SINN whose end-sequence does not contain any free t-variables into a proof figure ending with the same end-sequence in the system with the constructive ω-rule. In proving the consistency of SINN in [6, Ch. 2] we defined reductions on a proof-figure ending with $\longrightarrow$. This notion however can easily be extended to proof-figures whose end-sequences have no free t-variables. For the definition of 'reduction' and some related notions, cf. [6, Ch. 2, 3], [6, Ch. 1, 5] and [6, Ch. 3, 6].

For an o.d. α and a natural number m, we let $<\alpha, m>$ be short for $(0, 0, \alpha) \# O^{(m)}$, where $O^{(i)}$ is defined by $O^{(o)} = O$ and $O^{(i+1)} = O^{(i)} \# O$.

We use the same assignment of an o.d. to a sequence of a proof-figure as in [6, Ch. 2, 6], and define the o.d. of a proof-figure P to be $<\alpha, m>$, where α is the o.d. assigned to its end-sequence and m is the number of free t-variables in its end piece (denoted o(P)).

When we define a reduction we sometimes use an expression like 'take the bottommost inference satisfying certain conditions'. In many cases, there are many bottommost such inferences so that it is not uniquely determined. In those cases we understand that actually a Gödel number is given to each inference and we take the bottommost such inference whose Gödel number is the smallest.

THEOREM 1. There exists a $\prec$-recursive function f such that, for every proof-figure P ending with SINN to a sequence $\Gamma \longrightarrow \Delta$ which does not contain any free t-variable, $f(\ulcorner P \urcorner)$ is a Gödel number of an ω-proof ending with $\Gamma \longrightarrow \Delta$ which has no cut and no mathematical induction.

Proof. Let P be a proof-figure of SINN whose end-sequence contains no free t-variables. We define reductions r(P) and q(i, P) for each $i < \omega$ and a transformation f(P) by transfinite induction on the o.d. of P.

1. The case where the end-piece of P contains an induction or explicit logical inference.

1.1. If the end-piece of P contains a free t-variable which is not used as an eigenvariable, r(P) is defined to be the proof-figure obtained from P by substituting 0 for each a free t-variable. Obviously $r(P) \prec P$. We define f(P) to be f(r(P)).

1.2. If 1.1 is not the case, let I be the bottommost induction or explicit logical inference. We consider several cases.

1.2.1. If I is an induction, let r(P) be the proof-figure obtained from P by applying VJ-Reduktion to I (cf. [6, Ch. 2, 8.3]), and let f(P) be f(r(P)).

1.2.2. The case where I is an explicit logical inference.

1.2.2.1. The case where I is not an $\forall$ right on a t-variable. Since all the cases are similarly treated (cf. [6, Ch. 3, 6.1]) we consider the case where I is an $\wedge$ left. Let P be

$$\begin{array}{c} \vdots \\ A, \Gamma \longrightarrow \Delta \\ \hline A \wedge B, \Gamma \longrightarrow \Delta \\ \vdots \\ \Gamma_0 \longrightarrow \Delta_0 \end{array} .$$

We define r(P) to be the proof-figure

$$\begin{array}{c} \vdots \\ A, \Gamma \longrightarrow \Delta \\ \hline \text{Some exchanges and a weakening} \\ \hline A \wedge B, \Gamma, A \longrightarrow \Delta \\ \vdots \\ \Gamma_0, A \longrightarrow \Delta \end{array} .$$

Since $r(P) \prec P$, f(r(P)) has been defined by the inductive hypothesis. We define f(P) to be the following proof-figure

$$\begin{array}{c} \vdots \quad f(r(P)) \\ \Gamma_0, A \longrightarrow \Delta_0 \\ \hline \text{Some exchanges} \\ \hline A, \Gamma_0 \longrightarrow \Delta_0 \\ \hline A \wedge B, \Gamma_0 \longrightarrow \Delta_0 \\ \hline \text{Some exchanges and a contraction} \\ \hline \Gamma_0 \longrightarrow \Delta_0 \end{array}$$

We shall refer to this figure as g(f(r(P)).

1.2.2.2. The case where I is an $\forall$ right on a t-variable. Let P be of the following form:

$$\begin{array}{c} \vdots \\ \dfrac{\Gamma \longrightarrow \Delta,\ A(a)}{\Gamma \longrightarrow \Delta,\ \forall x A(x)}\ \text{I} \\ \vdots \\ \Gamma_o \longrightarrow \Delta_o \end{array}$$

We consider the proof-figure (referred to as q(i, P))

$$\begin{array}{c} \Gamma \longrightarrow \Delta,\ A(i) \\ \hline \text{Some exchanges and a weakening} \\ \hline \Gamma \longrightarrow A(i),\ \Delta,\ \forall x A(x) \\ \vdots \\ \Gamma_o \longrightarrow A(i),\ \Delta, \end{array}$$

where the proof-figure to $\Gamma \longrightarrow \Delta,\ A(i)$ is obtained from the proof-figure ending with the upper sequence of I by substituting the numeral i for a. Obviously $q(i, P) \prec P$ for any numeral i. Thus f(q(i, P)) has been defined for each i. We define f(P) to be the proof-figure

$$\begin{array}{c} f(q(i,P)) \\ \vdots \\ \Gamma_o \longrightarrow A(i),\ \Delta_o \\ \hline \text{Some exchanges} \\ \hline \ldots\ldots \quad \Gamma_o \longrightarrow \Delta_o,\ A(i) \quad \ldots\ldots\ldots\ldots \text{ for each } i \\ \hline \Gamma_o \longrightarrow \Delta_o,\ \forall x A(x) \\ \hline \text{Some exchanges and a contraction} \\ \hline \Gamma_o \longrightarrow \Delta_o \end{array}\ \omega\text{-rule}$$

2. The case where the end-piece of P does not contain any induction or logical inference. We define r(P) to be the proof-figure obtained from P by applying the reductions in [6, Ch. 2, 8-10], retaining explicit weakenings and explicit logical beginning sequences. (Also cf. the proof of Theorem 1a in [2].) Since the end-sequence is unchanged by the reductions we define f(P) to be f(r(P)).

We sometimes identify many notions with their Gödel numbers, e.g., a proof-figure P sometimes means its Gödel number. Thus we can consider the functions r, q, g, f to be number-theoretic functions. We can obviously take r, q and g to be primitive recursive. Let P(a) be a primitive recursive predicate stating that a is a proof-figure of SINN whose end-sequence contains no free t-variable. Let P_0, P_1, P_2 and P_3 be defined by

$P_0(m) \underset{dfn.}{\longleftrightarrow} P(m)$ and one of the reductions 1.1, 1.2.1 and 2 is applied.

$P_1(m) \underset{dfn.}{\longleftrightarrow} P(m)$ and the end-piece of m contains an explicit logical inference other than an $\forall$ right on a t-variable to which the reduction applies.

$P_2(m) \underset{dfn.}{\longleftrightarrow} P(m)$ and the reduction will apply to an $\forall$ right on a t-variable in the end-piece of m.

$P_3(m) \underset{dfn.}{\longleftrightarrow} \neg (P_0(m) \vee P_1(m) \vee P_2(m))$.

Obviously P_0, P_1, P_2 and P_3 are primitive recursive and have the following properties:

$$\forall x \exists ! i P_i(x) \qquad (i = 0, 1, 2, 3),$$

$$P_0(m) \longrightarrow r(m) \prec m,$$

$$P_1(m) \longrightarrow r(m) \prec m,$$

$$P_2(m) \longrightarrow \forall n (q(n, m) \prec m).$$

We shall show that f is recursive (in fact $\prec$-recursive). Let

$$f(e, m) \simeq \begin{cases} \{e\}(r(m)) & \text{if } P_0(m), \\ g(\{e\}(r(m))) & \text{if } P_1(m), \\ 3.5^{S_1^2(c_0, e, m)} & \text{if } P_2(m), \\ m & \text{if } P_3(m), \end{cases}$$

where c_0 is the Gödel number of $\lambda nem \{e\}(q(n, m))$. By the recursion theorem there is a number c such that $f(c, m) \simeq \{c\}(m)$. Then define f by $f(m) \simeq \{c\}(m)$, i.e.,

$$f(m) \simeq \begin{cases} f(r(m)) & \text{if } P_0(m), \\ g(f(r(m))) & \text{if } P_1(m), \\ 3.5^{S_1^2(c_0, c, m)} & \text{if } P_2(m), \\ m & \text{otherwise.} \end{cases}$$

Thus f is partial recursive. By transfinite induction on $\prec$ we can show that f is totally defined. It is also easy to see that f is $\prec$-recursive, that $f(P)$ has the same end-sequence as P and that $f(P)$ has no cuts or mathematical inductions or free t-variables.

DEFINITION. A number-theoretic function $\psi(a_1, \ldots, a_n)$ is called provably recursive in SINN, if the following sequence is provable in SINN:

$$\longrightarrow \forall x_1 \ldots \forall x_n \exists y T_n(e, x_1, \ldots, x_n, y),$$

where T_n expresses Kleene's primitive recursive predicate T_n and e is a Gödel number of ψ.

As an application of our technique we can give an alternative proof of Theorem 1 for SINN in Kino [2].

THEOREM. Let ψ be a provably recursive function in SINN. Then we can find an o.d. s of S such that ψ is $\prec^s$-recursive where $\prec^s$ is $\prec$ restricted to arguments $\prec^s$.

Proof. (Alternative) Without loss of generality we may assume that ψ has one argument-place. Let e be a Gödel number of ψ such that the sequence $\longrightarrow \forall x \exists y T_1(e, x, y)$ is provable in SINN. Let P be a proof-figure ending with $\longrightarrow \exists y T_1(e, a, y)$ whose o.d. is σ. We define P_m to be the proof-figure obtained from P by substituting the numeral m for a. The process of obtaining P_m from P is primitive recursive. We spply to each P_m the transformation f. $f(P_m)$ is a proof-figure without cut. Since P does not contain any explicit $\forall$ right on a t-variable (which is the only inference which induces an application of the ω-rule in the transformation), it is easily proved by transfinite induction that $f(P_m)$ does not contain any application of the ω-rule. By checking the proof-figure $f(P_m)$, we can find primitive recursively a numeral n satisfying $T_1(e, m, n)$

from $f(P_m)$. Since $n = \psi(m)$ and f is $\prec^{\sigma}$-recursive by Theorem 1, we see that ψ is $\prec^{\sigma}$-recursive.

§2. Let us now consider a system obtained from SINN by introducing function symbols (which act like free function variables). For brevity we shall deal with only one function symbol $\underline{f}$ in the following. We shall consider a slight modification of SINN obtained by adding a single function symbol $\underline{f}$.

(1) We add the inference schema 'term-replacement' and the following schema:

$$\frac{\underline{f}(n) = m,\ \Gamma(m) \longrightarrow \Delta(m)}{\underline{f}(n) = m,\ \Gamma(\underline{f}(n)) \longrightarrow \Delta(\underline{f}(n))}, \quad \frac{\underline{f}(n) = m,\ \Gamma(\underline{f}(n)) \longrightarrow \Delta(\underline{f}(n))}{\underline{f}(n) = m,\ \Gamma(m) \longrightarrow \Delta(m)},$$

where m and n are numerals, and $\Gamma(a)$ and $\Delta(a)$ indicate some or all a occurring in Γ and Δ respectively.

(2) As beginning sequences we add sequences of the form

$$\underline{f}(n) = m_1,\ \underline{f}(n) = m_2 \longrightarrow \quad ,$$

where m_1 and m_2 are distinct numerals.

In this section we shall show, by using transfinite induction on ordinal diagrams, that we can transform a proof-figure in SINN (thus modified) whose end-sequence does not contain any free t-variable into a proof-figure ending with the same end-sequence in the system obtained from the system with the constructive ω-rule by adjoining to it the $\underline{f}$-ω-rule:

$$\frac{\underline{f}(m) = n,\ \Gamma \longrightarrow \Delta \text{ for each } n < \omega}{\Gamma \longrightarrow \Delta},$$

where m is a numeral.

This rule is introduced if there exists a recursive function ϕ such that $\phi(n)$ is a Gödel number of a proof-figure ending with $\underline{f}(m) = n$, $\Gamma \longrightarrow \Delta$ for each $n < \omega$: 3.7^e is assigned to the proof-figure ending with $\Gamma \longrightarrow \Delta$, where e is a Gödel number of ϕ.

We assign to a proof-figure in the modified SINN an o.d. like the one in [6, Ch. 2, 6 and 8.1.1] except the following.

(1) We assign to the lower sequence of an $\forall$ left on a t-variable

$$\frac{A(t),\ \Gamma \longrightarrow \Delta}{\forall x A(x),\ \Gamma \longrightarrow \Delta}$$

$(\omega; a, \sigma)$, where a is the number of f's in t and σ is the o.d. of the upper sequence.

(2) We assign to the lower sequence of an induction

$$\frac{A(a),\ \Gamma \longrightarrow \Delta,\ A(a+1)}{A(0),\ \Gamma \longrightarrow \Delta,\ A(t)}$$

$(\omega, a_0+a_1+2, \sigma)$, where a_0 is the grade of the induction formula, a_1 is the number of f's in t and σ is the o.d. of the upper sequence.

(3) We assign to the lower sequence of a term-replacement in the modified form the same o.d. assigned to the upper sequence.

(4) The o.d. of a proof-figure P is $\langle \alpha, n \rangle$, where α is the o.d. of the end-sequence of P and n is the number of free t-variables in its end-piece.

The transformation is defined in the same way as in §1. Let P be a proof-figure whose end-sequence contains no free t-variable. We define reductions r and q and transformation t by a transfinite induction on S.

1. The case where the end-piece of P contains an induction or explicit logical inference.

1.1. If the end-piece of P contains a free t-variable which is not used as an eigenvariable, r(P) is defined to be the proof-figure obtained from P by substituting 0 for such a free t-variable. Obviously $r(P) \prec P$. We define t(P) to be t(r(P)).

1.2. If 1.1 is not the case, let I be the bottommost induction or explicit inference. We consider several cases separately.

1.2.1. Let I be an induction

$$\vdots$$

$$\frac{A(a),\ \Gamma \longrightarrow \Delta,\ A(a+1)}{A(0),\ \Gamma \longrightarrow \Delta,\ A(s)}\ I$$

$$\vdots$$

$$\Gamma_o \longrightarrow \Delta_o \quad .$$

1.2.1.1. If s contains no function symbol $\underline{f}$, let r(P) be the figure obtained from P by applying VJ-Reduktion to I, and let t(P) be t(r(P)).

1.2.1.2. If s contains the function symbol $\underline{f}$, let $\underline{f}(m)$ be the innermost one (i.e., m is a numeral) and let $s(\underline{f}(m))$ denote s. We reduce P to a proof-figure

$$\vdots$$

$$\cfrac{\cfrac{\cfrac{\cfrac{A(a),\ \Gamma \longrightarrow \Delta,\ A(a+1)}{A(0),\ \Gamma \longrightarrow \Delta,\ A(s(n))}\ I'}{\underline{f}(m) = n,\ A(0),\ \Gamma \longrightarrow \Delta,\ A(s(n))}}{\underline{f}(m) = n,\ A(0),\ \Gamma \longrightarrow \Delta,\ A(s(\underline{f}(m)))}}{\cfrac{\text{Some exchanges}}{A(0),\ \Gamma,\ \underline{f}(m) = n \longrightarrow \Delta,\ A(s)}}$$

$$\vdots$$

$$\Gamma_o,\ \underline{f}(m) = n \longrightarrow \Delta_o$$

where n is any numeral; s(n) is obtained from s by replacing $\underline{f}(m)$ in s by n. We shall refer to this figure as q(n, P). Since the number of f's in s(n) is smaller than that in s (i.e., $s(\underline{f}(m))$), the o.d. of q(n, P) is smaller than that of P for each number n. t(P) is defined to be the proof-figure.

$$\frac{\begin{array}{c} \vdots \quad t(q(n, P)) \\ \underline{f}(m) = n,\ \Gamma_o \longrightarrow \Delta_o \quad \text{for each } n < \omega \end{array}}{\Gamma_o \longrightarrow \Delta_o}\ f\text{-}\omega\text{-rule}$$

1.2.2. The case where I is an $\forall$ right on a t-variable. Let P be of the following form:

$$\begin{array}{c} \vdots \\ \dfrac{\Gamma \longrightarrow \Delta,\ A(a)}{\Gamma \longrightarrow \Delta,\ \forall x A(x)}\ I \\ \vdots \\ \Gamma_o \longrightarrow \Delta_o \end{array} .$$

We consider the proof-figure (referred to as q(i, P))

$$\begin{array}{c} \vdots \\ \Gamma \longrightarrow \Delta,\ A(i) \\ \hline \text{Some exchanges and a weakening} \\ \hline \Gamma \longrightarrow A(i),\ \Delta,\ \forall x A(x) \\ \vdots \\ \Gamma_o \longrightarrow A(i),\ \Delta, \end{array} ,$$

where the proof-figure ending with $\Gamma \longrightarrow \Delta,\ A(i)$ is obtained from the proof-figure ending with the upper sequence of I by substituting the numeral i for a. Obviously $q(i, P) \prec P$ for any i. Thus t(q(i, P)) has been defined for each i. We define t(P) to be the proof-figure

$$\begin{array}{c} \left.\begin{array}{c} \vdots \\ \Gamma_o \longrightarrow A(i),\ \Delta_o \end{array}\right\} \quad t(q(i,P)) \\ \hline \text{Some exchanges} \\ \hline \Gamma_o \longrightarrow \Delta_o,\ A(i) \qquad \ldots\ldots\ \text{for each } i \\ \hline \Gamma_o \longrightarrow \Delta_o,\ \forall x A(x) \\ \hline \text{Some exchanges and a construction} \\ \hline \Gamma_o \longrightarrow \Delta_o \end{array} \ \omega\text{-rule} .$$

1.2.3. The case where I is not an induction or an introduction of an $\forall$ on a t-variable: Like 1.2.2.1 in §1.

2. The case where the end-piece of P does not contain any induction or logical inference. We define r(P) to be the proof-figure obtained from P by applying the reductions in [6, Ch. 2, 8-10], retaining explicit weakenings and explicit logical beginning sequences. Since the end-sequence is unchanged by those reductions we define t(P) to be t(r(P)).

By obvious modifications of the technique given in §1, we can prove that t(m) is $\prec$-recursive and that t(P) has the desired properties.

Chapter II. Provable well-orderings.

In this chapter we shall consider a property of provable well-orderings in SINN. Let S denote the system of ordinal diagrams used to prove the consistency of SINN. We proved in [6, Ch. 5] that the order-type of any provable well-ordering $\prec$ of natural numbers in SINN is less than the order-type of S. We also proved in the Appendix of [7] that there exists a Π_1^0 order-preserving one-one mapping of $\prec$ into a segment of (arithmetized) S, mentioning a conjecture by Kreisel that Π_1^0 can be replaced by recursive. The purpose of this chapter is to give two proofs of this conjecture, i.e.,

THEOREM. Let $\prec$ be any provable recursive well-ordering in SINN. Then there exists a recursive order-preserving mapping of $\prec$ into a segment of S.

We begin with explaining the basic idea. Kreisel has pointed out that Schütte's method immediately implies the theorem for arithmetic. (This is the basis of his conjecture quoted above.) We shall outline his proof.

We first apply Schütte's cut elimination to the given proof, and, as usual, get a proof-figure that can be mapped into a (proper initial) segment of $<_{\varepsilon_0}$. The cut free proof has the following form, when one uses function symbols as in [4], i.e., with only axioms of the form $f(n) = m \vee f(n) \neq m$ for numerals n and m.

The final formula at the node which we shall call $N_{< >}$, is $\exists x \neg(f(x+1) \prec f(x))$. Denoting the predecessors of a node N_c by $N_{c0}, N_{c1}, \ldots$, (where c is a sequence of integers $c_0, \ldots, c_k$) the formula at N_c is

$$f(0) = c_0 \wedge \ldots \wedge f(k) = c_k \longrightarrow \exists x \neg(f(x+1) \prec f(x)) .$$

A terminal node N_c is such that $\neg(c_k \prec c_{k-1} \wedge \ldots \wedge c_1 \prec c_0)$ holds. We want to use this proof tree (which can be mapped into the ordering) to map the order $\prec$ itself into the ε_0 ordering.

What we should remark is the following (cf. Cor. 3.32 and 3.341 in Kreisel [3]).

1. The consistency of infinite proof-trees in Schütte [4] is such that, starting with a finite proof in arithmetic, each proof tree can be mapped primitive recursively into the standard ε_o-ordering.

2. An ordering can be embedded recursively in its Brouwer-Kleene ordering (of descending sequences).

In our case, we carried out the Schütte cut elimination for our system in Chapter I, and 2 also holds. Howere 1 is not obvious in our case and it may be said that 1 is the theorem itself. Therefore we need some elaboration. Roughly our general ideal is this. Let P(a) be a fixed proof-figure ending with the sequence expressing that $\prec$ is well-ordered up to a. First we define proof-figures P_k for all numerals k by induction on k. To do this, let $n_o, n_1, \ldots, n_k$ be a reordering of 0, 1, ..., k with respect to $\prec$ and let k be n_j, i.e.,

$$n_o \prec n_1 \prec \ldots \prec n_{j-1} \prec k \prec n_{j+1} \prec \ldots \prec n_k .$$

If $k = n_k$, then we define P_k to be the proof-figure obtained from P(a) by replacing a by the numeral k. If $k \neq n_k$, $P_{n_{j+1}}$ has been defined by the inductive hypothesis. Then P_k is defined from $P_{n_{j+1}}$ by applying a reduction procedure so that $o(P_k) \prec o(P_{n_{j+1}})$, where o(P) is the o. d. of P. Then we define the recursive order-preserving map ψ by

$$\psi(k) = \psi(n_{j-1}) + o(P_k),$$

where + is ordinary sum (not natural sum) of o. d.'s which is easily defined recursively by using natural sum # (one of the primitive operations of o. d.'s). Since

$$\psi(n_{j+1}) = \psi(n_{j-1}) + o(P_{n_{j+1}})$$

is easily proved by mathematical induction on k, it is obvious that ψ is order-preserving.

§1. First Proof. Our first proof will be given along the lines of the proof of the theorem in Chapter 5 of [6], which is again along the lines of the unprovability of transfinite induction up to ε_0 in the first order arithmetic given by Gentzen [1]. To make this section self-contained to some extent we shall first outline the proof.

Let $a \prec b$ be a recursive predicate such that $\prec$ is a well-ordering of natural numbers. (For simplicity we shall consider that the domain of $\prec$ is the set of all natural numbers.) We formulate 'transfinite induction on $\prec$' TI($\prec$) for SINN (or any second order arithmetic formalized by G^1LC) by

$$(*) \qquad \forall\psi(\forall x(\forall y(y \prec x \vdash \psi[y]) \vdash \psi[x]) \vdash \forall x\psi[x]),$$

and call $\prec$ a provable recursive well-ordering of SINN if TI($\prec$) is provable in SINN. We define a TJ-proof-figure with respect to $\prec$ for SINN, generalizing Gentzen's idea in [1] (also cf. [5]):

A TJ-proof-figure with respect to $\prec$ for SINN (referred to as a TJ-proof-figure in the following) is a figure which is obtained from a proof-figure of SINN by modifying it as follows:

(1) The beginning sequences of SINN and the sequences of the following form (called a TJ-beginning sequence) are allowed as beginning sequences:

$$\forall x(x \prec t \vdash \mathcal{E}(x)) \longrightarrow \mathcal{E}(t),$$

where t is an arbitrary term and $\mathcal{E}$ is considered a fixed free f-variable for a particular purpose.

(2) The inference schema 'term-replacement' is added.

(3) The end-sequence is of the form

$$\longrightarrow \mathcal{E}(s_1), \ldots, \mathcal{E}(s_n),$$

where $s_1, \ldots, s_n$ are numerals. The minimum of $|s_1|_\prec, \ldots, |s_n|_\prec$ is called the end-number of the TJ-proof-figure, where $|s|_\prec$ denotes the order-type of s with respect to $\prec$.

Let $\prec$ be a provable well-ordering in SINN. Then the formula (*) is provable in SINN, and so is the sequence

$$\forall x(\forall y(y \prec x \vdash a[y]) \vdash a[x]) \longrightarrow a[a] .$$

Replacing a by $\mathcal{E}$ we see that

$$\forall x(\forall y(y \prec x \vdash \mathcal{E}(y)) \vdash \mathcal{E}(x)) \longrightarrow \mathcal{E}(a)$$

is provable. By an obvious modification we see that there is a TJ-proof-figure whose end-sequence is $\longrightarrow \mathcal{E}(a)$. We shall define the reduction method for a TJ-proof-figure following the reduction defined for the consistency proof of SINN, defining the o.d. of a TJ-proof-figure. As will be seen below, the o.d.'s assigned to TJ-proof-figures belong to S. Since the end-sequence of a TJ-proof-figure is not empty and we possibly have TJ-beginning sequences in a TJ-proof-figure, we need to define a new reduction technique. We call a TJ-proof-figure non-critical or critical, according as we can apply one of the reduction steps for the consistency proof of SINN to it or not. By the reduction defined below a TJ-proof-figure P is reduced to a TJ-proof-figure with the same end-number, if P is non-critical; or with a smaller end-number otherwise. (We can take an arbitrary ordinal less than the end-number of P as the new end-number. See below.) When we define the reduction with the above property we can prove

FUNDAMENTAL LEMMA. For any TJ-proof-figure, its end-number is not greater than (the order-type of) its o.d.

Proof. By transfinite induction on the o.d. of a TJ-proof-figure: Let P be a TJ-proof figure whose o.d. is a and end-number is σ. We assume as the inductive hypothesis that our lemma is true for any TJ-proof-figure whose o.d. is less than $|a|$ and show $\sigma \leq |a|$. If P is non-critical, P is reduced to a TJ-proof-figure P' with the same end-number σ and an o.d. $\beta < a$. By the inductive hypothesis $\sigma \leq |\beta|$, and hence $\sigma \leq |a|$. Let P be critical. If σ were greater than $|a|$, we could reduce P to a TJ-proof-figure whose end-number is $|a|$ and whose o.d. is less than a, contradicting the inductive hypothesis, q.e.d.

We now define the reduction of TJ-proof-figures. A TJ-proof-figure with degree is defined in the same way as in [6, Ch. 2, 4]. The notion of i-loader and the assignment of an o.d. to a TJ-proof-figure with degree is defined similarly: The o.d. of the TJ-beginning sequence is $(\omega, 0, (\omega, 0, (\omega, 0, (\omega, 0, (\omega, 0, 0)))))$. For the sake of convenience the formula in the right side of a TJ-beginning sequence will be referred to as a principal formula. In other words, by a principal formula we mean the principal formula of a logical inference or the principal formula of a TJ-beginning sequence.

We can follow the reductions given for the consistency proof of SINN up to the existence of a suitable cut, i.e., until we get a TJ-proof-figure P with degree of the following properties:

p1. The end-piece of P contains no free t-variable.

p2. The end-piece of P contains no induction.

p3. The end-piece of P contains no beginning sequence for equality.

p4. If the end-piece of P contains a weakening I, then any inference under I is a weakening.

(REMARK. Since the end-piece of a TJ-proof-figure is not empty, the end-sequence S' of the proof-figure obtained from P by eliminating weakenings in the end-piece of P satisfying p1-p3 may be different from the end-sequence of P. In this case we apply weakenings under S' so that the end-sequence becomes the same as the end-sequence of P.)

p5. The end-piece of P contains no logical beginning sequence.

We see easily

PROPOSITION 1. Let P be a TJ-proof-figure with degree satisfying p1-p5. Then P contains at least one logical inference or TJ-beginning sequence. Therefore the end-piece of P contains a principal formula at the boundary or a TJ-beginning sequence.

Let P be a TJ-proof-figure with degree satisfying p1-p5. By Proposition 1 the end-piece of P contains a principal formula at the boundary or a TJ-beginning sequence. We call a formula A in the end-piece of P a principal descendant or a principal TJ-descendant according as A is a descendant of a principal formula at the boundary or a descendant of the principal formula of a TJ-beginning sequence in the end-piece of P.

PROPOSITION 2. Let P be a TJ-proof-figure with degree satisfying p1-p5, and S a sequence in the end-piece of P. If S contains a formula with a logical symbol, then there exists a formula A in S or in a sequence above S such that A is a principal descendant or a principal TJ-descendant.

Proof. Suppose S contains a formula with a logical symbol. Then S is above the uppermost weakening in the end-piece. The property 'a sequence contains a logical symbol' is preserved up to (at least) one of the upper sequences of each inference in the end-piece (but not beyond the inference at the boundary), or a TJ-beginning sequence, when we follow the string upward to which S belongs.

PROPOSITION 3. Let P be a TJ-proof-figure with degree satisfying p1-p5 and not containing a suitable cut. Then its end-sequence contains a principal TJ-descendant.

Proof. It suffices to prove that the end-sequence of P contains a principal descendant or a principal TJ-descendant, since the end-sequence contains no logical symbol. Suppose not. Since the end-piece contains a principal descendant or a principal TJ-descendant by Proposition 1, let us consider the following property (P) concerning a cut in the end-piece of P. A cut I in the end-piece of P is said to have the property (P) if (at least) one of its upper sequences contains a principal descendant or a principal TJ-descendant but its lower sequence contains no such formula. Let I be an uppermost cut with the property (P) in the end-piece of P:

$$\frac{\Gamma \longrightarrow \Delta, D \quad D, \Pi \longrightarrow \Lambda}{\Gamma, \Pi \longrightarrow \Delta, \Lambda} I_o$$

Let S_o and S_1 be the left and right upper sequences of I_o, respectively. By our assumption one of the cut formulas is a principal descendant or a principal TJ-descendant. First suppose D in S_o have that property. If D contains a logical symbol, then S_1 also contains a formula with a logical symbol (i.e., D). Therefore (by Proposition 2) there is a formula A in S_1 or above it such that A is a principal descendant or a principal TJ-descend
If A is not in S_1, there must be a cut having the property (P) above I_o,

contradicting our choice of T. If A is in S_1, A must be D itself, which contradicts our assumption that P does not contain a suitable cut. Thus D must be of the form $\mathcal{E}(t)$. If S_1 contains a logical symbol, there exists a principal descendant or a principal TJ-descendant either in S_1 or above it. If it is in S_1, it cannot be D (since D is $\mathcal{E}(t)$ and is in the left side of a sequence, it cannot be a principal TJ-descendant), and so it must also appear in the lower sequence of I_o, contradicting our assumption that I_o has the property (P). This means such a formula is in a sequence above S but not in S itself, contradicting our assumption that I_o is the uppermost cut with the property (P). Thus S_1 cannot contain a formula with logical symbol. No logical inference at the boundary or TJ-beginning sequence in the end-piece is above S_1. Therefore it is impossible that S_1 contains $\mathcal{E}(t)$, that is, it is impossible that D in S_o is a principal TJ-descendant. Next, suppose that the cut formula D in S_1 is a principal descendant or principal TJ-descendant. As was seen above, D cannot be a principal TJ-descendant: D must contain a logical symbol. Hence there is a principal descendant or a principal TJ-descendant either in S_o or in a sequence above S_o. If such a formula is not in S_o, there must be a cut having the property (P) above S_o, which is a contradiction. Therefore D in S_o must have that property, since the lower sequence of I_o cannot contain such, again contradicting our assumption P does not contain a suitable cut, q.e.d.

Let P be a TJ-proof-figure satisfying p1-p5 and not containing a suitable cut (called a critical TJ-proof-figure). We define the notion of critical reduction. By Proposition 3 the end-sequence of P contains a principal TJ-descendant $\mathcal{E}(r)$. Let s be any number such that $|s|_{\prec}$ is less than the end-number of P. Then the sequence $\longrightarrow s \prec r$ is a mathematical beginning sequence. We replace the TJ-beginning sequence

$$\forall x(x \prec r \vdash \mathcal{E}(x)) \longrightarrow \mathcal{E}(r)$$

in P by an ordinary proof-figure

$$\dfrac{\dfrac{\dfrac{\dfrac{\longrightarrow s \lessdot r \qquad \dfrac{\mathcal{E}(s) \longrightarrow \mathcal{E}(s)}{\longrightarrow \mathcal{E}(s), \neg\mathcal{E}(s)}}{\longrightarrow \mathcal{E}(s),\ s \lessdot r \wedge \neg\mathcal{E}(s)}}{s \lessdot r \vdash \mathcal{E}(s) \longrightarrow \mathcal{E}(s)}}{\forall x(x \lessdot r \vdash \mathcal{E}(x)) \longrightarrow \mathcal{E}(s)}}{\forall x(x \lessdot r \vdash \mathcal{E}(x)) \longrightarrow \mathcal{E}(s), \mathcal{E}(r)}$$

where $A \vdash B$ is the abbreviation of $\neg(A \wedge \neg B)$. The o.d. of this proof-figure $(\omega, 0, (\omega, 0, (\omega, 0, 0 \# (\omega, 0, 0))))$ is less than that of a TJ-beginning sequence. By this replacement and obvious changes P becomes a TJ-proof-figure P' with degree whose end-sequence is

$$\longrightarrow \mathcal{E}(s), \mathcal{E}(s_1), \ldots, \mathcal{E}(s_n),$$

where $\longrightarrow \mathcal{E}(s_1), \ldots, \mathcal{E}(s_n)$ is the end-sequence of P, and such that the o.d. of P' is less than that of P and the end-number of P' is $|s|_{\lessdot}$. We shall refer to P' as the proof-figure obtained from P by an application of the critical reduction at s.

Adjoining this reduction to the previous reduction and applying the fundamental lemma we can prove the

THEOREM. The order-type of $\lessdot$ is less than the order-type of S.

As we mentioned at the beginning of this section, there is a TJ-proof-figure ending with $\longrightarrow \mathcal{E}(a)$. Let P(a) be such a proof-figure.

1. First we shall assign to each number k a TJ-proof-figure P_k by induction on k:

1.0. The end-number of P_k is $|k|_{\lessdot}$.

1.1. The case where $\forall n < k(n \lessdot k)$. We define P_k to be the proof-figure obtained from P(a) by replacing a by the numeral k throughout P(a).

1.2. The case where $\exists n < k(k \lessdot n)$. Let

$$(*) \qquad n_o \lessdot \ldots \lessdot n_{j-1} \lessdot n_j (= k) \lessdot n_{j+1} \lessdot \ldots \lessdot n_k$$

be the reordering of the numbers $\leq k$ with respect to $\lessdot$. Then we define P_k to be the proof-figure obtained from $P_{n_{j+1}}$ by applying a critical reduction at k.

As is clearly seen we can define P_k for each k by induction on k and this definition is recursive.

We now define a map $\psi(k)$ which will turn out to be an order-preserving recursive map by making use of P_k. We see easily

LEMMA 1. Ordinal sum + of o.d.'s is recursive.

LEMMA 2. If two o.d.'s α and β are connected (i.e., the last operation used to form α or β is not $\#$) and $\alpha <_0 \beta$, then $\alpha + \beta = \beta$.

We define $\psi(k)$ by induction on k assuming (*) by

$$\psi(k) = \psi(n_{j-1}) + o(P_k),$$

where $o(P)$ is the o.d. of P.

LEMMA 3. Let

$$m_0 \lessdot m_1 \lessdot \ldots \lessdot m_i$$

be the reordering of numbers $< i+1$. Then

$$\psi(m_{j+1}) = \psi(m_j) + o(P_{m_{j+1}}),$$

where $0 \leq j < i$.

Proof. By mathematical induction on i.

§2. Second Proof. Let $a \lessdot b$ be a recursive predicate such that $\lessdot$ is a well-ordering of natural numbers. We assume that (i) the order-type of $\lessdot$ is a limit ordinal, (ii) we know the first element, (iii) the function h(a, n) such that $|h(a,n)|_{\lessdot} = |a|_{\lessdot} + n$ is recursive and (iv) if $a \lessdot b$, we can tell recursively if $|a|_{\lessdot} + \omega \leq |b|_{\lessdot}$ or not. Let $N_{\lessdot}$ denote the set of numbers ordered by the well-ordering $\lessdot$. Since lexicographical order of $N_{\lessdot} \times \omega$ satisfies this condition and $N_{\lessdot} \times \omega$ is easily embedded in $N_{\lessdot}$, we may assume this without loss of generality. We adjoin to SINN a function symbol $\underline{f}$ and formulate the notion '$\lessdot$ is a provable recursive well-ordering in SINN' by the sequence $\longrightarrow \exists x \neg(\underline{f}(x+1) \lessdot \underline{f}(x))$ being provable in SINN, or equivalently the sequence $\forall x(\underline{f}(x+1) \lessdot \underline{f}(x)) \longrightarrow$ being provable in SINN.

REMARK. This notion can also be formulated, using a predicate variable, by

$\forall\psi$('ψ expresses a function' $\longrightarrow \exists x \neg(\psi'x+1 \lessdot \psi'x))$, i.e. $\forall\psi(\forall x \exists! y \psi[y,x] \longrightarrow \exists x \neg \forall y \forall z(\psi[y,x] \wedge \psi[z,x+1] \longrightarrow z \lessdot y))$ is provable. This is stronger than $\exists x \neg(f(x+1) \lessdot f(x))$.

Let $\lessdot$ be a provable recursive well-ordering in SINN. We shall consider the modification of SINN given in §2, Chapter I. We shall show the existence of a recursive map of $\lessdot$ into an initial segment of S. Let $(i+j)_{\lessdot}$ denote the number k such that $|k|_{\lessdot} = |i|_{\lessdot} + j$.

We shall use a similar method to the one in §1. To each k, we assign a proof-figure P_k, a numeral $i(k)$ and a list of formulas

$$\underline{f}(0) = \ell_o, \ \underline{f}(1) = \ell_1, \ \ldots, \ \underline{f}(i(k)) = \ell_{i(k)}$$

where $\ell_o, \ldots, \ell_{i(k)}$ are numerals for which holds

$$\ell_o \gtrdot \ell_1 \gtrdot \ldots \gtrdot \ell_{i(k)}.$$

Let

(*) $$n_o \lessdot n_1 \lessdot \ldots \lessdot n_{j-1} \lessdot n_j \ (= k) \lessdot n_{j+1} \lessdot \ldots \lessdot n_k$$

be the reordering of the numbers $\leq k$ with respect to $\lessdot$. We define a $\lessdot$ - $\prec$-order-preserving recursive map $\psi(k)$ by mathematical induction on k by

$$\psi(k) = \psi(n_{j+1}) + o(P_k),$$

where $o(P)$ is the o.d. of a proof-figure P and + is ordinal sum as in §1. What we need to do first is to define reductions of proof-figures starting from a fixed proof-figure P to the sequence

$$\forall x(\underline{f}(x+1) \lessdot \underline{f}(x)) \longrightarrow$$

in such a way that P_k is obtained from $P_{n_{j+1}}$ and $o(P_k) \prec o(P_{n_{j+1}})$.

1. The assignment of an o.d. to a proof-figure is like the one in §2 of Chapter I except that we define the o.d. of a proof-figure to be the o.d. of its end-sequence. Let P be an arbitrary but fixed proof-figure ending with $\forall x(\underline{f}(x+1) \lessdot \underline{f}(x)) \longrightarrow$, and let its o.d. be σ. We define recursively reductions of proof-figures starting from P as in §2, Chapter I. In the following Q is understood to be P or a proof-figure obtained

by successive applications of reductions. Without loss of generality we may assume

(1) Every free t-variable in Q is used as an eigenvariable.

(2) The eigenvariables in Q are pairwise distinct.

(3) The o.d. of Q is not greater than σ with respect to $\prec$.

The end-sequence of a proof-figure is not necessarily preserved by some of the reductions. A formula of the form $\underline{f}(m) = n$ or $f(\ell+1) \lessdot f(\ell)$ will be added to the left side of the end-sequence by some reductions. Thus the end-sequence of proof-figures we are concerned with are of the form

$$\underline{f}(m_1) = n_1, \ldots, \underline{f}(m_i) = n_i,$$
$$\underline{f}(\ell_1+1) \lessdot \underline{f}(\ell_1), \ldots, \underline{f}(\ell_j+1) \lessdot \underline{f}(\ell_j), \quad \forall x(\underline{f}(x+1) \lessdot \underline{f}(x)) \longrightarrow,$$

where $0 \leq i$, $0 \leq j$ and $m_1, \ldots, m_i, n_1, \ldots, n_i, \ell_1, \ldots, \ell_j$ are numerals. We shall define two kinds of reductions $r(Q)$ and $q(n, Q)$. The latter will be referred to as a q-reduction. By an application of a q-reduction to Q we shall get a set of proof-figures $q(n, Q) : n < \omega$. The end-sequence of $q(n, Q)$ is of the form $\Gamma, \underline{f}(m) = n \longrightarrow \Delta$, where the end-sequence of Q is $\Gamma \longrightarrow \Delta$, and m is determined by Q. We refer to m as the number introduced by the q-reduction and refer to $q(n, Q)$ as the result of the q-reduction at n. The reductions are defined as follows:

1.1. If the end-piece of Q contains an induction, apply the reduction defined in 1.2.1, §2 of Chapter I.

1.2. If the end-piece of Q does not contain any induction but an explicit logical inference (which must be an $\forall$ left on a t-variable), apply the following reduction, where we should remark that A(n) is $\underline{f}(n+1) \lessdot \underline{f}(n)$. Let I be the bottommost $\forall$ left on a t-variable:

$$\frac{A(s), \Gamma \longrightarrow \Delta}{\forall x A(x), \Gamma \longrightarrow \Delta} I$$
$$\vdots$$
$$\Gamma_0 \longrightarrow \Delta_0$$

1.2.1. If s contains no $\underline{f}$, then we may assume s is a numeral n. We apply the usual reduction to define $r(Q)$

$$\begin{array}{c} \vdots \\ A(n),\ \Gamma \longrightarrow \Delta \\ \hline \text{Some exchanges and a weakening} \\ \hline \forall x A(x),\ \Gamma,\ A(n) \longrightarrow \Delta \\ \vdots \\ \Gamma_0,\ A(n) \longrightarrow \Delta_0 \end{array}$$

1.2.2. If s contains the symbol $\underline{f}$, let $\underline{f}(m)$ be the innermost one and let $s(\underline{f}(m))$ denote s. We define $q(n, Q)$ to be

$$\begin{array}{c} \vdots \\ A(s(\underline{f}(m))),\ \Gamma \longrightarrow \Delta \\ \hline \underline{f}(m) = n,\ A(s(\underline{f}(m))),\ \Gamma \longrightarrow \Delta \\ \hline \underline{f}(m) = n,\ A(s(n)),\ \Gamma \longrightarrow \Delta \\ \hline \text{Some exchanges} \\ \hline A(s(n)),\ \Gamma,\ \underline{f}(m) = n \longrightarrow \Delta \\ \hline \forall x A(x),\ \Gamma,\ \underline{f}(m) = n \longrightarrow \Delta \\ \vdots \\ \Gamma_0,\ \underline{f}(m) = n \longrightarrow \Delta_0 \end{array}$$

where n is any numeral. (Since the number of $\underline{f}$'s in s(n) is smaller than that in s, the o.d. of q(n, Q) is smaller than that of Q for each n.)

1.3. If the end-piece of Q does not contain any induction or explicit logical inference, we apply to Q the same reduction as in 2, §2 in Chapter I.

2. We now define for each k by mathematical induction on k a proof-figure Q(k), a number i(k) and a list of formulas

(0) $\quad \underline{f}(0) = \ell_0,\ \underline{f}(1) = \ell_1,\ \ldots,\ \underline{f}(i(k)) = \ell_{i(k)}$,

where $\ell_0, \ldots, \ell_{i(k)}$ are numerals and for which $\ell_0 \succ \ell_1 \succ \ldots \succ \ell_{i(k)}$. When we define Q(k) we are concerned with a q-reduction. i(k) is the number introduced by the q-reduction. It should be noticed that the formulas of the form $\underline{f}(m) = n$ occurring in the left side of the end-sequence of Q(k) will consist of part of the list (0) for k in the following definition. Suppose the definition has been given for any $n < k$.

2.1. The case where $\forall n < k (n \prec k)$. Q(k) is defined to be the proof-figure obtained as the result of the first critical reduction at k in the series of reductions from P. i(k) is the number introduced at the q-reduction:

$$\underline{f}(0) = \ell_o,\ \underline{f}(1) = \ell_1,\ \ldots,\ \underline{f}(i(k)) = \ell_{i(k)},$$

where $\ell_{i(k)} = k$ and $\ell_i = (k + (i(k)-i))_{\lessdot}$ for each $i < i(k)$.

2.2. The case where $\exists n < k(k \lessdot n)$. Let

$$n_o \lessdot n_1 \lessdot \ldots \lessdot n_{j-1} \lessdot k \lessdot n_{j+1} \lessdot \ldots \lessdot n_k$$

be the reordering of 0, ..., k. Since $n_{j+1} < k$, the proof figure $Q(n_{j+1})$, the number $i(n_{j+1})$ and the list

(1) $\underline{f}(0) = \ell_o,\ \ldots,\ f(i(n_{j+1})) = \ell_{i(n_{j+1})}$

are already given. Let $\ell_{i(n_{j+1})}$ be $(i(n_{j+1})+i)_{\lessdot}$. Consider the first q-reduction R in the series of reductions of proof-figures after $Q(n_{j+1})$ and let m be the numeral introduced at the reduction.

2.2.1. The case where $m < i(n_{j+1})$. By our assumption $\underline{f}(m) = \ell_m$ is in the list (1). Let $P^1(k)$ be the result of the q-reduction R at ℓ_m and consider the first q-reduction after $P^1(k)$. Let m_1 be the number introduced at the reduction. Repeat the same process until we get $m_{\ell+1} > m_\ell$. (It is evident this situation must happen in at most m steps.) Then apply the following process 2.2.2. reading m_ℓ and $m_{\ell+1}$ for $i(n_{j+1})$ and m there, respectively.

2.2.2. The case where $m > i(n_{j+1})$. i(k) is defined to be m.

2.2.2.1. If $|\underline{f}(i(n_{j+1}))|_{\lessdot} - |k|_{\lessdot} \geq m-i(n_{j+1})$, Q(k) is defined to be the result of the first q-reduction after $Q(n_{j+1})$ at k. We define the list from (1) by adjoining

(2) $\underline{f}(i(n_{j+1})+1) = \ell_{i(n_{j+1})}+1,\ \ldots,\ \underline{f}(m) = k,$

where $\ell_i = (k+(m-i))_{\lessdot}$ for $i(i(n_{j+1})+1 \leq i < m)$.

2.2.2.2. If $|\underline{f}(i(n_{j+1}))|_{\lessdot} - |k|_{\lessdot} < m-i(n_{j+1})$, then the left side is a natural number, say k*. Let Q(k) be the result of the first q-reduction after $Q(i(n_{j+1}))$ at (k+k*) . We define the list from (1) by adjoining

(3) $\underline{f}(i(n_{j+1})+1) = \ell_{i(n_{j+1})} + 1,\ \ldots,\ \underline{f}(m) = (k+k^*)_{\lessdot}$,

where $\ell_i = (k+k^*)_{\lessdot} + (m-i))_{\lessdot}$ for $i(i(n_{j+1})+1 \leq i < m)$.

This completes the definition by induction on k.

We define $\psi(k)$ assuming (*) in case $\exists n < k(k \lessdot n)$ by induction on k by

$$\psi(k) = \psi(n_{j-1}) + o(Q(k)) .$$

The proof that $\psi(k)$ is recursive and $\lessdot$-$\prec$ order-preserving is as in §1.

Appendix

In Chapters I and II we dealt with the system SINN and its modification by the addition of a function symbol. Here we consider carrying out the above argument for the system with extended inductive definitions (abbr. EID) defined in [6, Ch. 4] with a slight modification: Symbols for pred. $A_o, A_1, A_2, \ldots$ are introduced. We allow the sequences of the following forms as beginning sequences as well as those given in [6, Ch. 4, 1.2]:

$$A_n(i, t, V) \longrightarrow G_n(i, t, V, \{x, y\}(A_n(x, y, V) \wedge x <^* i));$$

$$G_n(i, t, V, \{x, y\}(A_n(x, y, V) \wedge x <^* i)) \longrightarrow A_n(i, t, V),$$

where i is a numeral, $\longrightarrow I(i)$ is provable and $G_n(a, b, \alpha, \beta)$ are semi-isolated.

In Chapter I we showed that every proof-figure in SINN (with a function symbol f) whose end-sequence contains no free t-variable can be transformed into a cut- and induction-free ω-proof (with the f-ω-rule) with the same end-sequence modifying the reduction technique used to prove the consistency-proof of SINN. We can apply the analogous argument to EID (with a function symbol f) until we get a proof-figure P with the following properties to which the reduction [6, Ch. 4, 9.1] will apply:

(1) The end-piece of P does not contain any induction, logical inference, beginning sequence for equality, implicit weakening or implicit logical beginning sequence.

(2) The end-piece of P contains a beginning sequence for inductive definition

(*) $$I(s),\ A_n(s, t, V) \longrightarrow G_n(s, t, V, \{x, y\}(A_n(x, y) \wedge x <^* s))$$

or

$$I(s),\ G_n(s, t, V, \{x, y\}(A_n(x, y) \wedge x <^* s)) \longrightarrow A_n(s, t, V)$$

where s and t are numerals.

Our argument does not go through (3) when the beginning sequence appears in such a manner that $\longrightarrow I(s)$ is provable, $A_n(s, t, V)$ is explicit and $G_n(s, t, V, \{x, y\}(A_n(x, y, V) \wedge x <^* s))$ is implicit. This is why we need to formulate the theorem in the following form.

Let us consider the system with extended inductive definitions with the constructive ω-rule (and $\underline{f}$-ω-rule) corresponding to the system with extended inductive definitions (with the function symbol $\underline{f}$). A cut in the system is called uninteresting if it is one of the following forms: one of its upper sequences is a beginning sequence for inductive definition $I(s),\ A_n(s, t, V) \longrightarrow G_n(s, t, V, \{x, y\}(A_n(x, y, V) \wedge x <^* s))$ or the end-sequence of a proof-figure of the form

$$\frac{I(s),\ G_n(s, t, V, \{x, y\}(A_n(x, y, V) \wedge x <^* s)) \longrightarrow A_n(s, t, V)}{G_n(s, t, V, \{x, y\}(A_n(x, y, V) \wedge x <^* s)),\ I(s) \longrightarrow A_n(s, t, V)}$$

and its cut formula is $G_n(s, t, V, \{x, y\}(A_n(x, y, V) \wedge x <^* s))$. Then we have

THEOREM. There exists a $\prec$-recursive function f such that for every proof-figure P ending with a sequence $\Gamma \longrightarrow \Delta$ without any free t-variable, f(P) is an ω-proof ending with the same sequence which has no mathematical induction and has only uninteresting cuts in such a manner that if I is an uninteresting cut in f(P), inferences under I are only exchanges, contractions and uninteresting cuts.

Proof. We can apply an argument analogous to that in Chapter I and in [6, Ch. 4] until we get a proof-figure P satisfying the properties (1) and (2) above. If $A_n(s, t, V)$ in (2) is implicit we apply the reduction [6, Ch. 4, 9.1] to P to define r(P). f(P) is defined to be f(r(P)). Let

$A_n(s, t, V)$ in (2) be explicit. If $I(s) \longrightarrow$ is provable and $G_n(s, t, V, A_n^s)$, where A_n^s denotes $\{x, y\}(A_n(x, y, V) \wedge x <^* s)$, is explicit, leave the beginning sequence for inductive definition as it stands and go to the next step (as we did for an explicit logical beginning sequence in the end-piece). Suppose P does have the property (3) because of, say, (*). Then we define r(P) to be the proof-figure obtained from P by replacing (*) by the proof-figure

$$(*') \qquad \frac{G_n(s, t, V, A_n^s) \longrightarrow G_n(s, t, V, A_n^s)}{I(s),\ G_n(s, t, V, A_n^s) \longrightarrow G_n(s, t, V, A_n^s)}$$

and replacing all the related formulas to $A_n(s, t, V)$ in (*) by $G_n(s, t, V, A_n^s)$ (with obvious changes). Evidently $r(P) \prec P$. f(P) is defined to be

$$\text{uninteresting cut}\quad \dfrac{A_n(s, t, V) \longrightarrow G_n(s, t, V, A_n^s) \qquad \dfrac{\left.\begin{array}{c}\vdots\\ \tilde{\Gamma}_0 \longrightarrow \Delta_0\end{array}\right\}\ f(r(P))}{G_n(s, t, V, A_n^s),\ \tilde{\Gamma}_0^* \longrightarrow \Delta_0}\ \text{Some exchanges}}{\Gamma_0 \longrightarrow \Delta_0}$$

where $\tilde{\Gamma}_0^*$ is the list of sequences obtained from $\tilde{\Gamma}_0$ by deleting $G_n(s, t, V, A_n^s)$, which is the descendant of the $G_n(s, t, V, A_n^s)$ in the left of (*').

The other form of the beginning sequence for inductive definition can be treated similarly.

Since no $A_n(s, t, V)$ can be explicit in a TJ-proof-figure or a proof-figure ending with $\forall x(\underline{f}(x+1) \lessdot \underline{f}(x)) \longrightarrow$, we can extend the theorem in Chapter II to (the modified) EID (with a function symbol $\underline{f}$) in a straightforward manner.

References

(1) G. Gentzen, Beweisbarkeit und Unbeweisbarkeit von Anfangsfällen der transfiniten Induktion in der reinen Zahlentheorie, Math. Ann., 119 (1943), 140-161.

(2) A. Kino, On provably recursive functions and ordinal recursive functions, to appear in J. Math. Soc. Japan.

(3) G. Kreisel, Mathematical Logic, Lectures on Modern Mathematics, vol. III, 95-195, New York, 1965.

(4) K. Schütte, Beweistheorie, Springer, 1960.

(5) G. Takeuti, A remark on Gentzen's paper 'Beweisbarkeit und Unbeweisbarkeit von Anfangsfällen der transfiniten Induktion in der reinen Zahlentheorie', I, II, Proc. Japan Acad., 39 (1963), 263-269.

(6) G. Takeuti, Consistency proofs of subsystems of classical analysis, Ann. of Math., 86 (1967), 299-348.

(7) G. Takeuti and M. Yasugi, Reflection principles of subsystems of analysis, to appear.

Offsetdruck: Julius Beltz, Weinheim/Bergstr.